Numerical Methods
in Fluid Dynamics

SERIES IN THERMAL AND FLUIDS ENGINEERING

JAMES P. HARTNETT and THOMAS F. IRVINE, JR., Editors
JACK P. HOLMAN, Senior Consulting Editor

Cebeci and Bradshaw	• Momentum Transfer in Boundary Layers
Chang	• Control of Flow Separation: Energy Conservation, Operational Efficiency, and Safety
Chi	• Heat Pipe Theory and Practice: A Sourcebook
Eckert and Goldstein	• Measurements in Heat Transfer, 2nd edition
Edwards, Denny, and Mills	• Transfer Processes: An Introduction to Diffusion, Convection, and Radiation
Fitch and Surjaatmadja	• Introduction to Fluid Logic
Ginoux	• Two-Phase Flows and Heat Transfer with Application to Nuclear Reactor Design Problems
Hsu and Graham	• Transport Processes in Boiling and Two-Phase Systems, Including Near-Critical Fluids
Kreith and Kreider	• Principles of Solar Engineering
Lu	• Introduction to the Mechanics of Viscous Fluids
Moore and Sieverding	• Two-Phase Steam Flow in Turbines and Separators: Theory, Instrumentation, Engineering
Richards	• Measurement of Unsteady Fluid Dynamic Phenomena
Sparrow and Cess	• Radiation Heat Transfer, augmented edition
Tien and Lienhard	• Statistical Thermodynamics, revised printing
Wirz and Smolderen	• Numerical Methods in Fluid Dynamics

PROCEEDINGS

Keairns	• Fluidization Technology
Spalding and Afgan	• Heat Transfer and Turbulent Buoyant Convection: Studies and Applications for Natural Environment, Buildings, Engineering Systems
Zarić	• Thermal Effluent Disposal from Power Generation

Numerical Methods in Fluid Dynamics

Edited by

H. J. WIRZ
J. J. SMOLDEREN
von Karman Institute for Fluid Dynamics
Rhode-Saint-Genèse, Belgium

 A von Karman Institute Book

HEMISPHERE PUBLISHING CORPORATION
Washington   London

McGRAW–HILL BOOK COMPANY

New York	Johannesburg	Paris
St. Louis	London	São Paulo
San Francisco	Madrid	Singapore
Auckland	Mexico	Sydney
Bogotá	Montreal	Tokyo
Düsseldorf	New Delhi	Toronto
	Panama	

NUMERICAL METHODS IN FLUID DYNAMICS

1 2 3 4 5 6 7 8 9 0   K P K P   7 8 3 2 1 0 9 8

This book was set in Press Roman by Hemisphere Publishing Corporation. The editors were Edward Millman and Mary Phillips; the designer was Lilia Guerrero. The printer and binder was The Kingsport Press, Inc.

Library of Congress Cataloging in Publication Data

Main entry under title:

Numerical methods in fluid dynamics.

  (Series in thermal and fluids engineering)
  "A Von Karman Institute book."
  Includes bibliographical references and index.
  1. Fluid dynamics—Addresses, essays, lectures.
2. Numerical analysis—Addresses, essays, lectures.
I. Wirz, Hans Jochen. II. Smolderen, J. J.
QA911.N84   532'.05 015194   77-18145

ISBN 0-07-071120-8

Contents

Contributors

William F. Ballhaus, U.S. Army Air Mobility Research and Development Laboratory, Moffett Field, California

Oleg M. Belotserkovskii, Computing Center, Academy of Sciences, Moscow, USSR

Antony Jameson, Courant Institute of Mathematical Sciences, New York University, New York, New York

Werner Kraus, Messerschmitt-Bölkow-Blohm GmbH, Unternehmensbereich Flugzeuge, Munich, West Germany

Thomas J. Mueller, Department of Aerospace and Mechanical Engineering, University of Notre Dame, Notre Dame, Indiana

Wolfgang Schmidt, Dornier GmbH, Theoretical Aerodynamics Group, Friedrichshafen, West Germany

Preface

This publication is a collection of the notes for a lecture series on numerical methods in fluid dynamics, organized in 1976 by the Computational Fluid Dynamics Department of the von Karman Institute and attended by more than 100 scientists and engineers from several countries.

The material was presented by active specialists working in this field and comprises recent developments in transonic flow computations, the application of numerical methods to physiological flow problems (Navier-Stokes), the present status of panel methods, and recent progress achieved by finite element methods to solve fluid flow problems. In addition, an attempt was made for the first time to give a broad survey of the present status and future prospects of computational approaches to fluid dynamics by leading scientists from the United States and the Soviet Union.

We gratefully acknowledge the support of the Plans and Programmes Division of the Advisory Group for Aerospace Research and Development (AGARD) of NATO. Finally, we express our thanks to Mrs. N. Toubeau, librarian at VKI, who carefully assembled and checked the material.

H. J. Wirz
J. J. Smolderen

Numerical Methods in Fluid Dynamics

CHAPTER 1

Transonic Flow Calculations

Antony Jameson

1 FORMULATION OF TRANSONIC FLOW PROBLEMS

1.1 Introduction

Following the introduction of type-dependent differencing by Murman and Cole [1] in 1970, relaxation methods have been widely used to calculate transonic flows in recent years. The purpose of this chapter is to discuss the development of these methods and some realistic applications, such as the calculation of the flow past a blunt-nosed airfoil, or past a lifting swept wing.

The first question that needs to be considered in a transonic flow calculation is the choice of an appropriate mathematical model. Experience has shown that rather accurate predictions can be made for a number of flows of interest using the transonic potential flow equation, which may be derived from the Euler equations for inviscid compressible flow by introducing the assumption that the flow is irrotational. According to Crocco's theorem, an irrotational flow is isentropic. Thus it is consistent to replace shock waves by discontinuities across which the entropy is conserved. Since the entropy generated by a shock wave is proportional to the third power of the shock strength (measured by the jump in the Mach number), this approximation should not be a source of serious error when the Mach number of the normal component of the flow ahead of the shock wave is less than 1.3. Stronger shock waves would in any case tend to cause boundary-layer separation and buffeting, calling for an altogether more elaborate analysis, including the effects of shock-wave–boundary-layer interaction and unsteady flow.

This work was supported by NASA under Grants NGR 33-016-167 and NGR 33-016-201. The computations were performed at the ERDA Mathematics and Computing Laboratory, New York University, under Contract E(11-1)-3077.

1

Once the choice of a mathematical model has been settled, two main elements can be recognized in the numerical procedure for actually computing a solution:

1 The construction of a discrete approximation that converges to the solution of the continuous problem in the limit as the mesh width is reduced to zero
2 The solution of the resulting set of nonlinear difference equations by a convergent iterative scheme

Each of these elements will be examined at some length in the following sections. The choice of an appropriate coordinate system can have a substantial influence on the accuracy of the discrete approximation. This question is discussed in Sec. 5, where a detailed description is given of several applications of the general method.

1.2   Exact Potential Flow Equation

It is convenient to develop the theory for two-dimensional calculations; the subsequent generalization to three-dimensional flows presents no particular difficulty. Consider the two-dimensional flow past a profile. Under the assumption of irrotational flow, we can introduce a velocity potential ϕ. Then we find that in smooth regions of the flow, ϕ satisfies the quasilinear equation

$$(a^2 - u^2)\phi_{xx} - 2uv\phi_{xy} + (a^2 - v^2)\phi_{yy} = 0 \tag{1}$$

where u and v are the velocity components

$$u = \phi_x \qquad v = \phi_y \tag{2}$$

and a is the local speed of sound. Given the ratio of specific heats γ and the stagnation speed of sound a_0, a can be determined from the energy relation

$$a^2 = a_0{}^2 - \frac{\gamma - 1}{2} q^2 \tag{3}$$

where q is the speed $\sqrt{u^2 + v^2}$. Equation (1) is hyperbolic when the local Mach number $M = q/a > 1$. On the profile the solution should satisfy the Neumann boundary condition

$$\frac{\partial \phi}{\partial n} = 0 \tag{4}$$

where n is the normal direction. At infinity the flow approaches a uniform stream with a Mach number M_∞. The density ρ and pressure p can be determined by the relations

$$\rho^{\gamma-1} = M_\infty^2 a^2 \tag{5}$$

and

$$p = \frac{\rho^\gamma}{\gamma M_\infty^2} \tag{6}$$

In the absence of discontinuities, Eq. (1) implies the conservation of both mass and momentum. Multiplied by ρ/a^2, it can be reduced to the equation for conservation of mass:

$$\frac{\partial}{\partial x}(\rho u) + \frac{\partial}{\partial y}(\rho v) = 0 \tag{7}$$

Multiplied by $\rho u/a^2$ or $\rho v/a^2$, on the other hand, it can be reduced to the equations for conservation of the x or y components of momentum. In an isentropic flow in which the energy is conserved it is not possible, however, to conserve both mass and momentum across a discontinuity. The approximation to a shock wave preferred here is one in which the mass is conserved. The momentum deficiency then provides an estimate of the wave drag [2]. Thus the desired solution should have the properties that ϕ is continuous, and that the velocity components are piecewise continuous, satisfying the conservation law (7) at points where the flow is smooth, together with the jump condition

$$[\rho v] - \frac{dy}{dx}[\rho u] = 0 \tag{8}$$

across a discontinuity, where [ ] denotes the jump, and dy/dx is the slope of the discontinuity. That is to say, ϕ should be a weak solution [3] of the conservation law (7) satisfying the condition that

$$\iint \rho(\phi_x w_x + \phi_y w_y)\, dx\, dy = 0 \tag{9}$$

for any smooth test function w, which vanishes in the far field. In a finite region the conservation law (7) is equivalent to the Bateman variational principle that

$$I = \iint p\, dx\, dy \tag{10}$$

is stationary [4].

1.3   Transonic Small-Disturbance Equation

A useful simplification is provided by small-disturbance theory. Suppose that the profile is given in the form $y = \tau f(x)$, and τ is small. If we expand the solution in powers of τ under the assumption that $1 - M_\infty^2 \sim \tau^{2/3}$ and retain only the leading term, we obtain the transonic small-disturbance equation [5]. Let K be the similarity constant $(1 - M_\infty^2)/\tau^{2/3}$. Then a typical form is

$$A\phi_{xx} + \phi_{yy} = 0 \tag{11}$$

where

$$A = K - (\gamma + 1)\phi_x \tag{12}$$

In this equation the y coordinate has been scaled by the factor $\tau^{1/3}$, and ϕ is the disturbance potential, scaled by the factor $\tau^{-2/3}$. The Neumann boundary condition is now transferred to the axis:

$$\phi_y = \frac{df}{dx} \quad \text{at } y = 0 \tag{13}$$

If the small-disturbance equation is written in the conservation form

$$\frac{\partial}{\partial x}\left(K\phi_x - \frac{\gamma + 1}{2}\phi_x^2\right) + \phi_{yy} = 0 \tag{14}$$

then the corresponding jump condition

$$[\phi_y] - \frac{dy}{dx}\left[K\phi_x - \frac{\gamma + 1}{2}\phi_x^2\right] = 0$$

yields a consistent approximation to shock waves [5].

1.4   Formulation of the Numerical Method

The general method to be described stems from the idea introduced by Murman and Cole [1], and subsequently improved by Murman [6], of using type-dependent differencing, with central difference formulas in the subsonic zone, where the governing equation is elliptic, and upwind difference formulas in the supersonic zone, where it is hyperbolic. The resulting directional bias in the numerical scheme corresponds to the upwind region of dependence of the flow in the supersonic zone. If we consider the transonic flow past a profile with fore and aft symmetry such as an ellipse, the desired solution of the potential flow equation is not symmetric. Instead it exhibits a smooth acceleration over the front half of the profile, followed by a discontinuous compression through a shock wave. In the absence of a directional bias in the numerical scheme, the fore and aft symmetry would be preserved in any solution that could be obtained, resulting in the appearance of improper discontinuities.

Since the quasilinear form (1) does not distinguish between conservation of mass and momentum, difference approximations to (1) will not necessarily yield solutions that satisfy the jump condition (8) unless shock waves are detected and special difference formulas are used in their vicinity. If we treat the conservation law (7), on the other hand, and preserve the conservation form in the difference approximation, we can ensure that the proper jump condition is satisfied. Similarly we can obtain proper solutions of the small-disturbance equation by treating it in the conservation form (14).

The general method of constructing a difference approximation to a conservation law of the form

$$f_x + g_y = 0$$

is to preserve the flux balance in each cell, as illustrated in Fig. 1. This leads to a scheme of the form

$$\frac{F_{i+1/2,\,j} - F_{i-1/2,\,j}}{\Delta x} + \frac{G_{i,\,j+1/2} - G_{i,\,j-1/2}}{\Delta y} = 0 \tag{15}$$

where F and G should converge to f and g in the limit as the mesh width tends to zero. Suppose, for example, that (15) represents the conservation law (7). Then on multiplying by a test function w_{ij} and summing by parts, there results an approximation to the integral (9). Thus the condition for a proper weak solution is satisfied [7]. Some latitude is allowed in the definitions of F and G, since it is only necessary that $F = f + 0(\Delta x)$ and

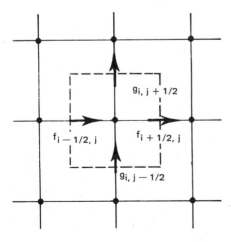

FIG. 1 Flux balance for difference scheme in conservation form.

$G = g + 0(\Delta x)$. In constructing a difference approximation we can therefore introduce an artificial viscosity of the form

$$\frac{\partial P}{\partial x} + \frac{\partial Q}{\partial y}$$

provided that P and Q are of order Δx. Then the difference scheme is an approximation to the modified conservation law

$$\frac{\partial}{\partial x}(f + P) + \frac{\partial}{\partial y}(g + Q) = 0$$

which reduces to the original conservation law in the limit as the mesh width tends to zero.

This formulation provides a guideline for constructing type-dependent difference schemes in conservation form. The dominant term in the discretization error introduced by the upwind differencing can be regarded as an artificial viscosity. We can, however, turn this idea around. Instead of using a switch in the difference scheme to introduce an artificial viscosity, we can explicitly add an artificial viscosity, which produces an upwind bias in the difference scheme at supersonic points. Suppose that we have a central difference approximation to the differential equation in conservation form. Then the conservation form will be preserved as long as the added viscosity is also in conservation form. The effect of the viscosity is simply to alter the conserved quantities by terms proportional to the mesh width Δx, which vanish in the limit as the mesh width approaches zero, with the result that the proper jump conditions must be satisfied. By including a switching function in the viscosity to make it vanish in the subsonic zone, we can continue to obtain the sharp representation of shock waves that results from switching the difference scheme.

There remains the problem of finding a convergent iterative scheme for solving the nonlinear difference equations that result from the discretization. Suppose that in the $(n + 1)$st cycle the residual R_{ij} at the point $i\,\Delta x, j\,\Delta y$ is evaluated by inserting the result $\phi_{ij}^{(n)}$ of the nth cycle in the difference approximation. Then the correction $C_{ij} = \phi_{ij}^{(n+1)} - \phi_{ij}^{(n)}$ is to be calculated by solving an equation of the form

$$NC + \sigma R = 0 \qquad (16)$$

where N is a discrete linear operator, and σ is a scaling function. In a relaxation method, N is restricted to a lower-triangular or block-triangular form so that the elements of C can be determined sequentially. In the analysis of such a scheme, it is helpful to introduce a time-dependent analogy. The vector R is an approximation to $L\phi$, where L is the operator appearing in the differential equation. If we consider C as representing $\Delta t\, \phi_t$, where t is an artificial time coordinate, and N is an approximation to a differential operator $(1/\Delta x)F$, then Eq. (16) is an approximation to

$$F\phi_t + \sigma \frac{\Delta x}{\Delta t} L\phi = 0 \qquad (17)$$

Thus we should choose N so that this is a convergent time-dependent process.

With this approach, the formulation of a relaxation method for solving a transonic flow is reduced to three main steps:

1 Construct a central difference approximation to the differential equation.
2 Add a numerical viscosity to produce the desired directional bias in the hyperbolic region.
3 Add time-dependent terms to embed the steady-state equation in a convergent time-dependent process.

Methods constructed along these lines have proved extremely reliable. Their main shortcoming is a rather slow rate of convergence. In order to speed up the convergence, we can extend the class of permissible operators N. Some possibilities are examined in Sec. 4.

2 SOLUTION OF THE TRANSONIC SMALL-DISTURBANCE EQUATION

2.1 Murman Difference Scheme

The basic ideas can conveniently be illustrated by considering the solution of the transonic small-disturbance equation. The treatment of the small-disturbance equation is simplified by the fact that the characteristics are locally symmetric about the x direction. Thus the desired directional bias can be introduced simply by switching to upwind differencing in the x direction at all supersonic points. To preserve the conservation form, some care must be exercised in the method of switching. Let p_{ij} be a central-difference approximation to the x derivatives at the point $i\,\Delta x, j\,\Delta y$:

$$
\begin{aligned}
p_{ij} &= K \frac{\phi_{i+1,\,j} - \phi_{ij} - (\phi_{ij} - \phi_{i-1,\,j})}{\Delta x^2} \\
&\quad - (\gamma + 1) \frac{(\phi_{i+1,\,j} - \phi_{ij})^2 - (\phi_{ij} - \phi_{i-1,\,j})^2}{2\Delta x^3} \qquad (18) \\
&= A_{ij} \frac{\phi_{i+1,\,j} - 2\phi_{ij} + \phi_{i-1,\,j}}{\Delta x^2}
\end{aligned}
$$

where

$$A_{ij} = K - (\gamma + 1)\frac{\phi_{i+1,j} - \phi_{i-1,j}}{2\Delta x} \tag{19}$$

Also let q_{ij} be a central-difference approximation to ϕ_{yy}:

$$q_{ij} = \frac{\phi_{i,j+1} - 2\phi_{ij} + \phi_{i,j-1}}{\Delta y^2} \tag{20}$$

Define a switching function μ with the value unity at supersonic points and zero at subsonic points:

$$\mu_{ij} = \begin{cases} 0 & \text{if } A_{ij} > 0 \\ 1 & \text{if } A_{ij} < 0 \end{cases} \tag{21}$$

Then the original scheme of Murman and Cole [1] can be written as

$$p_{ij} + q_{ij} - \mu_{ij}(p_{ij} - p_{i-1,j}) = 0 \tag{22}$$

Let

$$P = \Delta x \frac{\partial}{\partial x}\left(K\phi_x - \frac{\gamma+1}{2}\phi_x{}^2\right)$$

$$= \Delta x\, A\phi_{xx}$$

where A is the nonlinear coefficient defined in (12). Then the added terms are an approximation to

$$-\mu\frac{\partial P}{\partial x} = -\mu\,\Delta x\, A\phi_{xxx}$$

This may be regarded as an artificial viscosity of order Δx which is added at all points of the supersonic zone. Since the coefficient $-A$ of $\phi_{xxx} = u_{xx}$ is positive in the supersonic zone, it can be seen that the artificial viscosity includes a term similar to the viscous terms in the Navier-Stokes equation.

Since μ is not constant, the artificial viscosity is not in conservation form, with the result that the difference scheme does not satisfy the conditions stated in Sec. 1.4 for the discrete approximation to converge to a weak solution satisfying the proper jump conditions. To correct this, all that is required is to recast the artificial viscosity in a divergence form as $\partial(\mu P)/\partial x$. This leads to Murman's fully conservative scheme [6]

$$p_{ij} + q_{ij} - \mu_{ij}p_{ij} + \mu_{i-1,j}p_{i-1,j} = 0 \tag{23}$$

At points where the flow enters and leaves the supersonic zone, μ_{ij} and $\mu_{i-1,j}$ have different values, leading to special parabolic and shock-point equations

$$q_{ij} = 0$$

and

$$p_{ij} + p_{i-1,j} + q_{ij} = 0$$

With the introduction of these special operators, it can be verified by directly summing the difference equations at all points of the flow field that the correct jump conditions are satisfied across an oblique shock wave [6].

2.2 Solution of the Difference Equations by Relaxation

The nonlinear difference equations (18)–(21) and (22) or (23) may be solved by a generalization of the line relaxation method for elliptic equations. At each point we calculate the coefficient A_{ij} and the residual R_{ij} by substituting the result $\phi_{ij}^{(n)}$ of the previous cycle in the difference equations. Then we set $\phi_{ij}^{(n+1)} = \phi_{ij}^{(n)} + C_{ij}$, where the correction C_{ij} is determined by solving the linear equations

$$\frac{C_{i,j+1} - 2C_{ij} + C_{i,j-1}}{\Delta y^2} + (1 - \mu_{ij})A_{ij}\frac{-(2/\omega)\,C_{ij} + C_{i-1,j}}{\Delta x^2}$$

$$+ \mu_{i-1,j}A_{i-1,j}\frac{C_{ij} - 2C_{i-1,j} + C_{i-2,j}}{\Delta x^2} + R_{ij} = 0 \qquad (24)$$

on each successive vertical line. In these equations ω is the overrelaxation factor for subsonic points, with a value in the range $1 \le \omega \le 2$. In a typical line relaxation scheme for an elliptic equation, provisional values $\tilde{\phi}_{ij}$ are determined on the line $x = i\,\Delta x$ by solving the difference equations with the latest available values $\phi_{i-1,j}^{(n+1)}$ and $\phi_{i+1,j}^{(n)}$ inserted at points on the adjacent lines. Then new values $\phi_{ij}^{(n+1)}$ are determined by the formula

$$\phi_{ij}^{(n+1)} = \phi_{ij}^{(n)} + \omega(\tilde{\phi}_{ij} - \phi_{ij}^{(n)})$$

By eliminating $\tilde{\phi}_{ij}$, we can write the difference equations in terms of $\phi_{ij}^{(n+1)}$ and $\phi_{ij}^{(n)}$. Then it can be seen that ϕ_{yy} would be represented by $(1/\omega)\,\delta_y^2\phi^{(n+1)} + (1 - 1/\omega)\delta_y^2\phi^{(n)}$ in such a process, where δ_y^2 denotes the second central-difference operator. The appropriate procedure for treating the upwind difference formulas in the supersonic zone, however, is to march in the flow direction, so that the values $\phi_{ij}^{(n+1)}$ on each new column can be calculated from the values $\phi_{i-2,j}^{(n+1)}$ and $\phi_{i-2,j}^{(n+1)}$ already determined on the previous columns. This implies that ϕ_{yy} should be represented by $\delta_y^2 \phi^{(n+1)}$ in the supersonic zone, leading to a discontinuity at the sonic line. The correction formula (24) is derived by modifying this process to remove this discontinuity. New values $\phi_{ij}^{(n+1)}$ are used instead of provisional values $\tilde{\phi}_{ij}$ to evaluate ϕ_{yy} at both supersonic and subsonic points. At supersonic points, ϕ_{xx} is also evaluated using new values. At subsonic points, ϕ_{xx} is evaluated from $\phi_{i-1,j}^{(n+1)}$, $\phi_{i+1,j}^{(n)}$, and a linear combination of $\phi_{ij}^{(n+1)}$ and $\phi_{ij}^{(n)}$ equivalent to $\tilde{\phi}_{ij}$. In the subsonic zone the scheme acts like a line relaxation scheme, with a comparable rate of convergence. In the supersonic zone it is equivalent to a marching scheme, once the coefficients A_{ij} have been evaluated. Since the supersonic difference scheme is implicit, no limit is imposed on the step length Δx as A_{ij} approaches zero near the sonic line.

2.3   One-Dimensional Flow in a Channel—Nonunique Solutions of the Difference Equations

Some of the properties of the Murman difference formulas are clarified by considering a uniform flow in a parallel channel. Then $\phi_{yy} = 0$, and with a suitable normalization $K = 0$, so that the equation reduces to

$$ -\frac{\partial}{\partial x}\left(\frac{\phi_x^2}{2}\right) = 0 $$

with ϕ and ϕ_x given at $x = 0$, and ϕ given at $x = L$. Since ϕ_x^2 is constant ϕ_x simply reverses sign at a jump. Provided we enforce the entropy condition that ϕ_x decreases through a jump, there is a unique solution with a single jump whenever $\phi_x(0) > 0$ and $\phi(0) + L\phi_x(0) \geq \phi(L) \geq \phi(0) - L\phi_x(0)$.

Let $u_{i+1/2} = (\phi_{i+1} - \phi_i)/\Delta x$ and $u_i = (u_{i+1/2} + u_{i-1/2})/2$. Then the fully conservative difference equations can be written as

Elliptic:	$u_{i+1/2}^2 = u_{i-1/2}^2$	when $u_i \leq 0$	$u_{i-1} \leq 0$		(a)
Hyperbolic:	$u_{i-1/2}^2 = u_{i-3/2}^2$	when $u_i > 0$	$u_{i-1} > 0$		(b)
Shock point:	$u_{i+1/2}^2 = u_{i-3/2}^2$	when $u_i \leq 0$	$u_{i-1} > 0$		(c)
Parabolic:	$0 = 0$	when $u_i > 0$	$u_{i-1} \leq 0$		(d)

These admit the correct solution, illustrated in Fig. 2a with a constant slope on the two sides of the shock. The shock-point operator allows a single link with an intermediate slope, corresponding to the shock lying in the middle of a mesh cell.

The nonconservative difference scheme omits the shock-point operator,

Shock point

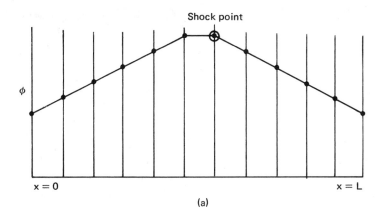

x = 0   x = L

(a)

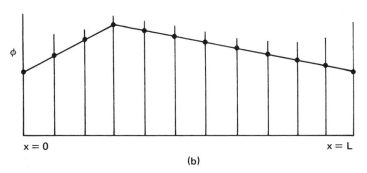

x = 0   x = L

(b)

FIG. 2 One-dimensional flow in a channel.

with the result that it admits solutions of the type illustrated in Fig. 2*b*, with the shock too far forward and the downstream velocity too close to the sonic speed (zero with the present normalization). The direct switch in the difference scheme from (*b*) to (*a*) allows a break in the slope as long as the downstream slope is negative. The magnitude of the downstream slope cannot exceed the magnitude of the upstream slope, however, because then $u_{i-1} < 0$, so that the elliptic operator would be used at the point $(i-1)\,\Delta x$. Thus the nonconservative scheme enforces the weakened shock condition

$$\phi_i - \phi_{i-2} > \phi_i - \phi_{i+2} > 0$$

which allows solutions ranging from the point at which the downstream velocity is barely subsonic up to the point at which the shock strength is correct. When the downstream velocity is too close to sonic speed, there is

an increase in the mass flow. Thus the nonconservative scheme may introduce a source at the shock wave.

The fully conservative difference equations also admit, however, various
improper solutions. Figure 3a illustrates a sawtooth solution with u^2 constant
everywhere except in one cell ahead of a shock point. Figure 3b illustrates
another improper solution in which the shock is too far forward. At the
last interior point, there is then an expansion shock that is admitted by the
parabolic operator. Since the difference equations have more than one root,
we must depend on the iterative scheme to find the desired root. The scheme
should ideally be designed so that the correct solution is stable under a
small perturbation, and improper solutions are unstable. Via a scheme
similar to (24), the instability of the sawtooth solution has been confirmed
in numerical experiments. The solutions with an expansion shock at the
downstream boundary are stable, on the other hand, if the compression
shock is too far forward by more than the width of a mesh cell. Thus
there is a continuous range of stable improper solutions, while the correct
solution is an isolated stable equilibrium point.

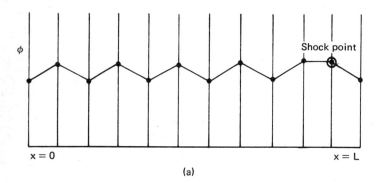

(a)

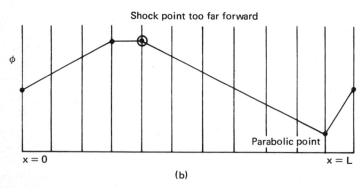

(b)

FIG. 3 One-dimensional flow in a channel. (a) Sawtooth solution; (b) solution
with downstream parabolic point.

3 SOLUTION OF THE EXACT POTENTIAL FLOW EQUATION

3.1 Difference Schemes for the Exact Potential Flow Equation in Quasilinear Form

It is less easy to construct difference approximations to the potential flow equation with a correct directional bias, because the upwind direction is not known in advance. If, however, the supersonic flow is confined to a bubble above the profile, it may be possible to use a coordinate system in which the x coordinate is more or less aligned with the flow in the supersonic zone. For this purpose we can use a conformal mapping to make the profile coincide with an x coordinate line (see Sec. 5). A simple difference approximation to the quasilinear form (1) can then be constructed in the following manner. The velocity components u and v are evaluated throughout the flow field by central-difference formulas, and the speed of sound is determined by Eq. (3). Then at subsonic points we use central-difference formulas for ϕ_{xx}, ϕ_{xy}, and ϕ_{yy}, while at supersonic points we switch to upwind difference formulas for ϕ_{xx} and ϕ_{xy}. The upwind difference formulas can be regarded as approximations to $\phi_{xx} - \Delta x\, \phi_{xxx}$ and $\phi_{xy} - (\Delta x/2)\phi_{xxy}$. Thus they introduce an effective artificial viscosity

$$\Delta x\,[(u^2 - a^2)\phi_{xxx} + uv\phi_{xxy}] = \Delta x\,[(u^2 - a^2)u_{xx} + uvv_{xx}]$$

When the flow is not perfectly aligned with the x coordinate, there exist supersonic points at which $u^2 < a^2 < u^2 + v^2$. One characteristic lies ahead of the y coordinate line at such a point, so that the difference scheme does not have the correct region of dependence (see Fig. 4). Also the artificial viscosity $\Delta x\,(u^2 - a^2)\phi_{xxx}$ introduced by the upwind difference formula for ϕ_{xx} is then negative. Despite this fact, schemes of this type have proved quite satisfactory in practice for flows with supersonic zones of moderate size [8,9].

The treatment of flows with large supersonic zones in a curvilinear coordinate system suited to the geometry of the problem requires the use of a more elaborate difference scheme, in which the direction of upwind differencing is independent of the coordinate system, and is instead rotated to conform with the local flow direction [10,11]. To illustrate the construction of such a scheme, consider the potential flow equation (1) in Cartesian coordinates. The required rotation of the upwind differencing at any particular point can be accomplished by introducing an auxiliary Cartesian coordinate system that is locally aligned with the flow at that point. If s

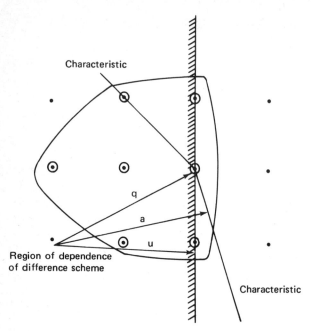

FIG. 4   Simple difference scheme.

and n denote the local streamwise and normal directions, then Eq. (1) becomes

$$(a^2 - q^2)\phi_{ss} + a^2\phi_{nn} = 0 \tag{25}$$

Since u/q and v/q are the local direction cosines, ϕ_{ss} and ϕ_{nn} can be expressed in the original coordinate system as

$$\phi_{ss} = \frac{1}{q^2}\left(u^2\phi_{xx} + 2uv\phi_{xy} + v^2\phi_{yy}\right) \tag{26}$$

and

$$\phi_{nn} = \frac{1}{q^2}\left(v^2\phi_{xx} - 2uv\phi_{xy} + u^2\phi_{yy}\right) \tag{27}$$

Then, at subsonic points, central-difference formulas are used for both ϕ_{ss} and ϕ_{nn}. At supersonic points, central-difference formulas are used for ϕ_{nn}, but upwind difference formulas are used for the second derivatives contributing to ϕ_{ss}, as illustrated in Fig. 5. At a supersonic point at which $u > 0$ and $v > 0$, for example, ϕ_{ss} is constructed from the formulas

$$\phi_{xx} = \frac{\phi_{ij} - 2\phi_{i-1,j} + \phi_{i-2,j}}{\Delta x^2}$$

$$\phi_{xy} = \frac{\phi_{ij} - \phi_{i-1,j} - \phi_{i,j-1} + \phi_{i-1,j-1}}{\Delta x \, \Delta y} \tag{28}$$

$$\phi_{yy} = \frac{\phi_{ij} - 2\phi_{i,j-1} + \phi_{i,j-2}}{\Delta y^2}$$

It can be seen that the scheme reduces to a form similar to the scheme of Murman and Cole for the small-disturbance equation if either $u = 0$ or $v = 0$. The upwind difference formulas can be regarded as approximations to $\phi_{xx} - \Delta\phi_{xxx}$, $\phi_{xy} - (\Delta x/2)\phi_{xxy} - (\Delta y/2)\phi_{xyy}$, and $\phi_{yy} - \Delta y\phi_{yyy}$. Thus at supersonic points the scheme introduces an effective artificial viscosity

$$\left(1 - \frac{a^2}{q^2}\right)[\Delta x \, (u^2 u_{xx} + uvv_{xx}) + \Delta y \, (uvu_{yy} + v^2 v_{yy})] \tag{29}$$

which is symmetric in x and y.

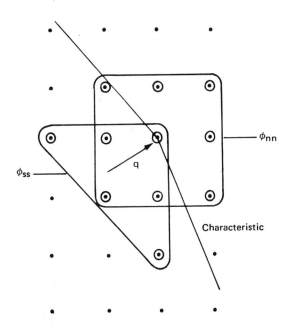

FIG. 5 Rotated difference scheme.

3.2 Difference Schemes for the Exact Potential Flow Equation in Conservation Form

In the construction of a discrete approximation to the conservation form (7) of the potential flow equation, it is convenient to accomplish the switch to upwind differencing by the explicit addition of an artificial viscosity in the manner proposed in Sec. 1.4. Thus we solve an equation of the form

$$S_{ij} + T_{ij} = 0 \tag{30}$$

where S_{ij} is a central-difference approximation to the left-hand side of Eq. (7), and T_{ij} is the artificial viscosity, which is constructed as an approximation to an expression in divergence form $\partial P/\partial x + \partial Q/\partial y$, where P and Q are appropriate quantities with a magnitude proportional to the mesh width. The central-difference approximation is constructed in the natural manner as

$$S_{ij} = \frac{(\rho u)_{i+1/2,\, j} - (\rho u)_{i-1/2,\, j}}{\Delta x} + \frac{(\rho v)_{i,\, j+1/2} - (\rho v)_{i,\, j-1/2}}{\Delta y} \tag{31}$$

Consider first the case in which the flow in the supersonic zone is aligned with the x coordinate, so that it is sufficient to restrict the upwind differencing to the x derivatives. In a smooth region of the flow, the first term of S_{ij} is an approximation to

$$\frac{\partial}{\partial x}(\rho u) = \rho\left(1 - \frac{u^2}{a^2}\right)\phi_{xx} - \frac{\rho u v}{a^2}\,\phi_{xy}$$

We wish to construct T_{ij} so that ϕ_{xx} is effectively represented by an upwind difference formula when $u > a$. Define the switching function

$$\mu = \min\left[0, \rho\left(1 - \frac{u^2}{a^2}\right)\right] \tag{32}$$

Then set

$$T_{ij} = \frac{P_{i+1/2,\, j} - P_{i-1/2,\, j}}{\Delta x} \tag{33}$$

where

$$P_{i+1/2,\, j} = -\frac{\mu_{ij}}{\Delta x}\left[\phi_{i+1,\, j} - 2\phi_{ij} + \phi_{i-1,\, j} - \varepsilon(\phi_{ij} - 2\phi_{i-1,\, j} + \phi_{i-2,\, j})\right] \tag{34}$$

The added terms are an approximation to $\partial P/\partial x$, where

$$P = -\mu[(1 - \varepsilon)\,\Delta x\,\phi_{xx} + \varepsilon\,\Delta x^2\,\phi_{xxx}]$$

Thus, if $\varepsilon = 0$, the scheme is first-order accurate; but if

$$\varepsilon = 1 - \lambda \, \Delta x$$

and λ is a constant, the scheme is second-order accurate. Also when $\varepsilon = 0$ the viscosity cancels the term $\rho(1 - u^2/a^2)\phi_{xx}$ and replaces it by its value at the adjacent upwind point.

In this scheme the switch to upwind differencing is introduced smoothly because the coefficient $\mu \to 0$ as $u \to a$. If the first term in S_{ij} were simply replaced by the upwind difference formula

$$\frac{(\rho u)_{i-1/2, j} - (\rho u)_{i-3/2, j}}{\Delta x}$$

the switch would be less smooth because there would also be a sudden change in the representation of the term $(\rho uv/a^2)\phi_{xy}$, which does not necessarily vanish when $u = a$. A scheme of this type proved to be unstable in numerical tests.

The treatment of flows that are not well aligned with the coordinate system requires the use of a difference scheme in which the upwind bias conforms to the local flow direction. The desired bias can be obtained by modeling the added terms T_{ij} on the artificial viscosity of the rotated difference scheme for the quasilinear form described in the previous section. Since Eq. (7) is equivalent to Eq. (1) multiplied by ρ/a^2, P and Q should be chosen so that $\partial P/\partial x + \partial Q/\partial y$ contains terms similar to Eq. (29) multiplied by ρ/a^2. The following scheme has proved successful [12,13]. Let μ be a switching function that vanishes in the subsonic zone:

$$\mu = \max\left[0, \left(1 - \frac{a^2}{q^2}\right)\right] \tag{35}$$

Then P and Q are defined as approximations to

$$-\mu[(1 - \varepsilon)|u| \, \Delta x \, \rho_x + \varepsilon u \, \Delta x^2 \, \rho_{xx}]$$

and

$$-\mu[(1 - \varepsilon)|v| \, \Delta y \, \rho_y + \varepsilon v \, \Delta y^2 \, \rho_{yy}]$$

where the parameter ε controls the accuracy in the same way as in the simple scheme. If $\varepsilon = 0$, the scheme is first-order accurate, and at a supersonic point where $u > 0$ and $v > 0$, P then approximates

$$-\Delta x \left(1 - \frac{a^2}{q^2}\right) u \rho_x = \Delta x \, \frac{\rho}{a^2} \left(1 - \frac{a^2}{q^2}\right)(u^2 u_x + uv v_x)$$

When this formula and the corresponding formula for Q are inserted in $\partial P/\partial x + \partial Q/\partial y$, it can be verified that the terms containing the highest derivatives of ϕ are the same as those in Eq. (29) multiplied by ρ/a^2. In the construction of P and Q, the derivatives of ρ are represented by upwind difference formulas. Thus the formula for the viscosity finally becomes

$$T_{ij} = \frac{P_{i+1/2, j} - P_{i-1/2, j}}{\Delta x} + \frac{Q_{i, j+1/2} - Q_{i, j-1/2}}{\Delta y} \qquad (36)$$

where if $u_{i+1/2, j} > 0$, then

$$P_{i+1/2, j} = u_{i+1/2, j}\mu_{ij}[\rho_{i+1/2, j} - \rho_{i-1/2, j} - \varepsilon(\rho_{i-1/2, j} - \rho_{i-3/2, j})] \quad (37a)$$

and if $u_{i+1/2, j} < 0$, then

$$P_{i+1/2, j} = u_{i+1/2, j}\mu_{i+1, j}[\rho_{i+1/2, j} - \rho_{i+3/2, j} - \varepsilon(\rho_{i+3/2, j} - \rho_{i+5/2, j})] \quad (37b)$$

while $Q_{i, j+1/2}$ is defined by a similar formula.

Formulas (31) and (37) call for the evaluation of the velocity components and density at the midpoint of each mesh interval. The precise method by which this is accomplished has been found to have little influence on the result. One method is to evaluate the velocity components by formulas such as

$$u_{i+1/2, j} = \frac{\phi_{i+1, j} - \phi_{ij}}{\Delta x} \qquad (38a)$$

and

$$v_{i+1/2, j} = \frac{\phi_{i+1, j+1} + \phi_{i, j+1} - (\phi_{i+1, j-1} + \phi_{i, j-1})}{4\Delta y} \qquad (38b)$$

Then the density $\rho_{i+1/2, j}$ is determined from Eq. (5). With this scheme two densities $\rho_{i+1/2, j}$ and $\rho_{i, j+1/2}$ have to be evaluated for each mesh point.

An alternative method is to determine first the density $\rho_{i+1/2, j+1/2}$ at the center of each cell. For this purpose the squared velocity components are evaluated as

$$u^2_{i+1/2, j+1/2} = \tfrac{1}{2}(u^2_{i+1/2, j} + u^2_{i+1/2, j+1}) \qquad (39a)$$

and

$$v^2_{i+1/2, j+1/2} = \tfrac{1}{2}(v^2_{i, j+1/2} + v^2_{i+1, j+1/2}) \qquad (39b)$$

The densities $\rho_{i+1/2, j}$ and $\rho_{i, j+1/2}$ are then formed by averaging as

$$\rho_{i+1/2, j} = \tfrac{1}{2}(\rho_{i+1/2, j+1/2} + \rho_{i+1/2, j-1/2}) \qquad (40a)$$

and

$$\rho_{i, j+1/2} = \tfrac{1}{2}(\rho_{i+1/2, j+1/2} + \rho_{i-1/2, j+1/2}) \qquad (40b)$$

This method has the advantage that Eq. (5), which requires the evaluation of a fractional power, has to be used to calculate only one density $\rho_{i+1/2, j+1/2}$ for each mesh point. In the case of a flow in a finite region, it results in a formula for S_{ij}, which could also be obtained from the Bateman variational principle. Let the integral (10) be approximated as

$$I = \sum_i \sum_j p_{i+1/2, j+1/2} \, \Delta x \, \Delta y$$

where $p_{i+1/2, j+1/2}$ is determined from $\rho_{i+1/2, j+1/2}$ by Eq. (6). Then $\partial I/\partial \phi_{ij} = S_{ij}$, where the averaged densities are used in the formula (26) for S_{ij}, and the condition that I is stationary is expressed by $S_{ij} = 0$.

If the profile coincides with a coordinate line, the Neumann boundary condition (3) can be treated in a simple manner suggested by Murman (private communication). Equation (3) is equivalent to requiring that $\rho v = 0$ at the boundary. The second term of Eq. (31) is therefore replaced by $(2/\Delta y)(\rho v)_{i, j+1/2}$, corresponding to the use of a one-sided difference formula for $\partial(\rho v)/\partial y$. When the densities are calculated by Eqs. (39) and (40), this is again just the formula which would be obtained from the variational principle.

3.3 Analysis of the Relaxation Method

The nonconservative rotated difference scheme and the difference schemes in conservation form lead to difference equations that are not amenable to solution by marching in the supersonic zone, and a rather careful analysis is needed to ensure the convergence of the iterative scheme. For this purpose it is convenient to introduce the time-dependent analogy proposed in Sec. 1.4. Thus we regard the iterative scheme as an approximation to the artificial time-dependent equation (17). It was shown by Garabedian [14] that this method can be used to estimate the optimum relaxation factor for an elliptic problem.

To illustrate the application of the method, consider the standard difference scheme for Laplace's equation. Typically, in a point overrelaxation scheme, a provisional value $\tilde{\phi}_{ij}$ is calculated by solving

$$\frac{\phi_{i-1, j}^{(n+1)} - 2\tilde{\phi}_{ij} + \phi_{i+1, j}^{(n)}}{\Delta x^2} + \frac{\phi_{i, j-1}^{(n+1)} - 2\tilde{\phi}_{ij} + \phi_{i, j+1}^{(n)}}{\Delta y^2} = 0$$

Then the new value $\phi_{ij}^{(n+1)}$ is determined by the formula

$$\phi_{ij}^{(n+1)} = \phi_{ij}^{(n)} + \omega(\tilde{\phi}_{ij} - \phi_{ij}^{(n)})$$

where ω is the overrelaxation factor. Eliminating $\tilde{\phi}_{ij}$, this is equivalent to calculating the correction C_{ij} by solving

$$\tau_1(C_{ij} - C_{i-1,j}) + \tau_2(C_{ij} - C_{i,j-1}) + \tau_3 C_{ij} = R_{ij} \qquad (41)$$

where R_{ij} is the residual, and

$$\tau_1 = \frac{1}{\Delta x^2} \qquad \tau_2 = \frac{1}{\Delta y^2}$$

$$\tau_3 = \left(\frac{2}{\omega} - 1\right)\left(\frac{1}{\Delta x^2} + \frac{1}{\Delta y^2}\right)$$

Equation (41) is an approximation to the wave equation

$$\tau_1 \Delta t \, \Delta x \, \phi_{xt} + \tau_2 \Delta t \, \Delta y \, \phi_{yt} + \tau_3 \Delta t \, \phi_t = \phi_{xx} + \phi_{yy}$$

This is damped if $\tau_3 > 0$, and to maximize the rate of convergence the relaxation factor ω should be chosen to give an optimal amount of damping.

In the case of a rectangular region with periodic boundary conditions, it is easily verified by a von Neumann test that schemes with

$$\tau_1 \geq \frac{1}{\Delta x^2} \qquad \tau_2 \geq \frac{1}{\Delta y^2} \qquad \tau_3 > 0$$

will converge. Suppose that

$$\phi^{(n)} = G^n \, e^{ipx} \, e^{iqy}$$

where G is the growth factor. Then substituting this expression in Eq. (41) gives

$$G = -\frac{(2 - \tau_1 \Delta x^2)A + (2 - \tau_2 \Delta y^2)B - iC}{\tau_1 \Delta x^2 \, A + \tau_2 \Delta y^2 \, B + \tau_3 + iC}$$

where $A = \dfrac{1 - \cos p \, \Delta x}{\Delta x^2} \geq 0 \qquad B = \dfrac{1 - \cos q \, \Delta y}{\Delta y^2} \geq 0$

and $C = \tau_1 \sin p \, \Delta x + \tau_2 \sin q \, \Delta y$

If we consider the potential flow equation (1) at a subsonic point, these considerations suggest that the scheme (41), where the residual R_{ij} is evaluated from the difference approximation described in Sec. 3.1, will converge if

$$\tau_1 \geq \frac{a^2 - u^2}{\Delta x^2} \qquad \tau_2 \geq \frac{a^2 - v^2}{\Delta y^2} \qquad \tau_3 > 0$$

Similarly the scheme

$$\tau_1(C_{ij} - C_{i-1,j}) - \tau_2(C_{i,j+1} - 2C_{ij} + C_{i,j-1}) + \tau_3 C_{ij} = R_{ij} \qquad (42)$$

which requires the simultaneous solution of the corrections on each vertical line, can be expected to converge if

$$\tau_1 \geq \frac{a^2 - u^2}{\Delta x^2} \qquad \tau_2 = \frac{a^2 - v^2}{\Delta y^2} \qquad \tau_3 > 0$$

At supersonic points, schemes similar to (41) or (42) are not necessarily convergent [10]. If we introduce a locally aligned Cartesian coordinate system and divide through by a^2, the general form of the equivalent time-dependent equation is

$$(M^2 - 1)\phi_{ss} - \phi_{nn} + 2\alpha\phi_{st} + 2\beta\phi_{nt} + \gamma\phi_t = 0 \qquad (43)$$

where M is the local Mach number, and s and n are the streamwise and normal directions. The coefficients α, β, and γ depend on the coefficients of the elements of C on the left-hand side of (41) or (42). The substitution

$$T = t - \frac{\alpha s}{M^2 - 1} + \beta n$$

reduces this equation to the diagonal form

$$(M^2 - 1)\phi_{ss} - \phi_{nn} - \left(\frac{\alpha^2}{M^2 - 1} - \beta^2\right)\phi_{TT} + \gamma\phi_T = 0$$

Since the coefficients of ϕ_{nn} and ϕ_{ss} have opposite signs when $M > 1$, T cannot be the timelike direction at a supersonic point. Instead, either s or n is timelike, depending on the sign of the coefficient of ϕ_{TT}. Since s is the timelike direction of the steady-state problem, it ought also to be the timelike direction of the unsteady problem. Thus, when M > 1, the relaxation scheme should be designed so that α and β satisfy the compatibility condition

$$\alpha > \beta\sqrt{M^2 - 1} \tag{44}$$

The characteristics of the unsteady equation (43) satisfy

$$(M^2 - 1)(t^2 + 2\beta nt) - 2\alpha st + (\beta s - \alpha n)^2 = 0 \tag{45}$$

Thus the characteristic cone touches the *s-n* plane. As long as condition (44) holds with $\alpha > 0$ and $\beta > 0$, it slants upstream in the reverse time direction, as illustrated in Fig. 6. To ensure that the iterative scheme has the proper region of dependence, the flow field should be swept in a direction such that the updated region always includes the upwind line of tangency between the characteristic cone and the *s-n* plane.

A von Neumann test of local stability at points in the supersonic zone leads to an additional guideline. Suppose that Eq. (1) is approximated at the mesh point $i\,\Delta x$, $j\,\Delta y$ by the general difference equation

$$\sum_{r,\,s} \left(a_{rs}\,\phi^{(n)}_{i+r,\,j+s} - b_{rs}\,\phi^{(n+1)}_{i+r,\,j+s} \right) = 0$$

Treating the equation as if it were a linear equation with constant coefficients, substitute

$$\phi^{(n)} = G^n\,e^{ipx}\,e^{iqy}$$

Then the growth factor is found to be

$$G = \frac{\sum_{r,\,s} a_{rs}\,e^{i(r\xi + s\eta)}}{\sum_{r,\,s} b_{rs}\,e^{i(r\xi + s\eta)}}$$

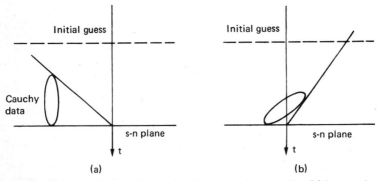

FIG. 6 Characteristic cone of equivalent time-dependent equation. (a) Supersonic; (b) subsonic.

where $$\xi = p\,\Delta x \qquad \eta = q\,\Delta y$$

For small values of ξ and η, G may be expanded as

$$G = \frac{A_{00} + A_{10}\,i\xi + A_{01}i\eta + A_{20}\,\xi^2 + A_{11}\xi\eta + A_{02}\,\eta^2 + \cdots}{B_{00} + B_{10}\,i\xi + B_{01}i\eta + B_{20}\,\xi^2 + B_{11}\xi\eta + B_{02}\,\eta^2 + \cdots}$$

where consistency with the steady-state differential equation (1) requires that

$$A_{00} = B_{00} \qquad\qquad A_{10} = B_{10} \qquad\qquad A_{01} = B_{01}$$

$$B_{20} - A_{20} = \frac{a^2 - u^2}{\Delta x^2} \qquad B_{11} - A_{11} = \frac{-2uv}{\Delta x\,\Delta y} \qquad B_{02} - A_{02} = \frac{a^2 - v^2}{\Delta y^2}$$

Since ξ and η can be chosen so that

$$(a^2 - u^2)\frac{\xi^2}{\Delta x^2} - \frac{2uv\xi\eta}{\Delta x\,\Delta y} + (a^2 - v^2)\frac{\eta^2}{\Delta y^2}$$

is either positive or negative whenever the flow is supersonic, $|G|$ can certainly exceed unity unless

$$A_{00} = B_{00} = 0$$

It follows that

$$\sum_{r,\,s} a_{rs} = \sum_{r,\,s} b_{rs} = 0 \tag{46}$$

If we interpret the difference scheme as the representation of a time-dependent process, this means that the coefficient of ϕ_t should be zero at supersonic points, reflecting the fact that t is not a timelike direction. The mechanism of convergence in the supersonic zone can be inferred from Fig. 6. An equation of the form of (43) with constant coefficients reaches a steady state because with advancing time its characteristic cone (45) eventually ceases to intersect the initial time plane. Instead it intersects a surface containing the Cauchy data of the steady-state problem. The rate of convergence is determined by the backward inclination of the most retarded characteristic

$$t = \frac{2\alpha s}{M^2 - 1} \qquad n = -\frac{\beta}{\alpha}s$$

and is maximized by using the smallest permissible coefficient α for the term in ϕ_{st}. In the subsonic zone, on the other hand, the cone of dependence contains the t axis, and it is important to introduce damping to remove the influence of the initial data.

Relaxation schemes that are derived simply by substituting the latest available values of ϕ in the difference equations do not necessarily satisfy the compatibility condition (44) at supersonic points. The following line relaxation scheme has proved successful in the calculation of a wide range of flows, including flows with supersonic free streams. At subsonic points the correction C_{ij} is determined by solving an equation similar to (42). At supersonic points the equation for the correction is determined by substituting in the upwind difference formulas for ϕ_{ss} a combination of new and old values such that the operator N in Eq. (16) is diagonally dominant. If $u > 0$, ϕ_{xx} is represented by the formula

$$\frac{2\phi_{ij}^{(n+1)} - \phi_{ij}^{(n)} - 2\phi_{i-1,j}^{(n+1)} + \phi_{i-2,j}^{(n)}}{\Delta x^2}$$

This is an approximation to $\phi_{xx} - 2(\Delta t/\Delta x)\phi_{xt}$. A similar formula is used for ϕ_{yy}, while ϕ_{xy} is represented by the formula

$$\frac{2\phi_{ij}^{(n+1)} - \phi_{ij}^{(n)} - \phi_{i-1,j}^{(n+1)} - \phi_{i,j-1}^{(n+1)} + \phi_{i-1,j-1}^{(n)}}{\Delta x \, \Delta y}$$

which is an approximation to $\phi_{xy} - (\Delta t/\Delta y)\phi_{xt} - (\Delta t/\Delta x)\phi_{yt}$. Thus the approximation to $(M^2 - 1)\phi_{ss}$ introduces a term

$$2(M^2 - 1)\left(\frac{u}{q}\frac{\Delta t}{\Delta x} + \frac{v}{q}\frac{\Delta t}{\Delta y}\right)\phi_{st}$$

in the equivalent time-dependent equation. To make sure that (44) is satisfied when the local Mach number is close to unity, the coefficient of ϕ_{st} is further augmented by adding an upwind approximation to ϕ_{st} at both supersonic and subsonic points. If $u > 0$ and $v > 0$, this term is

$$\omega_s\left[\frac{u}{\Delta x}(C_{ij} - C_{i-1,j}) + \frac{v}{\Delta y}(C_{ij} - C_{i,j-1})\right]$$

where ω_s is a relaxation factor with a value ≥ 0. The best rate of convergence is obtained by using the smallest possible value of ω_s, and in fact it often suffices to take $\omega_s = 0$.

The same ideas can be applied to the solution of the difference equations that result from approximations to the conservation form (7). Since the conservation form is equivalent to the quasilinear form multiplied by ρ/a^2, we can produce a time-dependent process for the conservation form that converges at about the same rate as the process for the quasilinear form simply by multiplying the operator N by ρ/a^2 [12]. This procedure has the advantage that the iterative scheme does not have to be modified to reflect detailed variations in the difference equations. The same operator N can be used for all values of the viscosity parameter ε in Eq. (37), for example, since the dominant terms of the equivalent time-dependent equation are independent of ε.

4 ACCELERATED ITERATIVE METHODS

4.1 Extrapolation

While relaxation has proved a very reliable method of solving the difference equations, it often requires a very large number of iterations. Since the effective time step, when relaxation is regarded as an artificial time-dependent process, is proportional to the mesh width, the number of iterations increases as the mesh width is reduced. As the process approaches convergence (and in particular after the supersonic zone is frozen), the coefficients of the difference equations become almost constant from iteration to iteration, so that an analysis of the linearized equations that would result from freezing the coefficients can provide a useful estimate of the rate of convergence that can be expected of the iterative method.

Suppose that $\hat{L}$ denotes the resulting linearized operator. Then, using vector notation, the iteration (16) for the linearized equations can be written as

$$NC^{(n)} + \sigma\hat{L}\phi^{(n)} = 0 \tag{47}$$

where $C^{(n)}$ is the correction $\phi^{(n+1)} - \phi^{(n)}$. Let ϕ be the exact solution of the linearized equation so that

$$\hat{L}\phi = 0 \tag{48}$$

and let $e^{(n)}$ be the error $\phi^{(n)} - \phi$ after the nth cycle. Then

$$C^{(n)} = e^{(n+1)} - e^{(n)} \tag{49}$$

and, subtracting (48) from (47), we find that

$$e^{(n+1)} = Me^{(n)} \tag{50}$$

where $$M = I - \sigma N^{-1}\hat{L} \tag{51}$$

For convergence it is clear that every eigenvalue of M must lie inside the unit disc. Otherwise, if the initial error $e^{(0)}$ is an eigenvector corresponding to an eigenvalue λ with a magnitude greater than unity, we shall have $e^{(1)} = \lambda e^{(0)}$, $e^{(2)} = \lambda^2 e^{(0)}$, ..., so that the error will grow.

Suppose that M has distinct eigenvalues ordered so that $1 > |\lambda_1| > |\lambda_i|$ for $i \geq 2$, and let $v_1, v_2, \ldots, v_N$ be the corresponding eigenvectors. Then if the initial error is decomposed as the sum

$$e^{(0)} = \sum_{i=1}^{N} \alpha_i v_i$$

the components $\alpha_2 v_2$, $\alpha_3 v_3$, ..., $\alpha_N v_N$ will be more rapidly reduced than the leading component $\alpha_1 v_1$. Thus, after a sufficient number of iterations, the error will approach the form of an eigenvector corresponding to the dominant eigenvalue, with the result that

$$e^{(n+1)} \sim \lambda_1 e^{(n)} \sim \lambda_1^2 e^{(n-1)} \tag{52}$$

Since the spectral radius $|\lambda_1|$ is nearly unity in a relaxation process, the final rate of convergence is slow. We can take advantage of the behavior indicated by Eq. (52), however, to extrapolate the error. Such a procedure was proposed by Lyusternik [15] (see Forsythe and Wasow [16]). Substituting from (49) in (52), we find that

$$C^{(n)} \sim (\lambda_1 - 1)e^{(n)} \tag{53}$$

so that we can estimate $e^{(n)}$, and hence the solution, as

$$\phi = \phi^{(n)} - e^{(n)} = \phi^{(n)} + \frac{C^{(n)}}{1 - \lambda_1} \tag{54}$$

This suggests that the correction should be multiplied by $1/(1 - \lambda_1)$. It appears, therefore, that in situations where (52) holds we ought to extrapolate with a very large overrelaxation factor for a single step.

It is generally difficult to estimate the eigenvalues in advance. However, if a dominant eigenvalue does exist, a situation in which (52) holds can easily be detected because it then follows from (53) that

$$C^{(n+1)} \sim \lambda_1 C^{(n)}$$

Thus the ratio of successive corrections should be the same at every point in the field. During each relaxation sweep, therefore, we can take some suitably large sample of the points (not necessarily the whole field), and compare the correction at each of these points with the correction in the preceding sweep. If the standard deviation from the mean ratio $\bar{\lambda}$ of successive corrections is less than some tolerance, a dominant eigenvalue with the value $\bar{\lambda}$ is judged to exist, and an additional correction is added to the potential at each point. Since the correction $C^{(n)}$ has already been made, the additional correction should be

$$\left(\frac{1}{1-\bar{\lambda}}-1\right)C^{(n)}=\frac{\bar{\lambda}}{1-\bar{\lambda}}C^{(n)}$$

In practice it often proves best to limit the additional correction to some fraction of this amount, particularly if $\bar{\lambda}$ is close to unity.

The effectiveness of this procedure depends on the existence of a dominant eigenvalue. It is well known that in the solution by relaxation of simple model problems, such as Laplace's equation in a rectangle, the use of an optimum relaxation factor distributes the eigenvalues around a circle of fixed radius, so that there is no dominant eigenvalue [16]. In transonic flow calculations, however, experience has shown that an attempt to use a relaxation factor close enough to 2 to cause this to occur often leads to divergence. Extrapolation can then be useful by allowing the use of a safe relaxation factor without too severe a penalty. Some results of numerical experiments with extrapolation are given in Sec. 5. Cheng and Hafez have also considered extrapolation in the case when there is a complex conjugate pair of dominant eigenvalues [17].

4.2 Methods Using a Fast Poisson Solver

The iterative scheme will converge rapidly if the operator N in Eq. (16) is a good approximation to the operator used to evaluate the residual R. If N is restricted to a triangular form, however, as in a relaxation method, it cannot be a very good approximation. Thus we can expect to improve the rate of convergence by introducing more general forms for N, which allow the approximation to be improved but still produce a set of equations that can easily be solved at each cycle. In recent years fast direct methods have been developed for inverting the discrete Laplacian [18–20]. The potential flow equation can be scaled so that the Laplacian represents its linear part by dividing the quasilinear form (1) by a^2, or the conservation form (7) by ρ. This suggests the use of an iteration in which N is the Laplacian, so that at each cycle we have to solve a discrete Poisson's equation with the nonlinear terms as forcing terms. A scheme of this type was proposed for the small-disturbance equation by Martin and Lomax [21]. A similar procedure has also been used by Periaux for subsonic flow calculations using the finite-element method [22].

In order to estimate the rate of convergence that might be obtained, consider the linearized small-disturbance equation

$$(1 - M_\infty^2)\phi_{xx} + \phi_{yy} = 0 \tag{55}$$

with $M_\infty < 1$. If we take $\sigma = 1$, Eq. (16) becomes a central-difference approximation to

$$\phi_{xx}^{(n+1)} + \phi_{yy}^{(n+1)} = M_\infty^2 \phi_{xx}^n$$

where $\phi^{(n)}$ is the result of the nth cycle. In the case of a rectangular region with periodic boundary conditions, we can estimate the amplification factor G by setting

$$\phi^{(n)} = G^n e^{ipx} e^{iqy} \tag{56}$$

Then we find that

$$G = M_\infty^2 \frac{\sin^2(p\,\Delta x/2)}{\sin^2(p\,\Delta x/2) + (\Delta x/\Delta y)^2 \sin^2(q\,\Delta y/2)}$$

Thus the error should be reduced by a factor bounded by M_∞^2 at every cycle, independent of the mesh size, leading to very rapid convergence in subsonic flow.

A similar argument suggests that the Poisson scheme will not converge in supersonic flow. Consider Eq. (55) with $M_\infty > 1$, and suppose that an upwind difference formula is used to evaluate ϕ_{xx} in calculating the residual, while N is still a central-difference approximation to the Laplacian. Then the substitution (56) results in the formula

$$G = \frac{[(M_\infty^2 - 1)e^{-ip\,\Delta x} + 1]\sin^2(p\,\Delta x/2)}{\sin^2(p\,\Delta x/2) + (\Delta x/\Delta y)^2 \sin^2(q\,\Delta y/2)}$$

for the growth factor. If we consider a harmonic with a low frequency in y and a moderate frequency in x, we find that $|G|$ exceeds unity, indicating divergence. This conclusion is confirmed by an analysis that includes the effect of the boundary conditions [23].

One method of stabilizing the scheme for transonic flow calculations is to desymmetrize the operator N by adding an upwind approximation to $(\alpha/\Delta x)\,\partial/\partial x$, where α is a sufficiently large positive coefficient. As long as α is a function of x only, independent of the local type of the flow, the equations can still be solved by a fast method such as the Buneman algorithm [18]. Martin has reported good results for calculations in which the transonic small-disturbance theory was formulated as a system of first-order equations, and an iterative scheme of this type was used to obtain the solution of the resulting difference equations [24,25].

An alternative method is to use the Poisson scheme in combination with some other method designed to remove the errors from the supersonic zone. The relaxation method is effective for this purpose [13,23,26]. It turns out that if a sufficient number of relaxation sweeps is used after each Poisson step, the combined scheme converges in both the supersonic and subsonic zones, and usually at a rate much faster than can be obtained by using relaxation sweeps alone. In such a hybrid scheme, which can easily be applied to the transonic potential flow equation in either quasilinear or conservation form, the Poisson steps are the principal source of convergence in the subsonic zone, while the relaxation sweeps are dominant in the supersonic zone. The best number p of relaxation sweeps to be used after each Poisson step is most easily determined by numerical experiments. Typically, when the rotated difference schemes of Secs. 3.1 and 3.2 are used, the best rate of convergence is obtained with $p \sim 5\text{-}8$.

4.3 Other Fast Methods

Some other promising alternatives for accelerating the iterative scheme will be briefly mentioned here. One approach is to generate the matrix N in Eq. (16) as a product of factors

$$N = P_1 P_2 \cdots P_n$$

each of which is easily invertible. Thus it should be possible to make N a better approximation to the operator used to evaluate the residual R. The transonic small-disturbance equation can be solved by a rapidly convergent alternating-direction method constructed along these lines [27]. Ballhaus and Steger have also shown that the low-frequency unsteady small-disturbance equation is well suited to solution by alternating-direction methods [28].

Another method that may prove to be competitive is the multigrid method first proposed by Federenko [29,30] and subsequently developed by Brandt [31,32]. Brandt and South have recently applied the method to a model problem requiring the solution of the transonic small-disturbance equation and have obtained some excellent results [33].

5 APPLICATIONS

5.1 Survey of Applications

There have been numerous applications of the method of Murman and Cole for solving the transonic small-disturbance equation. The method was extended to lifting airfoils by Krupp and Murman [34], and Murman also studied wind-tunnel wall effects [35]. The simplifications of the small-disturbance theory offer the prospect of performing quite economical calculations of three-dimensional transonic flows, and much effort has been concentrated in this direction. In particular, computer programs have been developed by Lomax et al. [36]; Newman and Klunker [37]; Schmidt et al. [38]; Albone et al. [39]; and van de Vooren et al. [40]. These authors have used a variety of forms of the transonic small-disturbance equations with the aim of improving the accuracy of calculations of flows past swept wings. This question is carefully reviewed by van de Vooren et al. [40].

There has been a parallel effort to develop computer programs to calculate solutions of the exact potential flow equation. Particularly encouraging results have been obtained in applications to two-dimensional and axially symmetric flows [8–13,41–43], and it has also proved possible to calculate three-dimensional flows past isolated wings [44,45] with the current generation of computers such as the CDC 6600 or IBM 370.

The choice of a coordinate system is a particularly important factor in the development of a satisfactory computer program to calculate solutions of the exact potential flow equation. The representation of the Neumann boundary condition (4) can be simplified by introducing curvilinear coordinates, such that the surface of the body coincides with a coordinate surface. Curvilinear coordinates can also be used to bunch the mesh points in sensitive regions of the flow, such as the regions near the leading or trailing edge of the airfoil. These questions will be discussed here in connection with three applications of the general method of solving the exact potential flow equation described in Secs. 3 and 4:

1 Calculation of the flow over an airfoil
2 Axially symmetric nacelle calculations
3 Calculation of the flow past a swept wing

5.2   Airfoil Calculations Using a Mapping to a Circle

A favorable coordinate system for the treatment of a flow past a two-dimensional profile can be generated by mapping the exterior of the profile conformally onto the interior of a unit circle [8–10]. This idea was first introduced by Sells for subsonic flow calculations [46]. The introduction of polar coordinates r and θ in the circle leads to a regular and finite mesh in which the profile becomes the coordinate line $r = 1$. Thus the Neumann boundary condition can be simply represented in the manner indicated in Sec. 3. Also, since the Laplacian is invariant under a conformal transformation, we can use a fast solver for Poisson's equation in polar coordinates to perform the Poisson steps of the hybrid fast iterative method proposed in Sec. 4.2.

The far-field boundary condition has to be applied at $r = 0$, where the potential becomes infinite. This singularity can be removed by defining a reduced potential

$$G = \phi - \frac{\cos (\theta + \alpha)}{r} + E(\theta + \alpha) \qquad (57)$$

where $2\pi E$ is the circulation, and α is the angle of attack. Then G is finite and single-valued. The modulus of the mapping function also becomes infinite at $r = 0$, and the use of finite-difference formulas to represent derivatives of quantities depending on the mapping function can lead to large errors. Thus it is best to calculate the mapping to the exterior of the circle, and to perform an explicit inversion to the interior of the circle. Equation (1) then becomes

$$(a^2 - u^2)G_{\theta\theta} - 2uvrG_{r\theta} + (a^2 - v^2)r\frac{\partial}{\partial r}(rG_r)$$

$$- 2uv(G_\theta - E) + (u^2 - v^2)rG_r + (u^2 + v^2)\left(\frac{u}{r} H_\theta + vH_r\right) = 0 \qquad (58)$$

where H is the modulus of the derivative of the transformation to the exterior of the circle, and u and v are the velocity components in the θ and r directions:

$$u = \frac{r(G_\theta - E) - \sin (\theta + \alpha)}{H} \qquad v = \frac{r^2 G_r - \cos (\theta + \alpha)}{H} \qquad (59)$$

The conservation form (7) becomes

$$\frac{\partial}{\partial \theta}\left(\rho \frac{Hu}{r}\right) + r\frac{\partial}{\partial r}\left(\rho \frac{Hv}{r}\right) = 0 \qquad (60)$$

The Neumann boundary condition (4) reduces to

$$G_r = \cos (\theta + \alpha) \qquad \text{at } r = 1$$

while the far-field boundary condition becomes [9]

$$G = E\{\theta + \alpha - \arctan [\sqrt{1 - M_\infty^2} \tan (\theta + \alpha)]\} \qquad \text{at } r = 0$$

Finally the circulation constant E is determined by the Kutta condition, which requires the velocity to be finite at the trailing edge, where $H = 0$.

Thus ϕ_θ must also vanish, giving

$$E = G_\theta - \sin \alpha \qquad \text{at } r = 1, \theta = 0$$

The derivative of the mapping function can be represented as a power series. Let $z = x + iy$ and $\sigma = (1/r)e^{-i\theta}$ be corresponding points exterior to the profile and the unit circle. Then $H = |dz/d\sigma|$ where, if ε is the included angle at the trailing edge and we use a series with N terms, we can set

$$\frac{dz}{d\sigma} = \left(1 - \frac{1}{\sigma}\right)^{1 - \varepsilon/\pi} \exp \sum_{n=0}^{N} \frac{c_n}{\sigma^n} \tag{61}$$

This method of representing the transformation has the advantage that it allows a profile with an open tail to be mapped to a closed circle. Expanding (61), the coefficient of $1/\sigma$ is

$$\tilde{c} = \left(c_1 - 1 + \frac{\varepsilon}{\pi}\right) \exp\left(c_0\right) \tag{62}$$

Then according to the Cauchy integral theorem, integration of $dz/d\sigma$ around any closed curve exterior to the unit circle in the σ plane results in a fixed gap

$$z_2 - z_1 = 2\pi i \tilde{c} \tag{63}$$

This gap can be used to represent a wake of constant thickness.

The mapping coefficients can be calculated by a simple iterative procedure [47]. Let β and s be the tangent angle and arc length of the profile, and let

$$c_n = a_n - ib_n$$

Then, separating the real and imaginary parts of (61) when $r = 1$, we obtain

$$\log \frac{ds}{d\theta} - \left(1 - \frac{\varepsilon}{\pi}\right) \log \left(2 \sin \frac{\theta}{2}\right) = \sum_{n=0}^{N} (a_n \cos n\theta + b_n \sin n\theta) \tag{64}$$

and

$$\beta + \theta + \frac{\pi}{2} - \left(1 - \frac{\varepsilon}{\pi}\right) \frac{\theta - \pi}{2} = \sum_{n=0}^{N} (a_n \sin n\theta - b_n \cos n\theta) \tag{65}$$

$\beta(s)$ is known from the geometry of the profile. Given an estimate $s(\theta)$ of the arc length as a function of the angle θ in the circle plane, we can therefore calculate the Fourier coefficients a_n and b_n from Eq. (65), and hence we can construct the conjugate Fourier series (64). The resulting value of $ds/d\theta$ can be integrated to provide a new estimate $s(\theta)$, and the process can then be repeated. The iterations usually converge quite rapidly, provided that the closure condition (63) is used to freeze the coefficient c_1. It is convenient to use a series with K terms to represent the mapping function at $2K$ equally spaced mesh points around the circle. Following a suggestion of Ives (private communication), the Fourier coefficients can then be evaluated with the aid of the fast Fourier transform, with the result that the number of operations required to perform each iteration can be reduced to $O(K \log K)$.

A mapping method similar to this was proposed by Timman [48]. In its present form, the method has proved fast and accurate in numerous calculations [9], allowing the construction of the coordinate system in a time that is negligible compared with the time required for the subsequent calculation of the flow by relaxation.

5.3 Results of Airfoil Calculations

Some typical results of airfoil calculations are presented in Figs. 7–11. Each figure shows the surface pressure distribution and the lift, drag, and moment coefficients. These were calculated by integrating the surface pressure. The critical pressure at which the flow has sonic velocity is marked by a horizontal line on the axis. The calculations were all performed on a sequence of grids, with, successively, 64 × 16 cells (that is, 64 cells in the θ direction, and 16 cells in the r direction); 128 × 32 cells; and 256 × 64 cells. The interpolated result of the calculation on the previous grid was used to provide the initial guess for each of the calculations on the finer grids. This procedure can lead to substantial savings in the number of cycles required for convergence on the finer grids. The use of such a sequence of grids also serves the purpose of providing an indication of the sensitivity of the numerical results to the mesh width of the grid. It can be seen from the results that in practice the solution on the 128 × 32 grid would usually be sufficiently accurate.

Figures 7 and 8 show calculations in quasilinear and conservation form for the 64A410 airfoil. The viscosity parameter ε was set equal to zero in the calculation in .conservation form, giving first-order accuracy. The difference in the two results is similar to that reported by Murman when a nonconservative scheme for the small-disturbance equation was replaced by a conservative scheme [6]. The jump at a normal shock wave is consistently underestimated by calculations that do not use conservation form. This is in line with the one-dimensional analysis given in Sec. 2.3. The downstream Mach number is too close to unity, and the resulting increase in mass flow corresponds to the introduction of a source distribution over the surface of the shock wave.

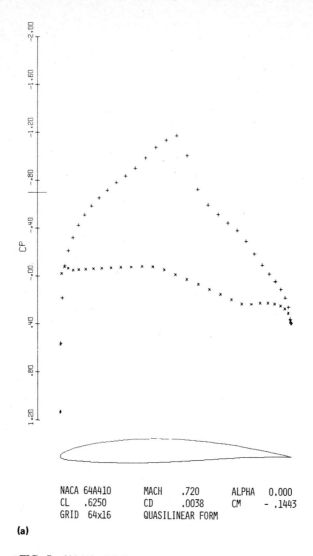

NACA 64A410	MACH .720	ALPHA 0.000
CL .6250	CD .0038	CM - .1443
GRID 64x16	QUASILINEAR FORM	

(a)

FIG. 7 64A410 airfoil, quasilinear form. (a) 64 × 16 grid; (b) 128 × 32 grid; (c) 256 × 64 grid.

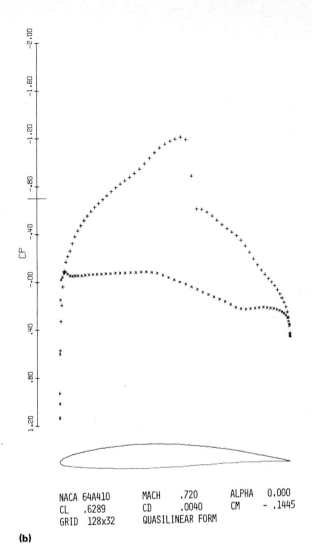

NACA 64A410   MACH   .720   ALPHA   0.000
CL .6289   CD   .0040   CM   - .1445
GRID 128x32   QUASILINEAR FORM

(b)

FIG. 7b

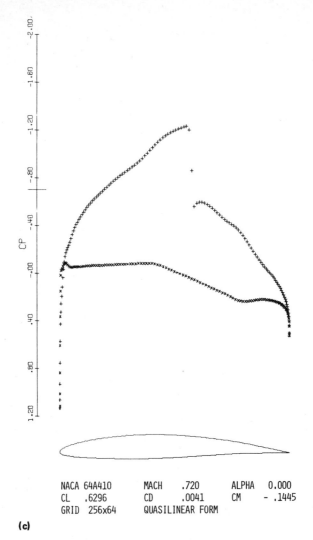

NACA 64A410   MACH   .720   ALPHA 0.000
CL   .6296   CD   .0041   CM   - .1445
GRID 256x64   QUASILINEAR FORM

(c)

FIG. 7c

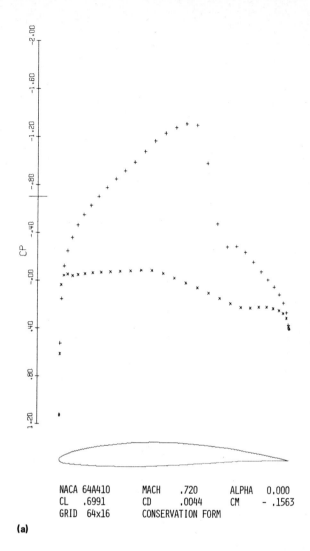

(a)

FIG. 8 64A410 airfoil, conservation form. (a) 64 × 16 grid; (b) 128 × 32 grid; (c) 256 × 64 grid.

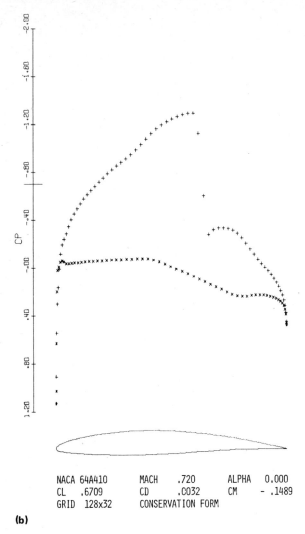

NACA 64A410	MACH .720	ALPHA 0.000
CL .6709	CD .C032	CM - .1489
GRID 128x32	CONSERVATION FORM	

(b)

FIG. 8b

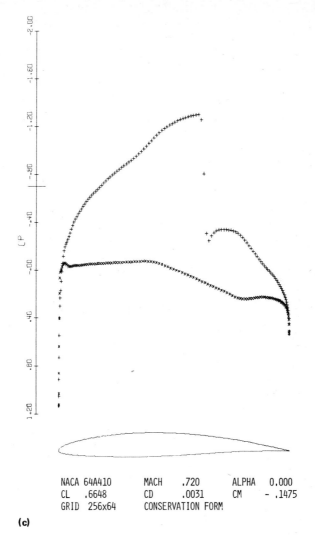

NACA 64A410   MACH   .720   ALPHA 0.000
CL   .6648   CD   .0031   CM   - .1475
GRID 256x64   CONSERVATION FORM

(c)

FIG. 8c

Figures 9 and 10 show a similar comparison for an NLR shock-free airfoil [49]. In this case the value $\varepsilon = 1$ was used for the viscosity parameter in the calculation in conservation form, giving second-order accuracy. In more difficult cases the use of a value of ε too close to unity can lead to divergence. It can be seen that even with $\varepsilon = 1$ the result of the calculation in conservation form exhibits an overexpansion on the coarse grid.

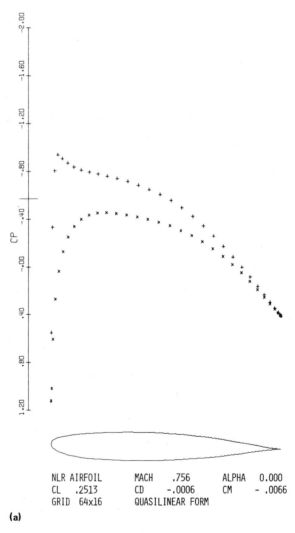

NLR AIRFOIL	MACH	.756	ALPHA	0.000
CL .2513	CD	-.0006	CM	- .0066
GRID 64x16	QUASILINEAR FORM			

(a)

FIG. 9 NLR shock-free airfoil, quasilinear form. (a) 64 × 16 grid; (b) 128 × 32 grid; (c) 256 × 64 grid.

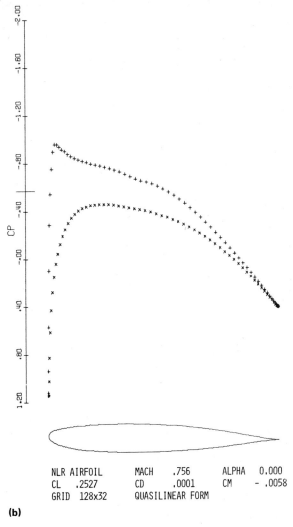

NLR AIRFOIL	MACH	.756	ALPHA	0.000
CL	.2527	CD	.0001	CM	- .0058
GRID	128x32	QUASILINEAR FORM		

(b)

FIG. 9b

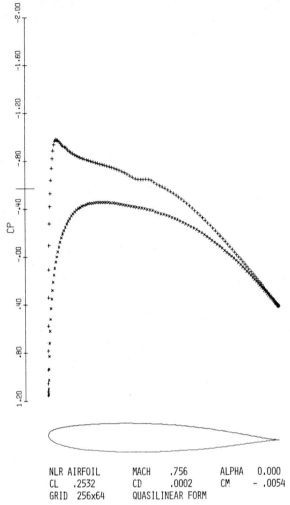

NLR AIRFOIL   MACH   .756   ALPHA   0.000
CL   .2532   CD   .0002   CM   - .0054
GRID 256x64   QUASILINEAR FORM

(c)

FIG. 9c

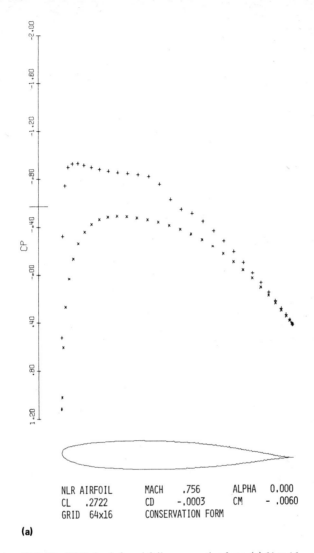

NLR AIRFOIL   MACH .756   ALPHA 0.000
CL .2722   CD -.0003   CM -.0060
GRID 64x16   CONSERVATION FORM

(a)

FIG. 10   NLR shock-free airfoil, conservation form. (a) 64 × 16 grid; (b) 128 × 32 grid; (c) 256 × 64 grid.

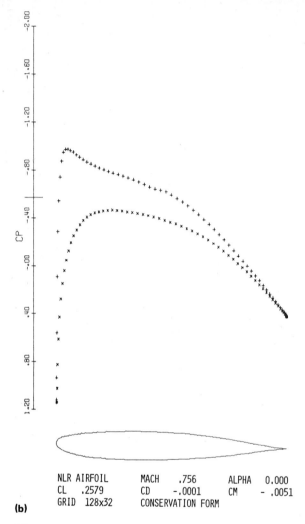

NLR AIRFOIL   MACH   .756   ALPHA 0.000
CL   .2579   CD   -.0001   CM   - .0051
(b)   GRID 128x32   CONSERVATION FORM

FIG. 10b

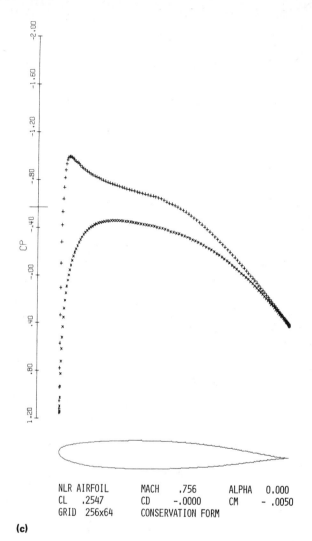

NLR AIRFOIL   MACH .756   ALPHA 0.000
CL   .2547   CD   -.0000   CM   -.0050
GRID 256x64   CONSERVATION FORM

(c)

FIG. 10c

This effect can be quite marked in some cases. As an example, Fig. 11 shows a result for an airfoil designed by the method of complex characteristics [50–52], the Korn airfoil, which has been widely tested to verify its ability to generate a shock-free transonic flow [53,54]. The calculation is for the design point, and the result shows a shock-free flow on the fine grid, yet on the coarse grid there is quite a strong shock wave. This calculation was performed with $\varepsilon = 0$, 0.5, and 0.75 on the successive grids.

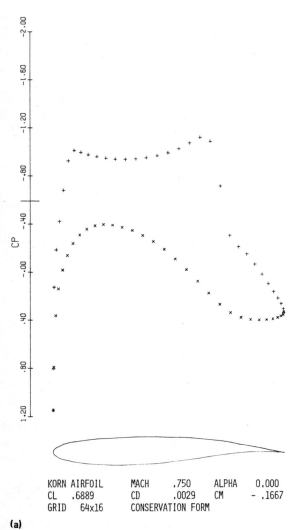

KORN AIRFOIL	MACH	.750	ALPHA	0.000
CL .6889	CD	.0029	CM	− .1667
GRID 64x16	CONSERVATION FORM			

(a)

FIG. 11 Korn airfoil, conservation form. (a) 64 × 16 grid; (b) 128 × 32 grid; (c) 256 × 64 grid.

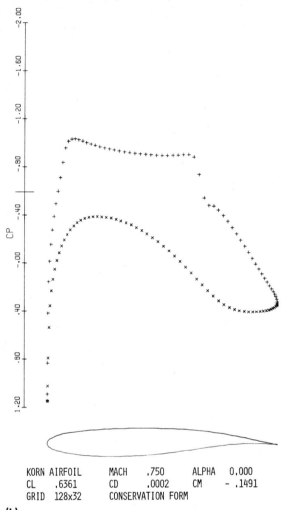

KORN AIRFOIL   MACH   .750   ALPHA   0.000
CL   .6361   CD   .0002   CM   - .1491
GRID   128x32   CONSERVATION FORM

(b)

FIG. 11b

The rate of convergence of the iterative method can be measured by a suitable norm of the vector of residuals R_{ij} appearing in Eq. (16), such as the largest absolute value of the residual at any point in the field, or the average absolute value of all the residuals. Other convenient measures of the rate of convergence are the largest absolute value of the correction C_{ij} at any point in the field, or the average absolute value of all the corrections. These have the disadvantage, however, that the size of the correction depends on the iterative method. If the residuals are normalized by multiplying by $\Delta\theta^2$, they are usually of the same order of magnitude as the corrections in a relaxation sweep. With this normalization, the error

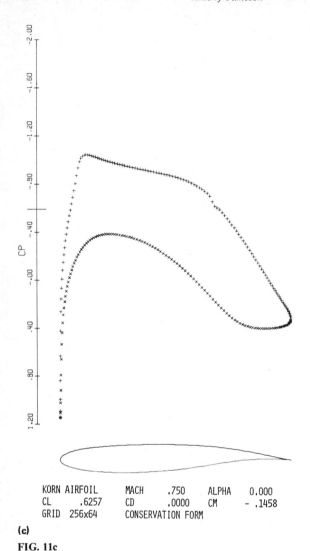

KORN AIRFOIL   MACH   .750   ALPHA   0.000
CL   .6257   CD   .0000   CM   - .1458
GRID   256x64   CONSERVATION FORM

(c)

FIG. 11c

resulting from incomplete convergence can be expected to be of the same order of magnitude as the discretization error when the residuals are reduced to $\sim \Delta\theta^{2+\alpha}$, where α is the order of accuracy of the difference scheme. If the difference scheme is second-order accurate, the residuals on a grid with 128 cells ought to be reduced to $\sim 10^{-8}$, for example.

When the solution is obtained by relaxation, rather a large number of cycles are needed to reduce the errors to acceptable levels. The errors can be reduced quite rapidly, however, with the aid of the hybrid fast iterative scheme described in Sec. 4.2. Some convergence histories are given in

Figs. 12–15. Each figure shows the decrease in the average absolute value of the residual against the amount of work performed by the computer, measured either by the number of relaxation cycles or, if the hybrid method was used, by the equivalent number of relaxation cycles. For this purpose each Poisson step was treated as the equivalent of two relaxation sweeps, since it was found that the Poisson steps, which were performed by the Buneman algorithm [18], typically required about the same amount of computer time as two relaxation sweeps. Each figure also shows the average reduction in the error per unit of work.

Figure 12 shows the convergence history for the calculation by relaxa-

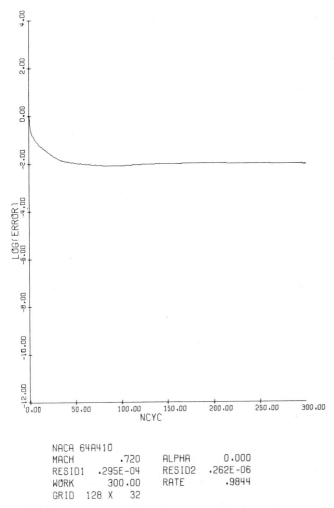

NACA 64A410
MACH   .720   ALPHA   0.000
RESID1 .295E-04   RESID2 .262E-06
WORK   300.00   RATE   .9844
GRID 128 X   32

FIG. 12 Convergence history for Fig. 7 (relaxation).

tion of the result of Fig. 7 on the 128×32 grid. The average reduction per cycle was 0.984. Figure 13 shows the convergence history for the same calculation using the hybrid method. Since the rotated difference scheme of Sec. 3.1 was used, each Poisson step was followed by six relegation sweeps. The average reduction of error per unit of work was 0.901. Even faster convergence can be obtained when the supersonic zone is small enough to permit the use of the simple unrotated difference scheme described at the beginning of Sec. 3.1, because then the relaxation method acts like a marching scheme in the supersonic zone, so that it almost entirely liquidates the errors in the supersonic zone, and only one relaxation sweep is needed after each Poisson step to stabilize the hybrid method. Figure 14 shows the con-

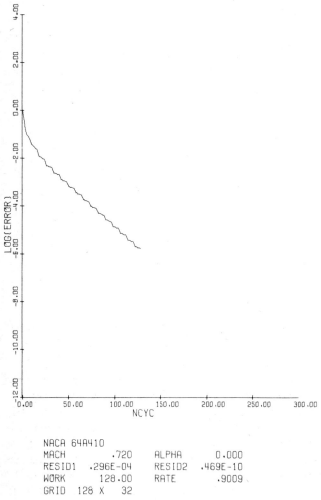

```
NACA 64A410
MACH          .720     ALPHA      0.000
RESID1   .296E-04      RESID2   .469E-10
WORK       128.00      RATE       .9009
GRID   128 X    32
```

FIG. 13   Convergence history for Fig. 7 (hybrid).

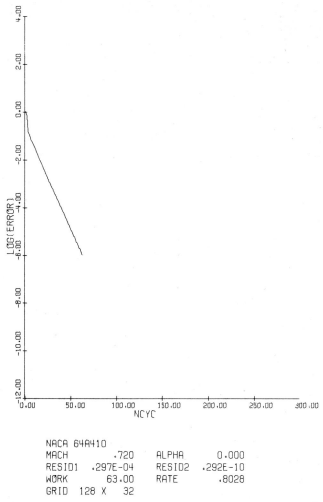

NACA 64A410
MACH   .720   ALPHA   0.000
RESID1   .297E-04   RESID2   .292E-10
WORK   63.00   RATE   .8028
GRID   128 X   32

FIG. 14 Convergence history for Fig. 7 (hybrid with simple difference).

vergence history for the calculation of the same case by the hybrid method with the simple difference scheme. An average reduction of error per unit of work of 0.803 was now obtained. Calculations in conservation form generally exhibit a slower rate of convergence, but a useful acceleration can still be achieved by introducing a Poisson solver. Figure 15 shows the convergence history for the calculation of the result of Fig. 8. In this case the hybrid method was used with a cycle of eight relaxation sweeps after a Poisson step, and a ripple can be observed over the length of each cycle. The average rate of reduction of error was 0.960.

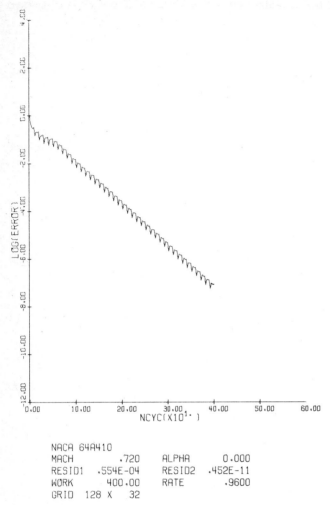

NACA 64A410
MACH   .720   ALPHA   0.000
RESID1   .554E-04   RESID2   .452E-11
WORK   400.00   RATE   .9600
GRID   128 X   32

FIG. 15   Convergence history for Fig. 8 (hybrid).

5.4   Boundary-Layer Correction

In order to simulate experimental results, it is important to allow for viscous effects in the boundary layer. It has been found that quite good agreement can often be obtained simply by adding the displacement thickness of the boundary layer to the profile [52,55–57]. This procedure is effective in the regime where the shock waves are not strong enough to separate the flow. Since the growth of the boundary layer, which is assumed to be turbulent downstream of a specified transition point, depends on the pressure distribution, the calculation of the boundary-layer correction is included in the iterative scheme. After a certain number of cycles of the flow calculation, the boundary-layer displacement thickness is calculated using the current estimate of the pressure distribution. The profile is then modified, and the mapping function is recalculated by the fast method of Sec. 5.2 before continuing the flow calculation. The whole process is repeated until the flow field and boundary layer have both converged. Results obtained by this procedure are compared with experimental data in Figs. 16 and 17. These calculations were performed by Frances Bauer, using conservation form on a 160×30 grid. Tunnel-wall effects were not included in the calculations, but in each case the angle of attack was adjusted so that the calculated lift coefficient matched the experimental lift coefficient.

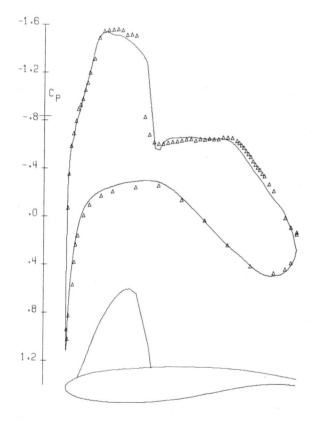

AIRFOIL 75-07-15

△ EXPERIMENT	MACH .687	ALPHA 4.09°	CL .809	CD .0170
— THEORY	MACH .687	ALPHA 2.66°	CL .809	CD .0137
CONSERVATION FORM	GRID 160x30	REYNOLDS NO. 20×10^6		

FIG. 16 75-07-15 airfoil, conservation form, with boundary-layer correction.

Figure 16 shows the flow past an airfoil designed by Garabedian to produce a shock-free flow at Mach 0.75 and a lift coefficient of 0.7 [52, p. 102]. The figure shows an off-design condition, with a shock wave that is quite well simulated by the calculation. The test was performed by Kacprzynski [58]. Figure 17 shows a supercritical airfoil designed by Whitcomb [59]. In this case the calculation required a Mach-number correction to move the shock wave forward to the location observed in the experiment. When the same case was calculated using the quasilinear form, the location of the shock wave was correctly predicted without a Mach-number correction, but the shock jump was less well simulated.

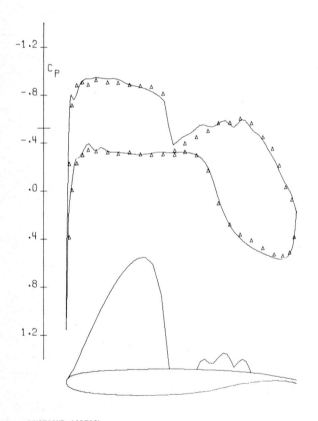

WHITCOMB AIRFOIL
△ EXPERIMENT   MACH .780   ALPHA 1.00   CL .576   CD .0098
—THEORY   MACH .770   ALPHA - .22   CL .576   CD .0084
CONSERVATION FORM   GRID 160x30   REYNOLDS NO. 8x10^6

FIG. 17 Whitcomb airfoil, conservation form, with boundary-layer correction.

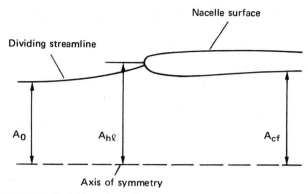

FIG. 18 Definition of nacelle geometry.

A definitive treatment of viscous effects would require a detailed analysis of wake effects near the trailing edge and of shock-wave–boundary-layer interaction. A start has been made in this area [60–63].

5.5 Nacelle Calculations

Type-dependent differencing has also proved an effective method for calculating inviscid transonic potential flow fields about axisymmetric inlet cowls [41–43]. Arlinger has obtained very satisfactory results [41] by a natural generalization of the method for treating airfoils described in Sec. 5.2. The flow field is mapped to a rectangular domain by a numerically calculated conformal transformation. Then a rotated difference scheme is used to introduce an upwind bias in the local flow direction at supersonic points.

The method of Caughey and Jameson [43] will be described here, as an illustration of an alternative procedure for generating a curvilinear coordinate system matched to the body. That is to introduce slightly nonorthogonal coordinates by performing a mild shearing of a simple conformal mapping generated by elementary functions. This avoids the need for a preliminary numerical calculation of the exact conformal mapping to the desired computational domain.

The geometry of the problem is illustrated in Fig. 18. The nacelle is treated as a semi-infinite body. The internal cross-sectional area at the engine compressor face, assumed to be far downstream, is denoted by A_{cf}, and the reference area of the nacelle, taken to be that defined by the leading-edge radius, is denoted by A_{hl}. The free stream is assumed to be uniform at infinity outside the nacelle, having a Mach number M_∞, and also uniform at the compressor face inside the nacelle, with a velocity

$1 + k$ times that of the free stream. The mass flow ratio is conventionally defined as

$$Q = \frac{A_0}{A_{hl}}$$

where A_0 is the capture area determined by the dividing streamline between the interior and exterior flows. Using the isentropic relations, the velocity ratio k can be related to the mass flow ratio Q by the implicit formula

$$k = Q \frac{A_{hl}}{A_{cf}} \left[1 - \frac{\gamma - 1}{2} M_\infty^{\,2}(2k + k^2) \right]^{-1/(\gamma - 1)} - 1$$

In cylindrical coordinates the equation for the velocity potential of an axially symmetric flow becomes

$$(a^2 - u^2)\phi_{xx} - 2uv\phi_{xy} + (a^2 - v^2)\phi_{yy} + a^2 \frac{v}{y} = 0 \qquad (66)$$

where y is the radial direction, and the velocity components u and v and the local speed of sound a are determined by Eqs. (2) and (3). On the axis $y = 0$, Eq. (66) is singular, and we must use the fact that $v \to 0 + y\phi_{yy} + O(y^2)$ to obtain the special form

$$(a^2 - u^2)\phi_{xx} + 2a^2\phi_{yy} = 0 \qquad (67)$$

Equations (66) and (67) are to be solved subject to the Neumann boundary condition (4) at the nacelle surface, and the conditions

$$u \to 1 \qquad \text{and} \qquad v \to 0 \qquad \text{as } x^2 + y^2 \to \infty \qquad (68)$$

outside the nacelle, and

$$u \to 1 + k \qquad v \to 0 \qquad (69)$$

inside the nacelle at the compressor face. The velocity potential ϕ is singular at infinity outside the nacelle. To remove this singularity, we define a reduced potential

$$G = \phi - x \qquad (70)$$

by subtracting the contribution of the uniform free stream.

The numerical calculation is performed on a rectangular domain, obtained from the original coordinates in the physical plane by a series of transformations. We denote the variable in the physical plane as

$$z = x + iy$$

where x and y are the axial and lateral variables, respectively, in any azimuthal plane, and the coordinates are scaled such that point $(1, \pi)$ lies just inside the leading edge of the nacelle. The region outside the nacelle surface is first mapped to an infinite strip of slowly varying width by the conformal transformation

$$z = Z - \exp(-Z) \tag{71}$$

If we denote the real and imaginary parts of Z by X and Y, respectively, then the width of the strip in the Y direction is nearly equal to π, while the limit as $X \to -\infty$ corresponds to infinity in the physical plane outside the nacelle, and the limit as $X \to +\infty$ corresponds to infinity in the physical plane inside the nacelle. (See Fig. 19.)

In order to provide a simple rectangular domain for the numerical calculation, the width of the strip is next made constant by the shearing transformation

$$\xi = X \qquad \eta = \frac{Y}{S(X)} \tag{72}$$

where $S(X)$ is the width of the strip as a function of X, which can be calculated from the nacelle coordinates and the transformation according to Eq. (71). If we set $T(X) = 1/S(X)$, the equation for the reduced potential in this coordinate system becomes

$$(a^2 - U^2)G_{\xi\xi} + (2\eta_X a^2 - 2U\overline{V})G_{\xi\eta} + [a^2(\eta_X^2 + T^2) - \overline{V}^2]G_{\eta\eta}$$
$$+ [\eta_{XX}(a^2 - U^2) - 2UVT_X]G_\eta + P(U^2 - V^2) + 2UVQ \tag{73}$$
$$- \frac{U^2 + V^2}{h}[(R^2 + P)U + QV] - \frac{a^2 h}{Y + Q}[QU - (1 + P)V] = 0$$

where
$$U = \frac{G_\xi + (\eta/T)T_X G_\eta + 1 + P}{h}$$

$$V = \frac{TG_\eta + Q}{h} \tag{74}$$

and
$$\overline{V} = TV + U\eta_X \tag{75}$$

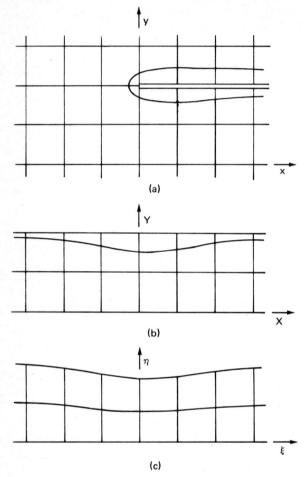

FIG. 19 Construction of coordinate system for nacelle calculation. (a) Cartesian coordinates; (b) coordinates in mapped
domain; (c) sheared coordinates.

Here P and Q are the real and imaginary parts of $\exp(-Z)$, respectively,
$R = \exp(-X)$, and

$$h^2 = \left| \frac{dz}{dZ} \right|^2 = 1 + 2P + R^2 \tag{76}$$

In this coordinate system the Neumann boundary condition (4) becomes

$$G_\eta = -\frac{Q - S_x(G_\xi + 1 + P)}{T(1 + S_x^2)} \tag{77}$$

to be applied along the line $\eta = 1$.

Two final transformations are performed before the equation is cast in finite-difference form for solution. First, to remove the exponential behavior as $X \to -\infty$, we define

$$\zeta = X - \exp(-X) \tag{78}$$

Then, to reduce the computational domain to a finite rectangle, we introduce the stretched coordinate $\overline{X}$ according to the formula

$$\zeta = \frac{B(\overline{X} - C)}{(1 - \overline{X}^2)^A} \tag{79}$$

whence $\overline{X} \to \pm 1$ corresponds to $X \to \pm \infty$, that is, infinity inside and outside the nacelle, respectively. A uniformly spaced finite-difference grid is set up in the $\overline{X}$-η plane spanning the region $-1 \le \overline{X} \le 1$, $0 \le \eta \le 1$, which corresponds to the entire physical plane exterior to the nacelle surface. The constants A, B, and C are chosen to concentrate the mesh points in regions of high gradients. The distribution of mesh points in the vicinity of the nacelle lip is shown for a typical case in Fig. 20 for a grid containing 1024 cells.

In this coordinate system the flow is not always aligned, even approximately, with one of the coordinate directions. Therefore a coordinate-invariant difference scheme of the type described in Sec. 3.1 is required in the supersonic zone to add an upwind bias in a direction parallel to the velocity vector. At a supersonic point the principal part of Eq. (66), consisting of the terms containing the second derivatives of G, is rewritten as

$$a^2 G_{nn} + (a^2 - q^2) G_{ss} = 0 \tag{80}$$

where n and s represent derivatives in the directions normal and parallel to the local velocity vector, and q is the magnitude of the velocity. In terms of the original coordinates,

$$G_{nn} = \frac{V^2}{q^2} G_{\xi\xi} + \frac{2}{q^2} (q^2 \eta_x - U\overline{V}) G_{\xi\eta} + \frac{1}{q^2} [q^2(\eta_x^2 + T^2) - \overline{V}^2] G_{\eta\eta}$$

$$G_{ss} = \frac{U^2}{q^2} G_{\xi\xi} + \frac{2U\overline{V}}{q^2} G_{\xi\eta} + \frac{V^2}{q^2} G_{\eta\eta} \tag{81}$$

Upwind difference formulas are then used to represent contributions to G_{ss}, while central-difference formulas are used to represent contributions to G_{nn}.

Alternative line relaxation schemes can be used to solve the difference equations. In the first scheme the equations for the correction at each point are solved simultaneously along lines of constant $\overline{X}$, sweeping the field to the right for lines intersecting the inner nacelle surface, and to the

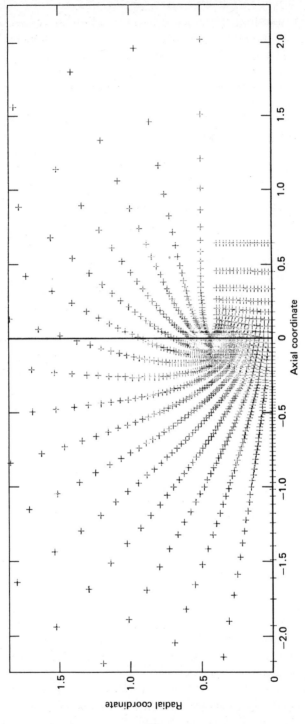

FIG. 20 Typical coordinate grid.

left for lines intersecting the outer nacelle surface, as illustrated in Fig. 21. This is necessary to avoid sweeping the field upstream in the supersonic zone, which is assumed to be near the nacelle surface. In the second scheme the equations are solved simultaneously along lines of constant η, and the field is swept from the axis of symmetry to the nacelle surface. It can be seen from the results of the numerical experiments reported in the next section that, in the absence of any accelerating device, the η-line scheme generally gives faster convergence. The $\overline{X}$-line scheme is the more effective, however, when a fast Poisson solver is introduced to accelerate the scheme in the manner proposed in Sec. 4.2.

5.6 Results of Nacelle Calculations

Some typical results of nacelle calculations by the method described in the last section are shown in Figs. 22–25. These calculations were performed on a grid with 128 cells in the $\overline{X}$ direction and 32 cells in the η direction. In some cases preliminary calculations were performed on 32×8 and 64×16 grids, and the interpolated results were used to provide starting data for calculations on finer grids.

Figures 22–24 show comparisons of calculated and measured pressure distributions on the outside of a cowl designed and tested by the Douglas Aircraft Company [64]. The three cases are for approximately the same mass flow ratio $Q = 0.700$. The agreement is fairly good despite the absence of a boundary-layer correction, except in the case with a free-stream Mach number of 0.851, in which there is a discrepancy in the region immediately behind the shock wave. The attenuated shock jump predicted by the calculation is typical of that encountered in airfoil calculations when a non-

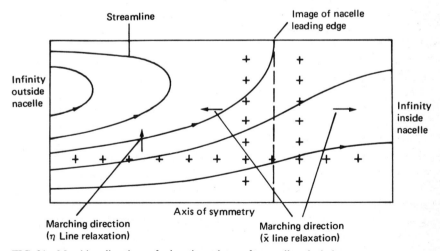

FIG. 21 Marching directions of relaxation schemes for nacelle calculation.

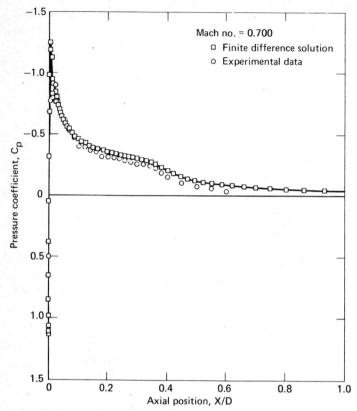

FIG. 22   Pressure distribution on cowl 36.

conservative difference scheme is used. Figure 25 shows a result for an NACA 1-85-100 cowl tested by Re [65], operating at a mass flow ratio of 0.8073. The agreement is encouraging.

Numerical experiments were performed to test the effectiveness of both the extrapolation method described in Sec. 4.1 and the hybrid method using a fast Poisson solver described in Sec. 4.2. The introduction of a fast Poisson solver is possible because the grid has a uniform spacing in the η direction. Figures 26–28 show some convergence histories in which the largest residual, normalized by multiplying through by $\Delta \overline{X}^2$, was used as the measure of error. Figures 26 and 27 show the improvement in the rate of convergence, which was obtained by using extrapolation on the 64×16 grid. The method proved less effective, however, on the 128×32 grid. Figure 28 shows the performance of the hybrid method on the 128×32 grid. As in the case of the airfoil calculations reported in Sec. 5.3, the Buneman algorithm was used to perform the Poisson steps, each of which required slightly less than three times the time required for a relaxation

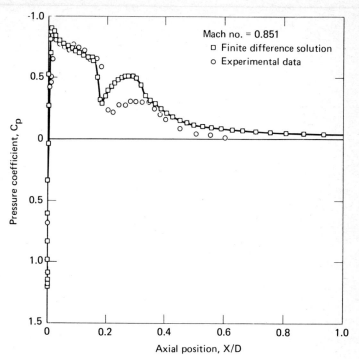

FIG. 23 Pressure distribution on cowl 36.

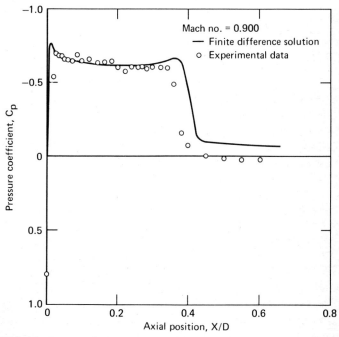

FIG. 24 Pressure distribution on cowl 36.

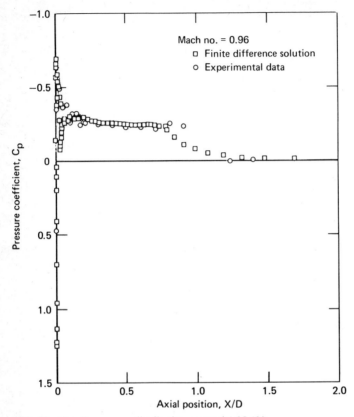

FIG. 25   Nacelle pressure distribution on cowl 1-85-100.

sweep. Each Poisson step was followed by five relaxation sweeps. It can be seen from Fig. 28 that the rates of convergence of both the $\overline{X}$-line and η-line relaxation schemes described in the last section decrease rapidly when the residuals are smaller than about 10^{-6}, whereas that of the hybrid method remains approximately constant.

5.7   Calculation of the Flow Past a Swept Wing

A computer program for calculating the potential flow past a swept wing has recently been developed by Jameson and Caughey [66]. It will be described in this section as a final example of the application of the general method of solving the exact potential flow equation.

It is now desired to solve the three-dimensional potential flow equation, which can be written in quasilinear form as

$$(a^2 - u^2)\phi_{xx} + (a^2 - v^2)\phi_{yy} + (a^2 - w^2)\phi_{zz} - 2uv\phi_{xy} - 2vw\phi_{yz}$$
$$- 2uw\phi_{xz} = 0 \quad (82)$$

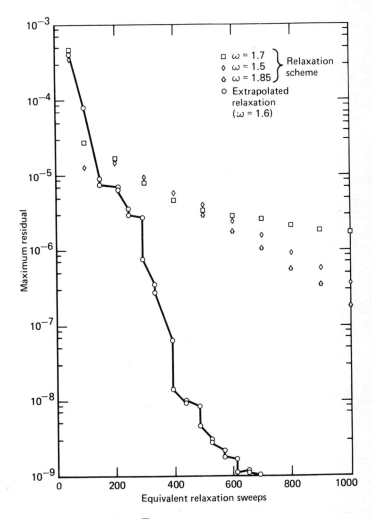

FIG. 26 Convergence of $\overline{X}$-line schemes (64 × 16 grid).

where u, v, and w are the velocity components

$$u = \phi_x \qquad v = \phi_y \qquad w = \phi_z \tag{83}$$

and a is the local speed of sound given by the relation (3). As in the case of the airfoil and nacelle equations, we remove the singularity at infinity in the velocity potential by introducing a reduced potential

$$G = \phi - x \cos \alpha - y \sin \alpha \tag{84}$$

where α is the angle of attack.

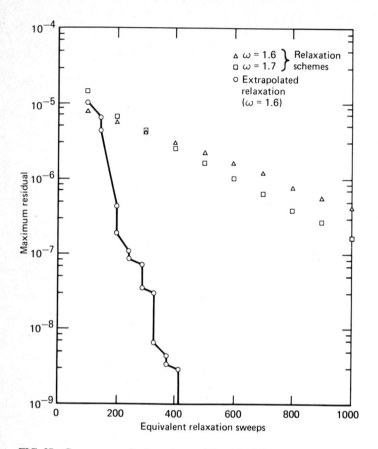

FIG. 27 Convergence of η-line schemes (64×16 grid).

In the case of a lifting flow, the velocity potential is discontinuous across the vortex sheet trailing behind the wing. Roll-up of the vortex sheet will be ignored: the conditions to be satisfied at the surface in which the vortex sheet is assumed to lie are that the jump Γ in the potential is constant along lines parallel to the free stream, and that the normal component of velocity is continuous through the sheet. At infinity the flow is undisturbed except in the Trefftz plane far downstream, where there will be a two-dimensional flow induced by the vortex sheet.

The construction of a satisfactory curvilinear coordinate system to suit the geometry of the configuration is one of the most difficult aspects of the three-dimensional problem. Here nonorthogonal coordinates will be generated by a sequence of elementary transformations in the same spirit as the method for treating nacelles described in Sec. 5.5. First we introduce para-

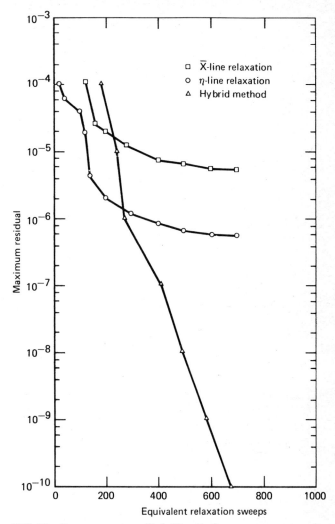

FIG. 28 Convergence rate of hybrid method.

bolic coordinates in planes containing the wing section by the square-root transformation

$$X_1 + iY_1 = \{x - x_0(z) + i[y - y_0(z)]\}^{1/2} \qquad Z_1 = z \qquad (85)$$

where z is the spanwise coordinate, and x_0 and y_0 define a singular line of the coordinate system located just inside the leading edge (see Figs. 29 and 30). The effect of this transformation is to unwrap the wing to form a shallow bump

$$Y_1 = S(X_1, Z_1) \qquad (86)$$

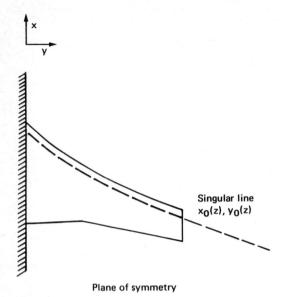

Plane of symmetry

FIG. 29   Configuration of swept wing.

Then we use a shearing transformation

$$X = X_1 \qquad Y = Y_1 - S(X_1, Z_1) \qquad Z = Z_1 \tag{87}$$

to map the wing surface to a coordinate surface. Finally, in order to obtain a finite computational domain, X, Y, and Z are replaced by stretched coordinates $\bar{X}$, $\bar{Y}$, and $\bar{Z}$. The stretching used in the present computer program is to set $X = \bar{X}$ in an inner domain $-\bar{X}_m \leq \bar{X} \leq \bar{X}_m$, and to set

$$X = \bar{X}_m + \frac{\bar{X} - \bar{X}_m}{\{1 - [(\bar{X} - \bar{X}_m)/(1 - \bar{X}_m)]^2\}^\alpha} \tag{88}$$

when $\bar{X} > \bar{X}_m$, with a corresponding formula when $\bar{X} < \bar{X}_m$, so that $X = \pm\infty$ when $\bar{X} = \pm 1$. Typically the parameter α has the value $1/2$. Similar stretchings are used for Y and Z.

The vortex sheet is assumed to coincide with the cut behind the singular line which is opened up by the square-root transformation (85). Thus a jump Γ is introduced in the potential between corresponding points representing the two sides of the vortex sheet. A complication is caused by the continuation of the cut beyond the wing. Points on the two sides of the cut must be identified as the same point in the physical space. Also, a special form of the equations must be used at points lying on the singular line beyond the wing. At these points the equation to be satisfied reduces to the two-dimensional Laplace equation in the X_1 and Y_1 coordinates. An

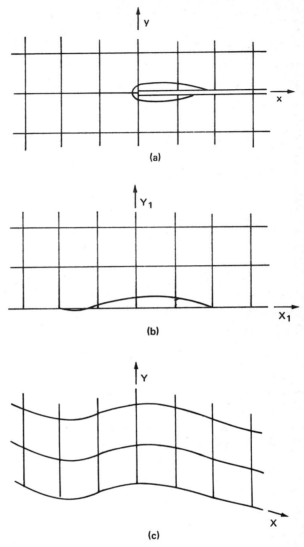

FIG. 30 Construction of coordinate system for swept-wing calculation. (a) Cartesian coordinates; (b) parabolic coordinates; (c) sheared coordinates.

advantage of the square-root transformation (85) is that it collapses the height of the disturbance due to the vortex sheet to zero in the parabolic coordinate system at points far downstream, where X_1 approaches infinity, with the result that the far-field boundary condition is simply

$$G = 0 \qquad\qquad (89)$$

The final equation for the reduced potential G contains numerous terms. In order to construct a rotated difference scheme with a proper upwind bias at supersonic points, it is only necessary, however, to consider the principal part, consisting of the terms containing the second derivatives of G. Suppose that Eq. (82) is written in the canonical form

$$(a^2 - q^2)\phi_{ss} + a^2(\Delta\phi - \phi_{ss}) = 0 \qquad (90)$$

where Δ is the Laplacian operator

$$\frac{\partial^2}{\partial x^2} + \frac{\partial^2}{\partial y^2} + \frac{\partial^2}{\partial z^2}$$

and ϕ_{ss} is the streamwise second derivative in a Cartesian coordinate system locally aligned with the flow

$$\phi_{ss} = \frac{1}{q^2}\left(u^2\phi_{xx} + v^2\phi_{yy} + w^2\phi_{zz} + 2uv\phi_{xy} + 2vw\phi_{yz} + 2uw\phi_{xz}\right) \quad (91)$$

Then at supersonic points we use upwind difference formulas for all second derivatives of G arising from the transformation of ϕ_{ss} into the curvilinear coordinate system, and central-difference formulas for all second derivatives of G arising from the transformation of $\Delta\phi - \phi_{ss}$. The formulation in terms of the Laplacian avoids the need to determine explicitly a pair of local coordinate directions normal to the stream direction.

The difference equations are solved by relaxation, with care taken to make sure that at supersonic points the equivalent time-dependent equation is compatible with the steady-state equation, as in the case of the two-dimensional theory of Sec. 3.3. If m and n are coordinates in a plane normal to the streamwise direction s, and M is the local Mach number q/a, the equivalent time-dependent equation can be written in the form

$$(M^2 - 1)\phi_{ss} - \phi_{mm} - \phi_{nn} + 2\alpha_1\phi_{st} + 2\alpha_2\,\phi_{mt} + 2\alpha_3\,\phi_{nt} + \gamma\phi_t = 0 \quad (92)$$

where the coefficients α_1, α_2, and α_3 depend on the split between new and old values used in the relaxation scheme. To make sure that this is a wave equation with s as the timelike direction, an analysis similar to that of Sec. 3.3 indicates that the difference formulas should be organized so that

$$\alpha_1 > \sqrt{(M^2 - 1)(\alpha_2{}^2 + \alpha_3{}^2)} \qquad (93)$$

Also, the balanced coefficient rule (46) should still be applied in the supersonic zone, corresponding to a zero coefficient of ϕ_t in Eq. (92). It is convenient to solve the equations for the correction to the potential simultaneously along lines, corresponding to a point relaxation process in two dimensions. Any of the coordinate lines can be used for this purpose, the choice being guided by the need to avoid advancing through the supersonic zone in a direction opposed to the flow. In practice it has been found convenient to divide each $\overline{X}$-$\overline{Y}$ plane into three strips, and to march toward the wing surface in each central strip, updating horizontal lines, and then outward in the left-hand and right-hand strips, updating vertical lines. (See Fig. 31.)

5.8 Results of Swept-Wing Calculations

Some results of swept-wing calculations are presented in Figs. 32 and 33. These were calculated on a grid with 144 cells in the chordwise X direction, 24 cells in the normal Y direction, and 32 cells in the spanwise Z direction, calling for the solution of the difference equations at 109,824 mesh points. In each case the interpolated result of a preliminary calculation on a $72 \times 12 \times 16$ grid was used to provide the starting guess: 200 cycles were used on the coarse grid, followed by 100 cycles on the fine grid. Such a calculation requires about 80 min on a CDC 6600 (which should be reduced to about 20 min on a CDC 7600).

Figure 32 shows the result of a calculation for a rather simple wing tested by ONERA, for which experimental data have been published [67]. It can be seen from Fig. 32c and d that agreement with the experimental data is quite good, despite the fact that no attempt was made to allow for viscous effects. As in the case of the two-dimensional calculations, the nonconservative difference scheme introduces a source distribution over the shock surfaces, causing a displacement of the streamlines and a forward

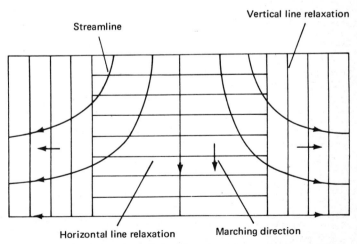

FIG. 31 Marching directions of relaxation scheme for swept-wing calculation.

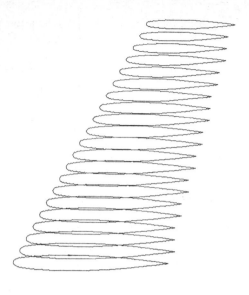

```
        VIEW OF WING

ONERA WING M6    L.E. SWEEP 30 DEG    ASPECT RATIO 3.8
MACH    .923    YAW    0.000    ALPHA  0.000
L/D    -.00     CL    -.0000    CD     .0246
(a)
```

FIG. 32 ONERA swept wing, 144 × 24 × 32 grid. (a) View of wing; (b) upper (left) and lower surface pressures; (c) and (d) theoretical versus experimental data.

shift in the location of the shock waves. Apparently this partially compensates for the absence of a correction for the displacement effect of the boundary layer.

To illustrate the geometric complexity of the configurations that can be treated by the program, Fig. 33 shows the results of a calculation for a wing designed and tested by the Douglas Aircraft Company [68]. The wing is a typical design for a long-range transport aircraft, with a sweepback of 35° at the leading edge, and a substantial change in the section between the root and tip. The test was of a wing-body combination. In the calculation the wing was extended to the plane of symmetry at the fuselage centerline. The calculation shows two shock waves over the inboard part of the wing. The forward shock wave originates from the leading edge at the wing root, and the aft shock wave is roughly normal to the free stream. The two shock waves merge at about the one-fourth-span point, forming a triangular shock pattern over the upper surface of the wing. The coalescence of the shock waves can be traced in Fig. 33c through *g*, in

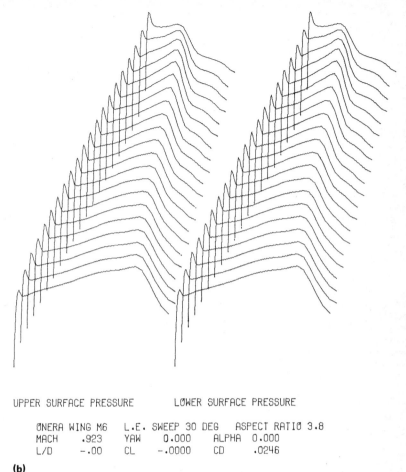

UPPER SURFACE PRESSURE   LOWER SURFACE PRESSURE

ONERA WING M6	L.E. SWEEP 30 DEG	ASPECT RATIO 3.8			
MACH	.923	YAW	0.000	ALPHA	0.000
L/D	-.00	CL	-.0000	CD	.0246

(b)

FIG. 32b

which the pressure distributions at a sequence of span stations over the inboard part of the wing are plotted separately. The convergence history of this calculation, measured by the largest residual, is shown in Fig. 34.

5.9 Remaining Problems

A number of problems remain to be solved. First there is the question of treating more complex three-dimensional configurations. Extensions of the method of Sec. 5.7 to wing–cylinder combinations and multibladed fans have been considered by Caughey and Jameson [69]. The construction of a suitable curvilinear coordinate system requires a certain amount of ingenuity and becomes more difficult with each increase in the geometric complexity

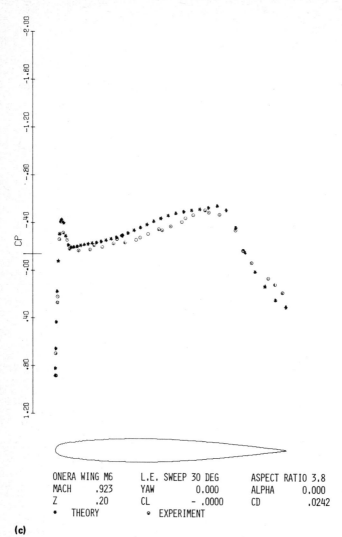

ONERA WING M6   L.E. SWEEP 30 DEG   ASPECT RATIO 3.8
MACH   .923   YAW   0.000   ALPHA   0.000
Z   .20   CL   - .0000   CD   .0242
   ✶ THEORY   ⊙ EXPERIMENT

(c)

FIG. 32c

of the configuration. An alternative is to use Cartesian coordinates; then we must attempt to obtain sufficient accuracy in the treatment of the Neumann boundary condition (4) by the use of interpolation formulas. This method has been explored for two-dimensional calculations by Carlson [70]. Another approach worth investigation is the finite-volume method [71], which has been used by Rizzi to solve the Euler equations [72]. The extension of the finite-element method [73] to treat equations of mixed type would also provide a way to circumvent the difficulties of treating complex configurations by difference methods.

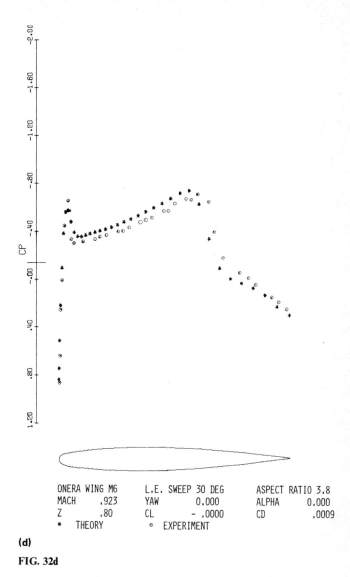

ONERA WING M6   L.E. SWEEP 30 DEG   ASPECT RATIO 3.8
MACH   .923   YAW   0.000   ALPHA   0.000
Z   .80   CL   - .0000   CD   .0009
*   THEORY   ⊚   EXPERIMENT

(d)

FIG. 32d

Certain features of type-dependent difference methods could usefully be improved. As long as the difference scheme is in conservation form, the solution of the difference equations should satisfy the proper jump conditions in the limit as the mesh width approaches zero. With the mesh widths realizable in practice, however, there is not enough resolution to provide a good representation of an oblique shock wave. The conservative difference schemes described in Sec. 3.2 also have the disadvantage of exhibiting rather large discretization errors on coarse grids. Thus the use of shock fitting [17,74–76] in conjunction with the quasilinear form would be an

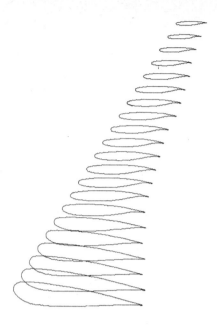

VIEW OF WING

DØUGLAS WING W2 (EXTENDED TØ CENTER LINE)
MACH   .819   YAW   0.000   ALPHA 0.000
L/D   23.86   CL   .5882   CD   .0247

(a)

FIG. 33 Douglas swept wing, 144 × 24 × 32 grid. (a) View
of wing; (b) upper (left) and lower surface pressures;
(c) through (g) pressure distributions at a sequence of span
stations.

attractive alternative, if a sufficiently reliable shock-fitting scheme could be
devised. This does not appear to be easy, however, in cases where two shock
waves coalesce, as in the flow over a swept wing illustrated in Fig. 33. If a
shock-fitting technique could be combined with a difference scheme having
a higher order of accuracy, it would have the additional advantage of
permitting the use of relatively coarse grids, as has been observed by
Moretti [75]. With the extension of fast iterative methods to three-
dimensional problems, it should then be possible to perform extremely
economical three-dimensional calculations.

Finally there remain problems requiring the use of a more complicated
mathematical model than the potential flow equation. These include the
treatment of viscous effects and the treatment of flows in which rotation is
important, such as flows containing strong shock waves, or flows in

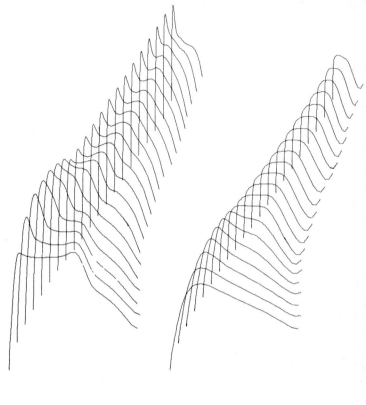

UPPER SURFACE PRESSURE   LOWER SURFACE PRESSURE

DOUGLAS WING W2 (EXTENDED TO CENTER LINE)
MACH   .819   YAW   0.000   ALPHA   0.000
L/D   23.86   CL   .5882   CD   .0247

(b)

FIG. 33b

turbomachinery. The device of integrating the time-dependent Euler equations until they reach a steady state has been widely used [72,76–80]. This is generally slow, however, and seemingly it ought to be possible to devise a more rapidly convergent method for solving the steady-state Euler equations with the aid of some of the ideas discussed in Secs. 3 and 4.

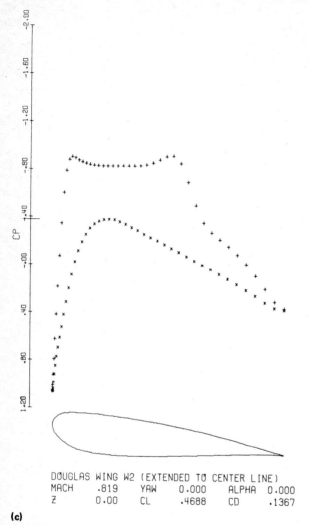

DOUGLAS WING W2 (EXTENDED TO CENTER LINE)
MACH   .819   YAW   0.000   ALPHA   0.000
Z   0.00   CL   .4688   CD   .1367

(c)

FIG. 33c

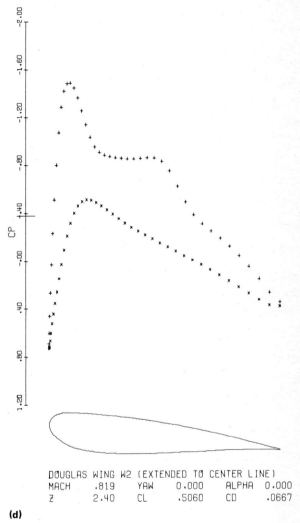

DOUGLAS WING W2 (EXTENDED TO CENTER LINE)
MACH   .819   YAW   0.000   ALPHA 0.000
Z   2.40   CL   .5060   CD   .0667

(d)

FIG. 33d

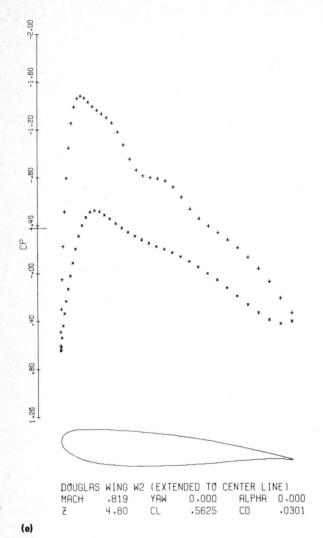

DØUGLAS WING W2 (EXTENDED TØ CENTER LINE)
MACH   .819   YAW   0.000   ALPHA   0.000
Z   4.80   CL   .5625   CD   .0301

(e)

FIG. 33e

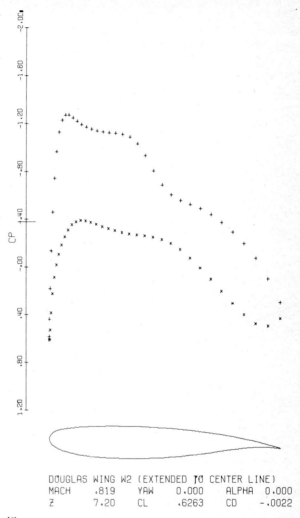

DOUGLAS WING W2 (EXTENDED TO CENTER LINE)
MACH   .819   YAW   0.000   ALPHA 0.000
Z   7.20   CL   .6263   CD   -.0022

(f)

FIG. 33f

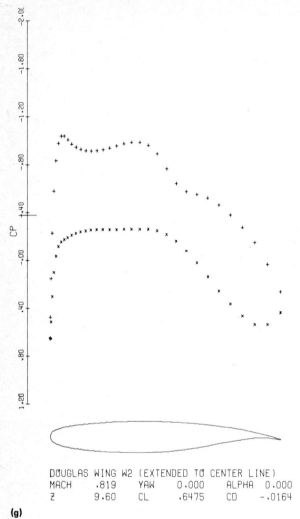

DØUGLAS WING W2 (EXTENDED TØ CENTER LINE)
MACH   .819   YAW   0.000   ALPHA 0.000
Z   9.60   CL   .6475   CD   -.0164

(g)

FIG. 33g

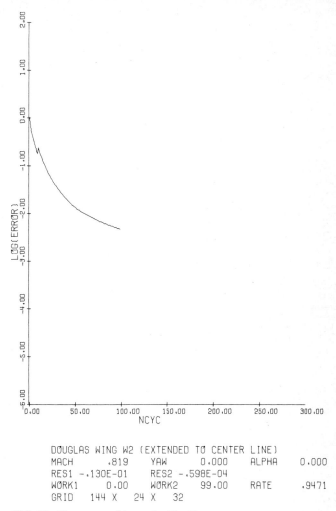

DOUGLAS WING W2 (EXTENDED TO CENTER LINE)
MACH   .819   YAW   0.000   ALPHA   0.000
RES1 -.130E-01   RES2 -.598E-04
WORK1   0.00   WORK2   99.00   RATE   .9471
GRID   144 X   24 X   32

FIG. 34   Convergence history for Fig. 33.

REFERENCES

1 Murman, E. M., J. D. Cole: Calculation of Plane Steady Transonic Flows, *AIAA J.*, 9: 114–121, 1971.
2 Steger, J. L., and B. S. Baldwin: Shock Waves and Drag in the Numerical Calculation of Isentropic Transonic Flow, NASA TN D-6997, 1972.
3 Lax, Peter D.: Weak Solutions of Nonlinear Hyperbolic Equations and their Numerical Computation, *Comm. Pure Appl. Math.*, 7: 159–193, 1954.
4 Bateman, H.: Notes on a Differential Equation which Occurs in the Two-dimensional Motion of a Compressible Fluid and the Associated Variational Problem, *Proc. Roy. Soc. London, Ser.* A, 125: 598–618, 1929.

5   Cole, Julian D.: Twenty Years of Transonic Flow, Boeing Scientific Research Laboratories report DL-82-0878, July, 1969.

6   Murman, E. M.: Analysis of Embedded Shock Waves Calculated by Relaxation Methods, *Proc. AIAA Conf. Comput. Fluid Dyn.*, Palm Springs, July, 1973, 27–40.

7   Lax, Peter, and Burton Wendroff: Systems of Conservation Laws, *Comm. Pure Appl. Math.*, 13: 217–237, 1960.

8   Garabedian, P. R., and D. G. Korn: Analysis of Transonic Airfoils, *Comm. Pure Appl. Math.*, 24: 841–851, 1972.

9   Jameson, Antony: Transonic Flow Calculations for Airfoils and Bodies of Revolution, Grumman aerodynamics report 391-71-1, December, 1971.

10 Jameson, Antony: Iterative Solution of Transonic Flows over Airfoils and Wings, Including Flows at Mach 1, *Comm. Pure Appl. Math.*, 27: 283–309, 1974.

11 South, J. C., and A. Jameson: Relaxation Solutions for Inviscid Axisymmetric Flow over Blunt or Pointed Bodies, *Proc. AIAA Conf. Comput. Fluid Dyn.*, Palm Springs, July, 1973, 8–17.

12 Jameson, Antony: Transonic Potential Flow Calculations Using Conservation Form, *Proc. Second AIAA Conf. Comput. Fluid Dyn.*, Hartford, June, 1975, 148–161.

13 Jameson, Antony: Numerical Computation of Transonic Flows with Shock Waves, Symposium Transsonicum II, Göttingen, September, 1975, 384–414.

14 Garabedian, P. R.: Estimation of the Relaxation Factor for Small Mesh Size, *Math. Tables Aids Comp.*, 10: 183–185, 1956.

15 Lyusternik, L. I.: Zamechania k chislemomu resheniyu kraevykh zadach uravnenia Laplasa i vychisleniyu sobstvennykh znachenii metodom setok, *Tr. Mat. Inst. Steklov*, 20: 1947.

16 Forsythe, George E., Wolfgang R. Wasow: "Finite Difference Methods for Partial Differential Equations," Wiley, New York, 1960.

17 Hafez, M. M., and H. K. Cheng: Convergence Acceleration and Shock Fitting for Transonic Flow Computations, AIAA paper 75-51, 1975.

18 Buneman, O.: A Compact Non-iterative Poisson Solver, Stanford University Institute for Plasma Research report 294, 1969.

19 Buzbee, B. L., G. H. Golub, and C. W. Nielsen: On Direct Methods of Solving Poisson's Equation, *SIAM J. Numer. Anal.*, 7: 627–656, 1970.

20 Fischer, D., G. Golub, O. Hald, C. Leiva, and O. Widlund: On Fourier Toeplitz Methods for Separable Elliptic Problems, *Math. Comp.*, 28: 349–368, 1974.

21 Martin, E. Dale, and Harvard Lomax: Rapid Finite Difference Computation of Subsonic and Transonic Aerodynamic Flows, AIAA paper 74-11, 1974.

22 Periaux, J.: Calcul tridimensionnel de fluides compressibles par la methode des elements finis, 10e Colloque d'Aerodynamique Appliquée, Lille, November, 1973.

23 Jameson, Antony: Accelerated Iteration Schemes for Transonic Flow Calculations Using Fast Poisson Solvers, New York University ERDA report COO-3077-82, March, 1975.

24 Martin, E. Dale: A Fast Semi-direct Method for Computing Transonic Aerodynamic Flows, *Proc. Second AIAA Conf. Comput. Fluid Dyn.*, Hartford, June, 1975, 162–174.

25 Martin, E. Dale: Advances in the Application of Fast Semi-direct Computational Methods in Transonic Flow, Symposium Transsonicum II, Göttingen, September, 1975, 431–438.

26 Jameson, Antony: Numerical Solution of Nonlinear Partial Differential Equations of Mixed Type, in B. Hubbard (ed.), "Numerical Solution of Partial Differential Equations III," SYNSPADE 1975, pp. 275–320, Academic, New York, 1976.

27 Jameson, Antony: Unpublished numerical experiments.

28 Ballhaus, W. F., and J. L. Steger: Implicit Approximate Factorization Schemes for the Low Frequency Transonic Equation, NASA TM X-72082, 1975.

29 Federenko, R. P.: A Relaxation Method for Solving Elliptic Difference Equations, *USSR Comput. Math. Math. Phys.*, 1: 1092–1096, 1962.

30 Federenko, R. P.: The Speed of Convergence of One Iterative Process, *USSR Comput. Math. Math. Phys.*, 4: 227–235, 1964.

31 Brandt, Achi: Multi-level Adaptive Technique (MLAT) for Fast Numerical Solution to Boundary Value Problems, *Proc. Third Int. Conf. Num. Methods Fluid Mech.*, Paris, 1972, vol. 1, pp. 82–89, Springer, New York, 1973.

32 Brandt, Achi: Multi-level Adaptive Techniques (MLAT), Part I, The Multi-grid Method, IBM Yorktown Heights report, to appear.

33 South, J. C., and A. Brandt: Application of the Multi-level Grid Method to Transonic Flows, Project SQUID Workshop on Transonic Flow Problems in Turbomachinery, Monterey, February, 1976.

34 Krupp, J. A., and E. M. Murman: Computation of Transonic Flows Past Lifting Airfoils and Slender Bodies, *AIAA J.*, 10: 880–886, 1972.

35 Murman, E. M.: Computation of Wall Effects in Ventilated Transonic Wind Tunnels, AIAA paper 72-1007, 1972.

36 Lomax, H., F. R. Bailey, and W. F. Ballhaus: On the Numerical Simulation of Three Dimensional Transonic Flow with Application to the C 141 Wing, NASA TN D-6933, 1973.

37 Klunker, E. B., and P. A. Neuman: Computation of Transonic Flow about Lifting Wing-Cylinder Combinations, *J. Aircr.*, 11: 254–256, 1974.

38 Schmidt, W., S. Rohlfs, and R. Vanino: Some Results Using Relaxation Methods for Two and Three-dimensional Transonic Flows, *Proc. Fourth Int. Conf. Numer. Methods Fluid Dyn.*, Boulder, June, 1974, pp. 364–372, Springer, New York, 1975.

39 Albone, C. M., M. G. Hall, and G. Joyce: Numerical Solution of Flows past Wing-Body Combinations, Symposium Transsonicum II, Göttingen, September, 1975.

40 van der Vooren, J., J. W. Slooff, G. H. Huizing, and A. van Essen: Remarks on the Suitability of Various Transonic Small Perturbation Equations to Describe Three-dimensional Transonic Flow: Examples of Computations Using a Fully Conservative Rotated Difference Scheme, Symposium Transsonicum II, Göttingen, September, 1975, 557–566.

41 Arlinger, B. G.: Calculation of Transonic Flow around Axisymmetric Inlets, AIAA paper 75-80, 1975.

42 Baker, T. J.: A Numerical Method to Compute Inviscid Transonic Flows around Axisymmetric Ducted Bodies, Symposium Transsonicum II, Göttingen, September, 1975, 495–506.

43 Caughey, D. A., and A. Jameson: Accelerated Iterative Calculation of Transonic Nacelle Flow Fields, AIAA paper 76-100, 1976.

44 Jameson, Antony: Three Dimensional Flows around Airfoils with Shocks, *Proc. IFIP Symp. Comput. Methods Appl. Sci. Eng.*, Versailles, December, 1973, part 2, pp. 185–212, Springer, New York, 1974.

45 Clapworthy, P. J., P. W. Duck, and K. W. Mangler: The Calculation of Steady Inviscid Flow around Non-lifting Bodies, Symposium Transsonicum II, Göttingen, September, 1975, 549–556.

46 Sells, C. C. L.: Plane Subcritical Flow past a Lifting Airfoil, *Proc. Roy. Soc., London*, 308A: 377–401, 1968.

47 James, R. M.: A New Look at Incompressible Airfoil Theory, McDonnell Douglas report J 0918, 1971.

48 Timman, R.: The Direct and Inverse Problem of Airfoil Theory: A Method to Obtain Numerical Solutions, Nat. Luchtv. Lab. Amsterdam report F 16, 1951.

49 Lock, R. C.: Test Cases for Numerical Methods in Two Dimensional Transonic Flow, AGARD report 575, 1970.

50 Garabedian, P. R., and D. G. Korn: Numerical Design of Transonic Airfoils, in E. Hubbard (ed.), "Numerical Solution of Partial Differential Equations II," SYNSPADE 1970, pp. 253–271, Academic, New York, 1971.

51   Bauer, F., P. Garabedian, and D. Korn: "Supercritical Wing Sections," Springer, New York, 1972.
52   Bauer, F., P. Garabedian, D. Korn, and A. Jameson: "Supercritical Wing Sections II," Springer, New York, 1975.
53   Kacprzynski, J. J.: A Second Series of Tests of the Shockless Lifting Airfoil No. 1, Project Report 5x5/0062, National Research Council of Canada, Ottawa, 1972.
54   Pollock, N., and B. D. Fairlie: An Experimental Investigation of Garabedian and Korn's Shockless Lifting Airfoil No. 1, Symposium Transsonicum II, Göttingen, September, 1975, 289–296.
55   Bauer, F., and P. Garabedian: Computer Simulation of Shock Wave Boundary Layer Interaction, Comm. Pure Appl. Math., 26: 659–665, 1973.
56   Bauer, F., and D. Korn: Computer Simulation of Transonic Flow past Airfoils with Boundary Layer Correction, Proc. Second AIAA Conf. Comput. Fluid Dyn., Hartford, June, 1975, 184–204.
57   Bavitz, Paul C.: An Analysis Method for Two-dimensional Transonic Viscous Flow, NASA TN D-7718, 1975.
58   Kacprzynski, J. J.: Wind Tunnel Test for Shockless Lifting Airfoil No. 2, Laboratory Technical Report LTR-HA-5X5/0067, National Research Council of Canada, Ottawa, 1973.
59   Whitcomb, R. T.: Review of NASA Supercritical Airfoils, Ninth International Congress on Aeronautical Sciences, Haifa, Israel, 1974.
60   Adamson, T. C., and A. Feo: Interaction between a Shock Wave and a Turbulent Boundary Layer in Transonic Flow, SIAM J. Appl. Math., 29: 121–145, 1975.
61   Adamson, T. C., and A. F. Messiter: Normal Shock Wave Turbulent Boundary Layer Interaction in Transonic Flow Near Separation, Project SQUID Workshop on Transonic Flow Problems in Turbomachinery, Monterey, February, 1976.
62   Melnik, R. E., and B. Grossman: Further Developments in an Analysis of the Interaction of a Weak Normal Shock Wave with a Turbulent Boundary Layer, Symposium Transsonicum II, Göttingen, September, 1975, 262–272.
63   Melnik, R. E., and R. Chow: Asymptotic Theory of Trailing Edge Flows, "Aerodynamic Analyses Requiring Advanced Computers," NASA SP 347, 1975, 177–249.
64   Ambrosiani, J.: Unpublished data of the Douglas Aircraft Company, Long Beach, California.
65   Re, Richard J.: An Investigation of Several NACA 1-Series Axisymmetric Inlets at Mach Numbers from 0.4 to 1.29, NASA TM X-2917, 1974.
66   Jameson, A., and D. A. Caughey: Numerical Calculation of the Transonic Flow Past a Swept Wing, New York University ERDA report COO-3077-140, June, 1977.
67   Monnerie, B., and F. Charpin: Essais de buffeting d'une aile en flèche transsonique, 10ᵉ Colloque d'Aerodynamique Appliquée, Lille, November, 1973.
68   Steckel, D., and J. Dahlin: Unpublished data of the Douglas Aircraft Company, Long Beach, California.
69   Caughey, D. A., and A. Jameson: Calculation of Transonic Potential Flow Fields about Complex Three-dimensional Configurations, Project SQUID Workshop on Transonic Flow Problems in Turbomachinery, Monterey, February, 1976.
70   Carlson, Leland A.: Transonic Airfoil Analysis and Design Using Cartesian Coordinates, Proc. Second AIAA Conf. Comput. Fluid Dyn., Hartford, June, 1975, 175–183.
71   MacCormack, R. W., and A. J. Paullay: The Influence of the Computational Mesh on Accuracy for Initial Value Problems with Discontinuous or Nonunique Solutions, Comput. Fluids, 2: 339–361, 1974.
72   Rizzi, Arthur: Transonic Solutions of the Euler Equations by the Finite Volume Method, Symposium Transsonicum II, Göttingen, September, 1975, 567–574.
73   Strang, G., and G. Fix: "Analysis of the Finite Element Method," Prentice-Hall, Englewood Cliffs, N.J., 1973.

74 Yu, N. J., and A. R. Seebass: Second Order Numerical Solutions of Transonic Flows over Airfoils with and without Shock Fitting, Symposium Transsonicum II, Göttingen, September, 1975, 449–456.

75 Moretti, Gino: On the Matter of Shock Fitting, *Proc. Fourth Int. Conf. Numer. Methods Fluid Dyn.*, Boulder, June, 1974, pp. 287–292, Springer, New York, 1975.

76 Grossman, B., and G. Moretti: Time Dependent Computation of Transonic Flows, AIAA paper 70-1322, 1970.

77 Magnus, R., and H. Yoshihara: Inviscid Transonic Flow over Airfoils, *AIAA J.*, 8: 2157–2162, 1970.

78 Oliver, D. A., and S. Panagiotis: Computational Aspects of the Prediction of Multi-dimensional Transonic Flows in Turbomachinery, "Aerodynamic Analyses Requiring Advanced Computers," NASA SP 347, 1975, 567–585.

79 Warming, R. F., and R. M. Beam: Upwind Second Order Difference Schemes and Applications in Unsteady Aerodynamic Flows, *Proc. Second AIAA Conf. Comput. Fluid Dyn.*, Hartford, June, 1975, 17–28.

80 Beam, R. M., and R. F. Warming: An Implicit Finite Difference Algorithm for Hyperbolic Systems in Conservation Law Form, *J. Comput. Phys.*, 22: 87–110, 1976.

CHAPTER 2

Application of Numerical Methods to Physiological Flows

Thomas J. Mueller

I give you thanks that I am fearfully,
wonderfully made   Psalm 139: 14

1   INTRODUCTION

1.1   General Remarks

It has been recognized for a long time that the flow of liquids and gases plays a vital role in the biological processes that occur in the body. In addition to the obvious, blood and air, some of the less obvious fluids of interest include urine, sweat and tears, and the synovial fluid in the joints. While air, urine, and sweat and tears are Newtonian, blood (under certain conditions) and the synovial fluid exhibit non-Newtonian behavior. Blood, a suspension, behaves as a Newtonian fluid for large shear rates, while it is highly non-Newtonian for small shear rates [1]. Synovial fluid is a viscoelastic fluid, and it appears that its elasticity is important during joint lubrication.

These physiological fluids are subjected to many types of motion in three-dimensional, usually distensible, passageways. For example, the pulsatile flow of blood in a healthy circulatory system is almost entirely laminar, although the peak Reynolds number is on the order of 10,000. The only exception to laminar blood flow appears to be the small "bursts" of tur-

It is a pleasure to thank my friends, colleagues, and former students Drs. F. N. Underwood, W. L. Oberkampf, and P. J. Roache for their valuable help and encouragement throughout this effort. My sincere thanks go to my co-investigators on the Heart Valve Project at Notre Dame, Professor J. R. Lloyd and Dr. E. H. MacDonell, for our many discussions of the physiological and experimental problems related to this work. I would also like to acknowledge the financial support of the American Heart Association, Indiana Affiliate, the National Science Foundation, and Dr. George Lea, Director of the Fluid Mechanics Division, and the Department of Aerospace and Mechanical Engineering.

bulence that have been detected in the aorta during a small fraction of each cycle. As a result of an occlusion or stenosis in the circulatory system (including stenosis of one or more of the heart valves), however, this normally laminar flow may become transitional or turbulent. While the flow of air in the respiratory system is also normally laminar, it may become turbulent during heavy breathing or as a result of an obstruction or during coughing, when the maximum Reynolds number may reach 50,000.

It appears that the human body has developed into a nearly optimum system. The norm of laminar flow maintains the resistance to flow at a minimum, ensuring the maximum life of the components, i.e., the heart, lungs, and accompanying distribution networks. Furthermore, the pulsatile nature of the flow in the elastic circulatory system first strains the heart and blood vessels and then relaxes them. This vascular exercise also contributes to long life.

In addition to the long interest in the flow of physiological fluids in their natural habit (i.e., *in vivo*), a more recent interest has developed in the flow of these same fluids in what might be referred to as unnatural habit (i.e., *in vitro*). The relatively recent success of the heart-lung machine during open heart surgery and the artificial kidney machine for those with kidney failure are two of the most striking examples of the *in vitro* flow of physiological fluids. Although these particular machines temporarily take over the functions of body organs, there are many other situations where an *in vitro* device can be used for determining the physical and chemical properties of physiological fluids or evaluating the performance of various prosthetic devices that ultimately may be implanted *in vivo*.

Most of the effort in bio-fluid mechanics has been directed toward problems related to the circulatory and respiratory systems; therefore, this chapter will be concerned with the numerical treatment of problems somehow related to either one or the other of these systems. In order to appreciate the complex geometry of these systems, one need only consult a text on physiology or anatomy. The principal elements of the arterial system are shown in Fig. 1 and Table 1, whereas a more detailed sketch of a single subsystem, the pulmonary arterial tree, is shown in Fig. 2. It is evident from these sketches that these blood vessels are tapered and branch in the flow direction. It is not surprising, therefore, that fully developed flow should be the exception—if it occurs at all—rather than the rule. The tracheobronchial tree has a similar geometry and is indicated in Fig. 3. There should be little doubt that simplifications of geometry have to be made in order to proceed.

Although the *in vivo* flows of blood in the circulatory system and air in the respiratory system take place in distensible vessels and are unsteady, it is often necessary to begin by studying steady flow in rigid vessels. Studies of this type are very useful and indicate the distributions of flow properties at a given instant in the unsteady flow cycle. Since our level of understanding of

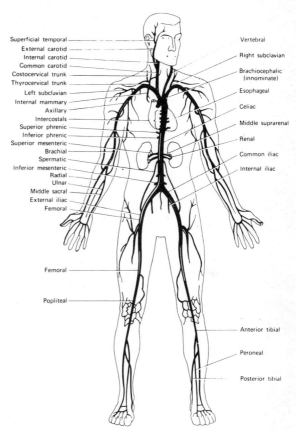

Superficial temporal
External carotid
Internal carotid
Common carotid
Costocervical trunk
Thyrocervical trunk
Left subclavian
Internal mammary
Axillary
Intercostals
Superior phrenic
Inferior phrenic
Superior mesenteric
Brachial
Spermatic
Inferior mesenteric
Radial
Ulnar
Middle sacral
External iliac
Femoral

Femoral

Popliteal

Vertebral
Right subclavian
Brachiocephalic (innominate)
Esophageal
Celiac
Middle suprarenal
Renal
Common iliac
Internal iliac

Anterior tibial
Peroneal
Posterior tibial

FIG. 1 Schematic of the arterial system. Taken from [2].

TABLE 1 Approximate dimensions and Reynolds numbers in human blood vessels[a]

Vessel	Diameter, cm	Length, cm	Wall thickness, cm	Average velocity, cm/s	Average Reynolds number	Maximum velocity, cm/s	Maximum Reynolds number
Aorta	2.5	50	0.2	48	3400	120	8500
Artery	0.4	50	0.1	45	500	90	1000
Arteriole	0.005	1	0.2	5	0.7	—	—
Capillary	0.0008	0.1	0.0001	0.1	0.002	—	—
Venule	0.002	0.2	0.0002	0.2	0.01	—	—
Vein	0.5	25	0.05	10	140	—	—
Vena cava	3.0	50	0.15	38	3300	—	—

[a] Blood viscosity and density taken as 3.5 cP and 1.0 g/cm^3, respectively. Data from [4–6].

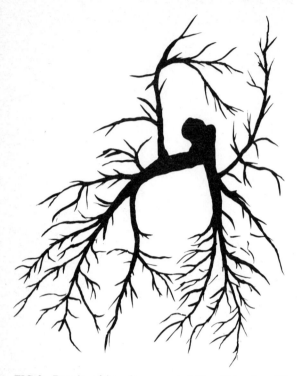

FIG. 2   Drawing of the pulmonary arterial tree. Taken from [3].

these extremely complex physiological flow problems is relatively poor, solutions to greatly simplified problems not only are very helpful, but also
usually provide the foundation for approaching the complex real problem.
The physiological flow problems that have been solved numerically up to
the present must be considered to be a foundation for the future. In this
spirit of building a foundation, therefore, the majority of numerical solutions
in this area have been for incompressible two-dimensional flow through
rather simple rigid geometries. Although a reasonable balance between unsteady and steady flow problems seems to have occurred, almost all investigators have assumed a Newtonian fluid and thus utilized the Navier-Stokes
equations. In the problems of interest here, extensive viscous separated flow
regions occur, and one must therefore look to the most complete mathematical model in fluid dynamics—the full Navier-Stokes equations.

   At present it is certainly possible, although very time consuming, to solve
bio-fluid-mechanics problems in three dimensions and/or with distensible
boundaries. Turbulence modeling will undoubtedly also be performed in the
near future in order to study specific problems, although most of the problems of present interest are laminar. Recent work on a constitutive equation
for blood [73], in addition to the use of the Casson type equation for suspen-

sions, indicates that we may soon be able to handle the non-Newtonian behavior of blood in a more satisfactory manner. However, we can learn much more in a shorter time by simplifying these complex problems as much as possible and by comparing our numerical results with the results of appropriate physical experiments. The solution of three-dimensional, turbulent, non-Newtonian problems will be greatly enhanced by establishing the methods and acquiring the experience of solving two-dimensional flows.

1.2 Bio-Fluid Mechanics of Separated Flows

Flow separation is common in most practical viscous-flow situations. In bio-fluid mechanics, it is common both *in vivo* and *in vitro*. *In vivo* physiological flow separation usually results from abberations of the circulatory system caused by atherosclerosis (especially near bifurcations), atheroma, heart valve and vessel stenosis, and aneurysms (particularly common in the aorta) as shown in Fig. 4. Separated flows are also introduced into the human circulatory system with such prosthetic cardiovascular devices as occluder heart valves, shown in Fig. 5, or assist pumps. Wherever blood is

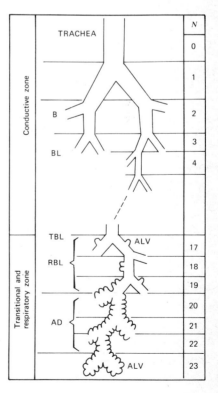

FIG. 3 Anatomical segments of the tracheobronchial tree. N = generation number; B = bronchi; BL = bronchioles; TBL = terminal bronchioles; RBL = respiratory bronchioles; AD = alveolar dusts; ALV = alveoli. Taken from [3].

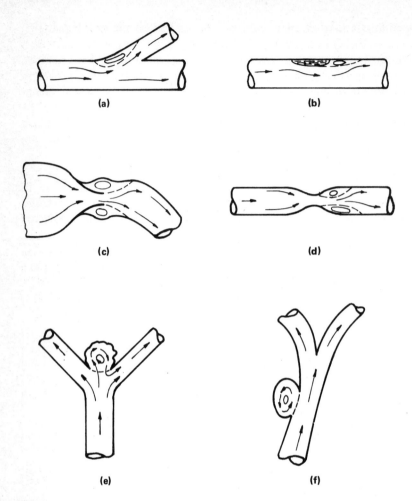

FIG. 4 Examples of *in vivo* separated flows in the circulatory system. (a) Blood-vessel bifurcation; (b) atheroma in blood vessel; (c) aortic heart-valve stenosis, i.e., valve fully open; (d) blood-vessel stenosis; (e) saccular aneurysm; (f) aortic aneurysm.

FIG. 5 Separated flows produced by fully open prosthetic aortic heart valves. (a) Disc prosthetic aortic heart valve; (b) tilting-disc prosthetic aortic heart valve.

pumped or circulated *in vitro*, such as in an oxygenator, dialyzer (see Fig. 6), blood analysis machine, or mock circulatory systems for the evaluation of prosthetic heart valves, the occurrence of this type of flow is also common. In these flows where blood is the working fluid, serious long-term mass-transfer problems arise (i.e., thrombus formation) in addition to the immediate performance loss owing to the increased resistance to flow.

Three important characteristics of separated flows illustrated in Fig. 7 are: (1) the stagnation regions in the immediate vicinity of separation and reattachment, (2) the low-velocity reverse flow along the wall, and (3) the free shear layer or mixing region between the mainstream and the recirculatory flow. Stagnation and low-velocity reverse-flow regions trap lipids, platelets, and other debris, which enhance thrombus formation. At the interface between the main flow and the recirculating flow there are large velocity gradients that produce strong shearing actions. The high local stresses may stretch and deform red blood cells so that they are more prone to damage when encountering the vessel wall farther downstream, or sufficient cell membrane injury may alter their deformability, making them susceptible to removal from the circulation by the filtering function of the spleen [8]. The free-shear-layer velocity profile also contains an inflection point that is known to be very unstable, thus enhancing the possible transition from laminar to turbulent flow. Since the fluid stresses in a turbulent flow are two orders of magnitude greater than for a laminar flow, the possibility of red cell damage or hemolysis is greatly increased. Release of the erythrocyte

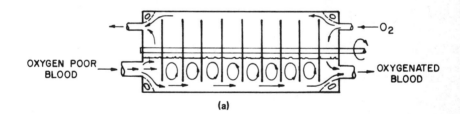

(a)

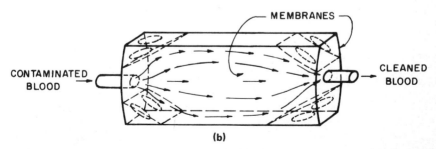

(b)

FIG. 6 Examples of *in vitro* separated blood flows. (a) Rotating-disc oxygenator; (b) membrane dialyzer (one section).

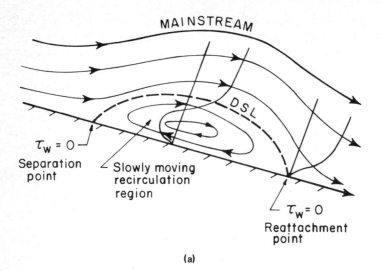

FIG. 7 Essential features of separated flows. (a) Separation resulting from an adverse pressure gradient; (b) separation resulting from abrupt change in geometry.

contents due to this damage can result in a variety of serious effects including anemia [9], long-term toxic effects of free hemoglobin [10], and the liberation of clotting factors [11]. Therefore the problems of erythrocyte damage and hemolysis and of thrombus formation, are related, especially when separated flow regions are present.

Recent experiments indicate that the transport of certain materials between the interluminal blood and the artery wall appears to be dependent upon the value of wall shear stress [12]. These experiments show that the distribution of certain early atheromatous lesions in humans, such as early plaques and fatty streaks, and the accumulation pattern of lipids in arteries of animals on high-lipid diets occur preferentially in arterial regions where the wall shear stress is expected to be low. There is also evidence that the development of advanced atherosclerotic plaques may occur preferentially in high-shear regions [13]. Although a complete description of this complex interaction of blood flow and artery wall is not yet available [14], it appears that the wall shear stress is an important factor.

It seems worthwhile from the fluid-dynamics point of view, therefore, to develop reliable numerical methods of identifying regions of very high shear and normal stresses in the flow (erythrocyte deformation or damage and/or hemolysis), regions of very low or very high shear stress at walls (atheromatous lesions), and the extent of separated- or reverse-flow regions (thrombosis). The favorable comparison of such numerical results with appropriate laboratory experiments will provide the confidence necessary to continue our attempt to solve, or at the very least to obtain a better understanding of, these complex physiological flow problems.

1.3 Scope of This Chapter

The purpose of this chapter is to describe the application of finite-difference numerical methods to the Navier-Stokes equations for selected physiological flow problems. It is not possible to cover here all the problems that have been completed or published, since this field has grown very rapidly in recent years. The problems discussed, however, may be considered to be typical of the large number of complex separated flows that occur in physiological situations. Comparisons of the numerical results with the results of appropriately designed physical experiments will be made whenever possible. The difficulties involved in obtaining accurate experimental data concerning the detailed velocity, shear, and pressure fields for such complicated flows leads us to fully appreciate the value of numerical experiments. Physical experiments at low Reynolds numbers are actually more difficult than their numerical counterparts. Producing a truly planar or axisymmetric separated flow in the laboratory is always difficult. In addition, one must face the equally difficult problem of measuring velocity and/or pressure in such low-Reynolds-number flows. When the flow is unsteady, these difficulties are compounded. The physiological problems reviewed include the axisymmetric prosthetic heart valve and planar and axisymmetric local vessel constriction.

2 THE NAVIER-STOKES EQUATIONS

The numerical treatment of the Navier-Stokes equations involves deriving the desired form of the equation, choosing a suitable mesh system, developing the finite-difference form of the equations and solution procedure, determining the stability of the resultant scheme, and examining the question of accuracy of the solution in view of computational costs as well as the use to be made of the results. This section presents a brief review of this procedure for the incompressible physiological flow problems to be discussed in detail later.

2.1   Incompressible Axisymmetric Flow

For the incompressible, laminar flow of a Newtonian fluid with constant viscosity in the absence of body forces, the Navier-Stokes equations and continuity equation in cylindrical coordinates (r, θ, z) for two-dimensional axisymmetric (i.e., $v_\theta = 0$ and $\partial/\partial\theta = 0$) flow may be expressed as

$$\frac{\partial v_z}{\partial t} + v_r \frac{\partial v_z}{\partial r} + v_z \frac{\partial v_z}{\partial z} = -\frac{1}{\rho}\frac{\partial P}{\partial z} + \nu\left(\frac{\partial^2 v_z}{\partial r^2} + \frac{\partial^2 v_z}{\partial z^2} + \frac{1}{r}\frac{\partial v_z}{\partial r}\right) \tag{1}$$

$$\frac{\partial v_r}{\partial t} + v_r \frac{\partial v_r}{\partial r} + v_z \frac{\partial v_r}{\partial z} = -\frac{1}{\rho}\frac{\partial P}{\partial r} + \nu\left(\frac{\partial^2 v_r}{\partial r^2} + \frac{\partial^2 v_r}{\partial z^2} + \frac{1}{r}\frac{\partial v_r}{\partial r} - \frac{v_r}{r^2}\right) \tag{2}$$

$$\frac{1}{r}\frac{\partial(r v_r)}{\partial r} + \frac{\partial v_z}{\partial z} = 0 \tag{3}$$

These equations are referred to as the primitive equations since they are written in terms of the primitive variables v_r, v_z, and P. If the pressure is of primary importance in a particular incompressible problem, the equations in this form plus a Poisson equation for pressure should be considered. Experience in solving these equations may also be useful if one is ultimately interested in three-dimensional flows. Solution procedures for solving the primitive equations in both two- and three-dimensional rectangular coordinates are discussed by many investigators (e.g., Piacsek [15] and Roache [16]).

By eliminating the pressure terms in Eqs. (1) and (2) by cross differentiation, applying Eq. (3) to this result, and then using the vorticity relation for this axisymmetric flow

$$\xi = \frac{\partial v_r}{\partial z} - \frac{\partial v_z}{\partial r} \tag{4}$$

the following equation, with the vorticity as dependent variable, is obtained:

$$\frac{\partial \xi}{\partial t} + v_r \frac{\partial \xi}{\partial r} + v_z \frac{\partial \xi}{\partial z} - \frac{v_r \xi}{r} = \nu\left(\frac{\partial^2 \xi}{\partial r^2} + \frac{\partial^2 \xi}{\partial z^2} + \frac{1}{r}\frac{\partial \xi}{\partial r} - \frac{\xi}{r^2}\right) \tag{5}$$

This form of the Navier-Stokes equations is referred to as the "vorticity transport" equation. If the slightly modified version of the continuity equation,

$$\xi\left[\frac{1}{r}\frac{\partial(rv_r)}{\partial r} + \frac{\partial v_z}{\partial z}\right] = \xi\frac{1}{r}\frac{\partial(rv_r)}{\partial r} + \xi\frac{\partial v_z}{\partial z} = 0 \tag{6}$$

is added to the left side of Eq. (5), then another form of the vorticity transport equation is obtained:

$$\frac{\partial\xi}{\partial t} + \frac{\partial(v_r\xi)}{\partial r} + \frac{\partial(v_z\xi)}{\partial z} = v\left(\frac{\partial^2\xi}{\partial r^2} + \frac{\partial^2\xi}{\partial z^2} + \frac{1}{r}\frac{\partial\xi}{\partial r} - \frac{\xi}{r^2}\right) \tag{7}$$

Introducing the incompressible stream function defined by

$$\frac{1}{r}\frac{\partial\psi}{\partial r} = v_z \qquad \frac{1}{r}\frac{\partial\psi}{\partial z} = -v_r \tag{8}$$

into the continuity equation (3) gives the following equation for vorticity in terms of stream function:

$$-\xi = \frac{1}{r}\left(\frac{\partial^2\psi}{\partial z^2} + \frac{\partial^2\psi}{\partial r^2} - \frac{1}{r}\frac{\partial\psi}{\partial r}\right) \tag{9}$$

Because the magnitudes of the stresses are often very important in the flow of physiological fluids, the components of the stress tensor for the incompressible axisymmetric flows studied are of increasing interest. These stresses, in terms of velocity gradients and viscosity, are

$$\tau_{rr} = -2\mu\frac{\partial v_r}{\partial r} \tag{10}$$

$$\tau_{\theta\theta} = -2\mu\frac{v_r}{r} \tag{11}$$

$$\tau_{zz} = -2\mu\frac{\partial v_z}{\partial z} \tag{12}$$

$$\tau_{zr} = \tau_{rz} = -\mu\left(\frac{\partial v_z}{\partial r} + \frac{\partial v_r}{\partial z}\right) \tag{13}$$

Although in most bio-fluid-dynamics problems the velocity and stress fields are of primary importance, there are cases in which the pressure field is also of interest. Recent results for the two-dimensional driven cavity problem indicate that convergence was more rapid with the ξ-ψ formulation than with the primitive formulation that includes the pressure [17]. This study found that the accuracy of the primitive-variable solution was very sensitive to the convergence tolerance used for the pressure solver.

If the Navier-Stokes equations are solved in primitive-variable form [i.e., Eqs. (1)–(3)], the determination of the pressure is part of the solution procedure and is thus available when a solution is achieved. If, however, the ξ-ψ form of the Navier-Stokes equations is solved [i.e., Eqs. (7) and (9)], then the pressure may be obtained after the ξ-ψ solution is obtained. This "pressure extraction" as the name suggests, is neither simple nor painless. A Poisson equation for pressure is formed from the primitive equations (1) and (2) and is then written in terms of the stream function. For axisymmetric flow this Poisson equation for the pressure is

$$\frac{\partial^2 P}{\partial z^2} + \frac{\partial^2 P}{\partial r^2} + \frac{1}{r}\frac{\partial P}{\partial r}$$

$$= \frac{2}{r^2}\left[\frac{\partial^2 \psi}{\partial z^2}\left(\frac{\partial^2 \psi}{\partial r^2} - \frac{1}{r}\frac{\partial \psi}{\partial r}\right) - \left(\frac{\partial \psi}{\partial z}\right)^2 + \frac{\partial^2 \psi}{\partial z\,\partial r}\left(\frac{1}{r}\frac{\partial \psi}{\partial r} - \frac{\partial^2 \psi}{\partial z\,\partial r}\right)\right] \quad (14)$$

A method of solution for this equation will be discussed later.

By substituting the vorticity given by Eq. (9) into Eq. (7), a single equation in terms of the stream function is obtained. This equation is called the "biharmonic equation" and may be expressed as

$$\frac{\partial}{\partial t}(D^2\psi) - \frac{1}{r}\frac{\partial(\psi, D^2\psi)}{\partial(r, z)} - \frac{2}{r^2}\frac{\partial\psi}{\partial z}D^2\psi = vD^4\psi \quad (15)$$

where $\qquad D^2 = \dfrac{\partial^2}{\partial r^2} - \dfrac{1}{r}\dfrac{\partial}{\partial r} + \dfrac{\partial^2}{\partial z^2} \qquad$ and $\qquad D^4\psi = D^2(D^2\psi) \quad (16)$

Although the biharmonic equation contains only one unknown ψ, it is a fourth-order nonlinear partial differential equation and is usually more difficult to solve. However, Roache [16] has suggested an efficient method of solving the steady planar biharmonic equation.

The above equations may be conveniently nondimensionalized using a reference radius R and a reference velocity U_0. For example, Eqs. (7) and (9) become

$$\frac{\partial\xi}{\partial t} + \frac{\partial(v_r\xi)}{\partial r} + \frac{\partial(v_z\xi)}{\partial z} = \frac{1}{\mathrm{Re}_R}\left(\frac{\partial^2\xi}{\partial r^2} + \frac{\partial^2\xi}{\partial z^2} + \frac{1}{r}\frac{\partial\xi}{\partial r} - \frac{\xi}{r^2}\right) \quad (17)$$

and

$$-\xi = \frac{1}{r}\left(\frac{\partial^2\psi}{\partial z^2} + \frac{\partial^2\psi}{\partial r^2} - \frac{1}{r}\frac{\partial\psi}{\partial r}\right) \tag{18}$$

where all quantities are nondimensionalized, and $\mathrm{Re}_R = RU_0/\nu$ is the Reynolds number.

Other minor variations of Eq. (17) have also been used together with Eq. (18). These forms differ only by the manner in which the diffusion terms are written. For example, if in Eq. (17)

$$\frac{1}{r}\frac{\partial}{\partial r}\left(r\frac{\partial\xi}{\partial r}\right)$$

is used instead of

$$\frac{\partial^2\xi}{\partial r^2} + \frac{1}{r}\frac{\partial\xi}{\partial r}$$

the following equivalent form is obtained:

$$\frac{\partial\xi}{\partial t} + \frac{\partial(v_r\xi)}{\partial r} + \frac{\partial(v_z\xi)}{\partial z} = \frac{1}{\mathrm{Re}_R}\left[\frac{1}{r}\frac{\partial}{\partial r}\left(r\frac{\partial\xi}{\partial r}\right) - \frac{\xi}{r^2} + \frac{\partial^2\xi}{\partial z^2}\right] \tag{19}$$

There does not appear to be any significant advantage in using one of these equations rather than the other, since neither is truly in conservative form. The physical interpretation of the conservative form of the equations for fluid dynamics has been discussed by many authors (e.g., [18,19]). If, for example, the total flux of vorticity is conserved for a finite volume in the region of interest, then the vorticity transport equation is said to be in conservative form. Equations (17) and (19) have previously been referred to, by inspection, as conservative forms by several investigators including this author. This misconception results from the natural transfer of vast experience with the planar equations to the much less frequently used axisymmetric equations. In other words, the identical procedure of adding a modified version of the continuity equation, Eq. (6), to the left side of the vorticity equation (5) is used in both the planar and axisymmetric cases. In both cases, this puts the convective terms, on the left side, into the conservative form without affecting the diffusive terms on the right side of the equation. In the planar case, this simple manipulation produces the conservative form desired. In the axisymmetric case, however, this is not true, since the term $(1/r)(\partial\xi/\partial r)$ is not in the proper form. This point has only recently been discussed by Oberkampf[20] for curvilinear coordinate systems. According to Oberkampf, for a second-order partial differential equation to be in conservative form, it must be written in the form

$$\sum_{i=1}^{m} \{\phi\}_{x_i}{}^i + \{c\} = \sum_{j,\,i=1}^{m} \{[A]^{ij}\{\phi\}_{x_i}\}_{x_j} \tag{20}$$

where $\{\phi\}$ and $\{\phi\}^i$ are n-component vectors of the second-order and first-order terms, respectively, subscripts indicate partial differentiation, $\{c\}$ is an n-component vector whose components are functions of the dependent and independent variables, and $[A]^{ij}$ is the i, j element of the sequence of $n \times n$ coefficient matrices. It is not assumed that $[A]^{ij} = [A]^{ji}$, so that variable viscosity can be considered in the Navier-Stokes equations. For the Navier-Stokes equations with constant viscosity, the nondimensional vorticity transport equation in conservative form becomes

$$\frac{\partial \xi}{\partial t} + \frac{\partial (v_r \xi)}{\partial r} + \frac{\partial (v_z \xi)}{\partial z} = \frac{1}{\mathrm{Re}_R}\left[\frac{\partial^2 \xi}{\partial r^2} + \frac{\partial^2 \xi}{\partial z^2} + \frac{\partial (\xi/r)}{\partial r} \right] \tag{21}$$

This can be obtained from Eq. (17) by the substitution

$$\frac{1}{r}\frac{\partial \xi}{\partial r} - \frac{\xi}{r^2} = \frac{\partial (\xi/r)}{\partial r}$$

Equation (21) may also be shown to possess the conservative property by examining its finite-difference analog as demonstrated by Roache [18] for the planar case.

In order to preserve the conservation property of the equation for stream function, Oberkampf and Goh [21] used the form

$$\frac{\partial^2 \psi}{\partial r^2} + \frac{\partial^2 \psi}{\partial z^2} - \frac{\partial (\psi/r)}{\partial r} - \frac{\psi}{r^2} = -r\xi \tag{22}$$

The relations between stream function and velocity components, Eqs. (8), were also written in conservative form as

$$\frac{\partial (\psi/r)}{\partial r} + \frac{\psi}{r^2} = v_z \tag{23a}$$

$$\frac{\partial (\psi/r)}{\partial z} = -v_r \tag{23b}$$

Using Eqs. (21), (22), (23a), and (23b), Oberkampf and Goh were able to satisfy the general conservative form given by Eq. (20).

The formulation of Eqs. (21)–(23) appears to be unique with these investigators. In order to determine any advantage that might be gained by this formulation, Oberkampf and Goh performed numerical experiments for the case of fully developed flow in a constant-diameter tube. They found that the truncation error in v_r was zero for the differenced conservative form, Eq. (23b), while the truncation error for the differenced nonconservative form, Eq. (8), was nonzero. Thus, better accuracy using the conservative form was demonstrated in this restrictive problem [21].

Vinokur [22] examined this principle of conservation for the inviscid gas dynamic equations in curvilinear coordinates. He distinguished between the strong conservation-law form when there are no undifferentiated terms, that is, $\{c\} = 0$ in Eq. (20), and the weak conservation-law form when such terms are present, as in Eqs. (22) and (23a). These undifferentiated terms act as sources that violate the conservation principle. It is evident that the strong conservation-law form of the equations cannot be achieved for axisymmetric flow [22].

Oberkampf and Goh [23] did not extract the pressure from their ξ-ψ solutions and therefore did not investigate the equation for pressure. However, the nonconservative form of Eq. (14) can be readily cast into the conservative form, namely

$$\left[\frac{\partial^2 P}{\partial z^2} + \frac{\partial^2 P}{\partial r^2} + \frac{\partial(P/r)}{\partial r} + \frac{P}{r^2}\right] = \frac{2}{r^2}\left\{\frac{\partial^2 \psi}{\partial z^2}\left[\frac{\partial^2 \psi}{\partial r^2} - \frac{\partial(\psi/r)}{\partial r} - \frac{\psi}{r^2}\right]\right.$$
$$\left. - \left(\frac{\partial \psi}{\partial z}\right)^2 + \frac{\partial^2 \psi}{\partial z\,\partial r}\left[\frac{\partial(\psi/r)}{\partial r} + \frac{\psi}{r^2} - \frac{\partial^2 \psi}{\partial z\,\partial r}\right]\right\} \qquad (24)$$

Whether there is any advantage in using this form of the pressure equation for a particular problem remains to be shown.

Although conservative does not necessarily imply accuracy [18], experience indicates that conservative systems usually produce more accurate results [24]. There are also cases, however, in which nonconservative forms produce more accurate numerical results than conservative forms [25]. In certain situations, the conservative form leads to instabilities that are not present when the nonconservative form is used (e.g. [26]). It will be shown later that excellent numerical results can be obtained using the nonconservative vorticity equation (17). It should have been pointed out in references 27 and 28 that it was the nonconservative form that was used.

2.2 Incompressible Planar Flow

The Navier-Stokes equations and the continuity equation in a rectangular Eulerian coordinate system (x, y, z) for two-dimensional (i.e., $v_z = 0$ and $\partial/\partial z = 0$) are

$$\frac{\partial v_x}{\partial t} + v_x\frac{\partial v_x}{\partial x} + v_y\frac{\partial v_x}{\partial y} = -\frac{P}{\rho} + v\left(\frac{\partial^2 v_x}{\partial x^2} + \frac{\partial^2 v_x}{\partial y^2}\right) \qquad (25)$$

$$\frac{\partial v_y}{\partial t} + v_x \frac{\partial v_y}{\partial x} + v_y \frac{\partial v_y}{\partial y} = -\frac{P}{\rho} + v\left(\frac{\partial^2 v_y}{\partial x^2} + \frac{\partial^2 v_y}{\partial y^2}\right) \qquad (26)$$

$$\frac{\partial v_x}{\partial x} + \frac{\partial v_y}{\partial y} = 0 \qquad (27)$$

The vorticity in such a two-dimensional flow can be written

$$\xi = \frac{\partial v_y}{\partial x} - \frac{\partial v_x}{\partial y} \qquad (28)$$

and the vorticity transport equation in conservative form

$$\frac{\partial \xi}{\partial t} + \frac{\partial(v_x \xi)}{\partial x} + \frac{\partial(v_y \xi)}{\partial y} = v\left(\frac{\partial^2 \xi}{\partial x^2} + \frac{\partial^2 \xi}{\partial y^2}\right) \qquad (29)$$

Using the incompressible planar stream function

$$\frac{\partial \psi}{\partial y} = v_x \qquad \frac{\partial \psi}{\partial x} = -v_y \qquad (30)$$

the following Poisson equation results:

$$-\xi = \frac{\partial^2 \psi}{\partial x^2} + \frac{\partial^2 \psi}{\partial y^2} \qquad (31)$$

The appropriate stresses in terms of velocity gradients and viscosity are

$$\tau_{xx} = -2\mu \frac{\partial v_x}{\partial x} \qquad (32)$$

$$\tau_{yy} = -2\mu \frac{\partial v_y}{\partial y} \qquad (33)$$

$$\tau_{xy} = \tau_{yx} = -\mu\left(\frac{\partial v_x}{\partial y} + \frac{\partial v_y}{\partial x}\right) \qquad (34)$$

The planar form of the Poisson equation for pressure is

$$\nabla^2 P = 2\left[\frac{\partial^2 \psi}{\partial x^2}\frac{\partial^2 \psi}{\partial y^2} - \left(\frac{\partial^2 \psi}{\partial x \, \partial y}\right)^2\right] \qquad (35)$$

and the biharmonic equation may be written as

$$\frac{\partial \nabla^2 \psi}{\partial t} + \frac{\partial \psi}{\partial y} \frac{\partial \nabla^2 \psi}{\partial x} - \frac{\partial \psi}{\partial x} \frac{\partial \nabla^2 \psi}{\partial y} = \nu \nabla^4 \psi \tag{36}$$

where
$$\nabla^2 = \frac{\partial^2}{\partial x^2} + \frac{\partial^2}{\partial y^2}$$

and
$$\nabla^4 \psi = \nabla^2 (\nabla^2 \psi).$$

If L and U_0 are the characteristic or reference length and velocity, then the nondimensional vorticity transport equation and Poisson equation for stream function become

$$\frac{\partial \xi}{\partial t} + \frac{\partial (v_x \xi)}{\partial x} + \frac{\partial (v_y \xi)}{\partial y} = \frac{1}{Re_L} \left(\frac{\partial^2 \xi}{\partial x^2} + \frac{\partial^2 \xi}{\partial y^2} \right) \tag{37}$$

$$-\xi = \frac{\partial^2 \psi}{\partial x^2} + \frac{\partial^2 \psi}{\partial y^2} \tag{38}$$

where all quantities are nondimensional and $Re_L = LU_0/\nu$ is the Reynolds number.

2.3   Mesh Systems and Geometric Considerations

A glance at the arterial and bronchial trees, as well as the types of local aberrations of these systems that are of interest (Fig. 4), will quickly indicate that the choice of mesh system for these irregularly shaped vessels usually determines the problem that can be solved. For example, there has been a good deal of interest in the flow through a locally constricted vessel. This problem is related to the local deposit or growth of plaques that restrict blood flow and is part of the overall problem of arteriosclerosis. Consider the symmetric local vessel constriction as shown in Fig. 8a. There are a variety of possible methods of proceeding with this type of problem. For those who prefer a rectangular mesh at any cost, a nonuniform rectangular mesh as illustrated in Fig. 8b (variable mesh spacing in one or both directions) may be developed, which will approximate the irregular wall shape. The finer the mesh, the more closely it will approximate the irregular geometry. This method also has the advantage of allowing an increase in the number of grid points in areas where gradients in the flow are high (i.e., near the constriction), and a decrease in the number of grid points in low-gradient regions (i.e., near the center of the channel and far upstream and downstream of the constriction). It is important to expand or contract the mesh slowly enough to limit truncation errors and thus not sacrifice absolute

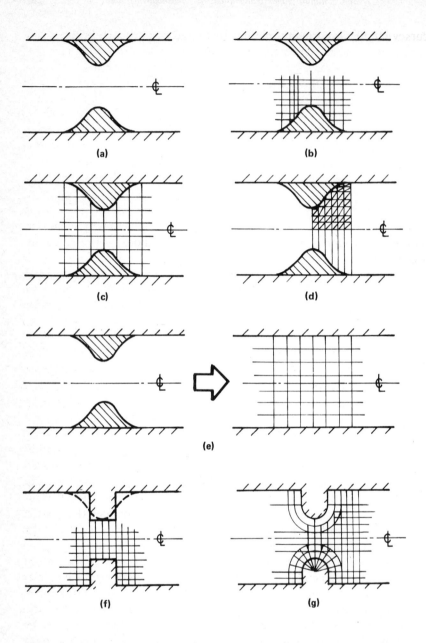

FIG. 8 Possible methods of handling arbitrarily shaped boundaries. (a) Symmetric local constriction; (b) nonuniform rectangular mesh; (c) overlay uniform rectangular mesh; (d) triangular finite-element mesh; (e) transform geometry to rectangular region and use uniform rectangular mesh; (f) simplify geometry to fit uniform rectangular mesh; (g) simplify geometry to fit hybrid mesh.

accuracy [18,19]. Tailoring the mesh in this way for a particular problem can produce good results; however, if a constriction with a different geometry becomes of interest, another such mesh must be constructed.

No one familiar with finite-difference methods uses the overlay of a uniform rectangular mesh for irregular boundaries as shown in Fig. 8c, because of the large truncation error associated with a grid point that is very close to the boundary. However, recent cases in which this method has been tried may be found (e.g., [30]).

One of the significant advantages of the finite-element method is the relative ease with which triangular elements may be used to fit an irregular boundary, as illustrated in Fig. 8d. However finite-element methods have not yet been demonstrated to produce accurate results for a practical viscous-flow problem. A good discussion concerning the many claims of finite-element methods versus finite-difference methods is presented by Roache [16].

One of the most efficient methods of dealing with irregularly shaped boundaries is to use a coordinate transformation as illustrated in Fig. 8e. This is especially advantageous when the stretching of only one irregular wall is necessary. Lee and Fung [31] considered an infinitely long cylindrical tube with bell-shaped local constriction (Fig. 8a) and used conformal mapping to transform this to an infinitely long constant-diameter tube (i.e., Fig. 8e). Numerical results were obtained for this single constriction for steady flow and $Re_R = 0$, 10, and 25. Oberkampf and Goh [23], using the same bell-shaped local constriction, transformed this geometry into a constant-diameter tube of finite length with a nonorthogonal transformation. Numerical results were obtained for both a single and a double constriction for steady flow and $Re_R = 10$ and 25. This appears to be the best method of dealing with such problems, although each transformation used in a numerical solution should be examined closely in order to determine if the mapping is one to one. A recent description of this method for conservative and nonconservative systems of first- and second-order partial differential equations using generalized mapping functions is given by Oberkampf [20].

Extensive results have been published by Thompson et al. [32,33] using a method of automatic numerical generation of a general curvilinear coordinate system with coordinate lines coincident with all boundaries of arbitrarily shaped bodies. There are no restrictions on the shape of the boundaries, and they may even be time-dependent. Although the problems studied by Thompson et al. have all been for external flow situations, this approach appears particularly attractive when more than one irregular boundary is present in the same problem. For example, the flow through a prosthetic heart valve in the aortic position as shown in Fig. 5 could be studied in this way. In addition to the irregular shape of the aorta and heart valve, one would like to take into account the distensibility of the aorta and the movement of the occluder as affected by the pulsatile flow.

There is always an important lesson to be learned by simplifying the irregular geometry to fit a uniform rectangular mesh while retaining the most important feature of the original problem. An extensive study of steady, oscillating, and pulsatile flows past a square constriction, illustrated in Fig. 8*f*, in planar flow has been studied by Cheng et al. [34–36]. A similar rectangular-cross-section orifice geometry has recently been studied for which a nonuniform rectangular mesh in both directions was used [37]. These works will be discussed in more detail later. If the simplification of the smooth local constriction that has sharp corners is deemed too drastic, the geometry can be approximated by an analytic function with the thought in mind of using a hybrid mesh. Figure 8*g* shows the same constriction with a semicircular protrusion. Here overlapping cylindrical-polar and square coordinate systems could be used. It is evident from the current literature that the mesh used in a particular problem depends on the complexity of the problem, the desired result, and the background and experience of the investigator.

2.4 Computational Methods

The principal difficulties in applying numerical methods to physiological flows of the circulatory and respiratory systems, excluding the distensible boundaries, include (1) approximating or modeling the irregular wall shape, (2) accounting for the non-Newtonian behavior of the physiological fluids where necessary, and (3) duplicating the pulsatility of the flow. The irregular wall shape can be handled in a number of different ways as discussed in the previous section. The non-Newtonian behavior of blood, where it is important, may be accounted for by using a Casson type of equation or one of the more recent empirical relationships, as mentioned in the introduction.

The inclusion of the pulsatility of the flow into the numerical procedure, whether it be for the present or for future calculations, should be considered before the computational method is chosen. This requirement will be important sooner or later in most of the problems we are interested in at this time, and it favors the use of explicit time-dependent computational methods. The time-dependent approach also seems particularly well suited to problems with separated flows, which often exhibit a time-dependent behavior, for example, oscillating wakes and vortex shedding. Another advantage of an explicit method of solution is that the mathematics and calculations are kept simple. Of course the principal disadvantage is that there is an upper limit on the size of the time step that can be used if the finite-difference system is to remain stable. This disadvantage can be overcome with reasonable care and at the expense of longer computer runs. Implicit techniques are currently very popular and efficient for problems with relatively simple boundary geometries when only the steady-flow solution is required. Both types of methods have been utilized for physiological flow problems.

Roache [38,16,39] has presented four semidirect finite-difference methods for solving the steady-state Navier-Stokes equations. These methods are neither time-dependent nor timelike in their iterations. They are based on recent advances in solving linear equations by noniterative or direct methods. These four semidirect methods are referred to as NOS, LAD, Split NOS, and BID. The NOS, LAD, and Split NOS methods have been applied to the planar vorticity transport equation for flow-through (i.e., developing channel flow) and recirculating-flow (i.e., driven-cavity) problems. For the driven-cavity problem the convergence rates of the NOS, LAD, and Split NOS methods are about the same, but better than time-dependent methods [16]. Even faster convergence rates were obtained with the BID method, which uses a form of the biharmonic equation. Another semidirect method called FOD has been suggested by Roache but not tested. Similar approaches have been used by Morihara and Cheng [40] and Lomax and Martin [41]. For physiological problems where steady-state solutions are sought and the geometry presents no serious difficulties, the use of one of these fast semidirect methods should be considered.

2.5   Treatment of Unsteady Viscous Flow

The time-dependent Navier-Stokes equations in the vorticity–stream-function form have been used successfully by several authors to include the effect of pulsatile flow (e.g., [36,42]). The vorticity transport equation and the equation for stream function [e.g., Eqs. (27) and (38) for planar flow] do not contain the pressure gradient, which is necessary if we are to prescribe a pulsatile flow. However, the pressure gradient can be introduced as a forcing function by means of the following nondimensional equation:

$$\frac{\partial v_x}{\partial t} = -\frac{\partial P}{\partial x} + \frac{\partial^2 v_x}{\partial y^2} \tag{39}$$

In order to make use of this simple relation, the inflow and outflow cross sections are chosen far enough upstream and downstream of the region of interest so that the flow is uniform at these sections. Since the intervening fluid is incompressible, the conditions at the inlet and outlet are identical and described by Eq. (39). Also, since v_y is zero at inlet and outlet, the vorticity is obtained by differentiating v_x with respect to y. The stream function is obtained by integrating v_x across the cross section.

The flow velocity (and thus the Reynolds number) varies with time in a pulsatile flow. Sometimes this difficulty is overcome by using the maximum value or the average value of Reynolds number over the cycle. If the detailed effects of the pulsatile flow cycle are desired, then the Reynolds number is not as appropriate a nondimensional parameter as it is for steady flow. Additional dimensionless parameters have been developed and used for unsteady viscous flows (e.g. [35]). The Stokes number is defined as $S = D^2(2\pi f/v)$, where D is the tube diameter, v the kinematic viscosity, and f the frequency. This characterization of the frequency may be thought of as the ratio of the temporal inertia force to the viscous force. The Karman number represents the ratio of the pressure-gradient force to the viscous force and is defined by

$$K = \frac{D^3}{\rho v^2} \left| -\frac{dP}{dx} \right|_{ampl}$$

where ρ is the density, and $\left| -dP/dx \right|$ is the maximum magnitude of the pressure gradient.

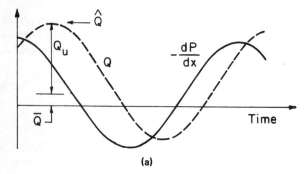

(a)

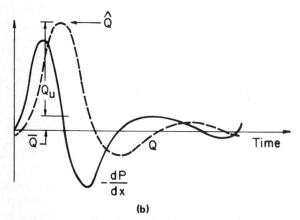

(b)

FIG. 9 Pulsatile forcing functions and resultant tube flow rates. (a) Simple pulsatile flow; (b) arterial pulsatile flow. Q = volumetric flow rate; $\overline{Q}$ = temporal average (steady) value of Q; Q_u = unsteady component of Q; $\hat{Q}$ = instantaneous peak value of Q; dP/dx = axial pressure gradient. Taken from [43].

The arterial pulsatile flow has been modeled by a steady component plus an oscillatory component, as shown in Fig. 9a. This type of flow in a constant-diameter tube is uniquely defined by the steady-flow value of K and the oscillatory-flow values of K and S [43]. The arterial pulsatile flow, however, is more complicated, as indicated in Fig. 9b. Harmonics of the basic oscillations are important and should be included in the specifications of this flow.

Pulsatile flows may be characterized by comparing the relative strengths of the oscillatory and mean flow components. The kinematic parameter λ is defined for this purpose. The unsteady flow ratio λ is the ratio of the amplitude of the unsteady flow component Q_u to the mean flow rate $\bar{Q}$ for the simple pulsatile flow of Fig. 9a. For the arterial pulsatile flow it is the ratio of the peak flow rate $\hat{Q}$ to the mean flow rate $\bar{Q}$ [43].

3 PROSTHETIC HEART VALVES

3.1 Description of the Physiological Problem

During the past 15 years the implantation of prosthetic heart valves has become a relatively common operation. Many different designs have been tried, with varying degrees of success. Today the most common types of valves are the disc valve, the ball valve, and the tilting-disc valve (see Fig. 5). Severe short- and long-term problems of hemolysis and thrombus formation were encountered with the early valve designs. While these problems have been reduced through different valve designs and materials, they are by no means under control with today's valves, as is evident in the recent medical literature [44,45]. Problems of severe hemolysis and massive thrombus formation continue to exist. The importance of understanding the mechanisms that cause these problems is evident. Prosthetic valves are widely used to replace those natural valves congenitally deformed or damaged by rheumatic heart disease, and additional life-saving uses are now being found.

Several studies have reported that erythrocytes taken from patients who have prosthetic valves have shorter survival times than normal when transfused into patients without prosthetic valves [46–48,8]. This would indicate that these cells had sufficient injury to their cell membranes so that their deformability was altered, making them susceptible to removal from the circulation by the filtering function of the spleen [49,8]. Many of these patients with chronic hemolysis require constant iron therapy, and some must be exposed to the hazards of multiple blood transfusions. Even with valves that exhibited low erythrocyte damage or destruction, thrombosis is a serious problem [50–53,47]. Thrombi that continue to form around the valve break away and move on downstream. Emboli have been found in coronary arteries [52,47,54], in the brain [51,54,55], and at various locations in the body [52,54–56]. Thrombus buildup can also inhibit the motion of the valve-occluding mechanisms and result in death [51,52,47]. Most patients must take anticoagulants to help control this problem.

From an engineering point of view, the problem of hemolysis and the formation of thrombi are related. High fluid shear stresses in the absence of walls can hemolyze erythrocytes if their value exceeds 50,000 dyn/cm^2 [57]. Shear stresses lethal to erythrocytes in the presence of walls appear to be much lower. According to Nevaril et al. [49], in the presence of walls shear stresses between 1500 and 3000 dyn/cm^2 can cause damage to the erythrocytes. Small increases in shear stress above 3000 dyn/cm^2 cause significant hemolysis [49]. Recently [58] it has been found that normal or elongational fluid stresses of the order of 10,000 dyn/cm^2 can rupture the red-cell membrane, resulting in significant bulk hemolysis. Other investigations have found the rate of hemolysis to be more a function of the fluid velocity [59,60], the material of the contact wall [61], and the temperature [62] than of the value of shear stress. It appears that the regions close to the walls of the valve (i.e., sewing rings, occluding mechanism, and support struts) are of greatest importance in hemolysis. This does not rule out the effect of fluid shear stresses or normal stresses in the absence of walls as being hemolytic, but casts doubt on their relative importance. Just what shear or normal stress in the absence of a wall is large enough to damage an erythrocyte is not completely known. Crushing of erythrocytes between the occluding mechanism and the sewing ring has also been considered to contribute to the hemolytic effects of prosthetic valves [63].

The blood flow through a natural aortic valve is smooth and able to wash the walls of the valve. This reduces the possibility of thrombus formation in the vicinity of the valve. Furthermore, there is negligible backflow during the closing of the natural aortic valve. For the prosthetic valve, however, the flow is greatly disturbed both upstream and downstream of the sewing ring, as well as downstream of the occluding mechanism. The primary flow pattern thus produced by the prosthesis is a separated or wake-type flow. In order to reduce these flow-induced effects, which are closely related to the occurrence and extent of separated-flow regions, shear-stress magnitudes, and mass-transfer distributions, a better understanding of the mechanisms involved is necessary. If accurate computational methods can be developed to determine the complex velocity, shear, and pressure fields for the prosthetic heart-valve problem, new valve designs may be evaluated faster and possibly more cheaply than by using laboratory experiments alone.

3.2   Description of the Numerical Problem

Interest in the numerical determination of heart-valve flows is very recent [64–71,27,28,42], and most of the results are therefore of a preliminary nature. Although the numerical origin of this problem dates back to about

1970, the first published information appeared in 1972. Greenfield and Kolff [64] solved the time-dependent planar Navier-Stokes equations for the flow through an idealized aortic-shaped geometry for both the ball and the disc-type prosthetic valves. No details of their numerical procedures were given. Their principal interest was to describe the use of computer graphics to display the results of these calculations. The authors emphasize that the results of such numerical calculations are only really useful if they can be presented in such a visual way that our understanding of the complex hemodynamic problem is enhanced. Two of their results are presented in Figs. 10 and 11. While these graphic results certainly add a great deal to our understanding of this problem, the authors presented no quantitative information about these calculations, not even the Reynolds number.

Other short papers related to leaflet and prosthetic heart valves appeared in 1972 [67,68]. Schuessler and Hung [67] studied the leaflet motion during the opening of an axisymmetric leaflet valve in an axisymmetric aortic-shaped geometry. However, they assumed that the viscous effects would be

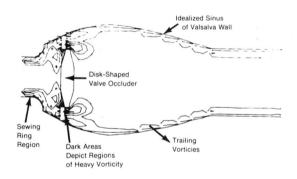

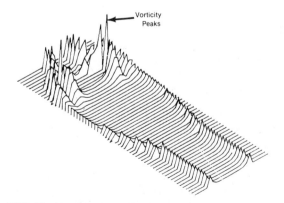

FIG. 10 Vorticity for a Kay-Shiley type heart valve in fully open position. Taken from [64].

small in this accelerating flow and thus solved the inviscid Euler equations. While these results of the leaflet motion are certainly useful, this approach does not add to our understanding of the viscous blood motion, which is of equal or even greater interest, since the problems of hemolysis and thrombus formation are closely related to viscous phenomena. A numerical and experimental study of the plane pulsatile flow through a bilaterally symmetric type valve with an eight-sided-polygon occluder shape was reported by Saklad and Moskowitz [68,69]. Computer solutions of the conservative Navier-Stokes equations in primitive-variable form for the fully open valve were obtained. The results [69] indicated that the mesh used was so coarse that smooth streamlines could not be obtained. An important contribution to the study of leaflet valves was presented by Peskin [70]. Plane two-dimensional mitral-valve closure was studied using the Navier-Stokes equations and an implicit scheme that allowed the fluid forces and elastic properties of the leaflets to interact in order to determine the motion of the leaflets. The leaflets, which are really immersed boundaries in the rectangular domain of interest, are replaced by a field of force that is defined on the mesh points.

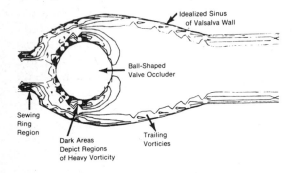

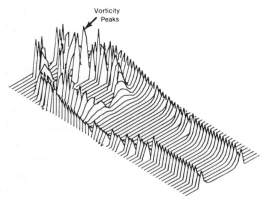

FIG. 11 Vorticity for a Starr-Edwards type heart valve in fully open position. Taken from [64].

This field of force has approximately the same effect on the fluid as the immersed boundary would have. This approach is closely related to the work of Viecelli [72,73].

In a study by F. N. Underwood that began in 1970 under the author's direction, calculations of the steady axisymmetric flow through a fully open model disc heart valve in a circular tube were made [65,66]. The smooth shapes of the sewing ring and disc were approximated by rectangular elements of similar dimensions in order to take advantage of the simplifying features of a square mesh. The time-dependent Navier-Stokes equations in the vorticity–stream-function formulation were solved for the asymptotic steady solution. Early numerical results were encouraging, although they did not agree well with the experimental data [65,55,74]. However a minor error was found in the computer program, which, when corrected, produced results that agreed very well with the data [27,28]. Krause and Giese [71] repeated the calculations for the model disc valve in the circular tube and found good agreement with our corrected results. Their work also presented the results of planar microchannel experiments on this valve shape by Lambert [75] at Aachen. An important additional use for numerical solutions of physiological problems was demonstrated by Krause and Giese. These solutions produce or contain the streamline patterns as well as the stress components throughout the flow field. One can, therefore, extract the stress history along any streamline of interest. This type of information is important in determining how long a red blood cell is exposed to a certain level of stress. It is not just the magnitude of the stress that may be important in erythrocyte damage and hemolysis, but the time of exposure as well. This type of data could also be used together with recently published approximate mathematical techniques to determine the changes in red blood cell shape as a result of the changing stress applied to the cell [76 and 77].

The numerical study of the model disc heart valve in the circular tube was continued, and a study of the model disc valve in an aortic-shaped geometry was added [28]. Experiments indicated, as one would expect, that the flow in the vicinity of the valve is greatly affected by the channel or wall shape. Thus, an axisymmetric aortic-shaped channel was used. Laboratory experiments were also initiated to check the numerical results [78].

A very recent study of the pulsatile flow through a planar idealized or model tilting-disc heart valve was presented by Sator [42]. The time-dependent planar Navier-Stokes equations in the vorticity–stream-function formulation [i.e., Eqs. (37) and (38)] were used. The pressure gradient was utilized as a forcing function to produce the pulsatile flow as described in Sec. 2.4, and a finite-element method was used. Results include velocity and shear-stress values for the case in which the Karman number is 1000 and the Stokes number is 10.

The following sections will describe the computational techniques used by Underwood and Mueller for the solution of the Navier-Stokes equations and the results for steady flow through an axisymmetric fully open disc-type heart valve in a circular tube and an axisymmetric aortic-shaped channel as shown in Fig. 12. The purpose of this research was not to duplicate physiological conditions but to develop diagnostic techniques that can be used together with *in vitro* physical experiments to evaluate existing or future prosthetic heart valve designs. Data concerning the hemolytic and thrombogenic potential of various valve designs can help in understanding the mechanisms involved and contribute to the evolution of better valve designs.

3.3   Model Disc Heart Valve in a Constant-Diameter Tube

The disc-type prosthetic heart valve geometry was modeled so that maximum advantage could be taken of a uniform rectangular mesh. This model valve was first studied in a constant-diameter tube to further simplify the geometry (see Fig. 12a). Since it was desired to study the worst case from the point of view of the extent of flow separation and fluid stresses, the fully open position was chosen. The valve occluder, that is, the disc, remains fully open for the largest portion of the cardiac cycle, and the maximum flow rate or Reynolds number occurs at this condition. In order to live within our computational means, the fluid was assumed to be Newtonian, and the flow

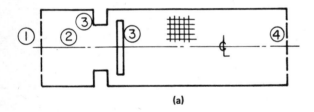

(a)

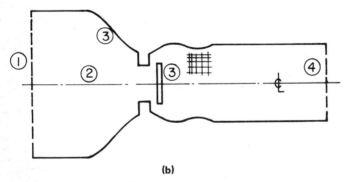

(b)

FIG. 12   Axisymmetric geometries used in the numerical studies. (a) Constand-diameter tube with model disc valve; (b) aortic-shaped channel with model disc valve.

axisymmetric. Equations (7) and (9) describe the form of the Navier-Stokes equations used. The solution of Eq. (14) for the pressure was also examined. Two uniform-mesh configurations were used: (1) the regular mesh containing about 7900 points (43 × 184), and (2) the extended mesh, which continued further in the downstream direction and had about 11,500 points (43 × 268). Both configurations had a nondimensional mesh spacing of $\Delta z = \Delta r = 0.0238$.

3.3.1 Finite-Difference Equations

The axisymmetric vorticity transport equation (7) is written in finite-difference form with the new vorticity at time $k + 1$ determined from values of vorticity and velocity at time k.

$$\zeta_{i,j}^{k+1} = \zeta_{i,j}^{k} - \Delta t \left(\frac{v_{ri,j}^{k} \zeta_{i,j}^{k} - v_{ri,j-1}^{k} \zeta_{i,j-1}^{k}}{h} + \frac{v_{zi,j}^{k} \zeta_{i,j}^{k} - v_{zi-1,j}^{k} \zeta_{i-1,j}^{k}}{h} \right)$$

$$+ \frac{\Delta t}{\mathrm{Re}_R} \left(\frac{\zeta_{i+1,j}^{k} - 2\zeta_{i,j}^{k} + \zeta_{i-1,j}^{k}}{h^2} + \frac{\zeta_{i,j+1}^{k} - 2\zeta_{i,j}^{k} + \zeta_{i,j-1}^{k}}{h^2} \right.$$

$$\left. + \frac{\zeta_{i,j+1}^{k} - \zeta_{i,j-1}^{k}}{2rh} - \frac{\zeta_{i,j}^{k}}{r^2} \right) \tag{40}$$

Centered space differences that are second-order accurate are used to represent the first and second derivatives of the diffusion terms. The convective or advective terms of Eq. (7) are approximated by using the first upwind differencing method. The second upwind differencing method, also known as the "donor cell" method, was also used for the advection terms. Both methods are statically stable, and both are transportive and conservative [18]. Upwind differencing schemes allow the effects of advection to be felt only in the flow direction, while the diffusion terms allow the disturbance to be felt in all directions. In the first upwind differencing scheme, backward differencing for the advection terms is used if all the velocities in the neighborhood of a typical point i are positive:

$$\frac{\Delta(v_z \xi)}{\Delta x} = \frac{(v_z \xi)_i - (v_z \xi)_{i-1}}{\Delta x} \tag{41}$$

If all the velocities are negative, forward differencing is used:

$$\frac{\Delta(v_z \xi)}{\Delta x} = \frac{(v_z \xi)_{i+1} - (v_z \xi)_i}{\Delta x} \tag{42}$$

If a flow reversal occurs, neither method can be used since an artificial source or sink occurs. A modification formulated by using the control-volume approach provides a conservative and transportive differencing method to handle this problem.

$$\frac{\Delta \xi}{\Delta t} = \frac{f_{i-1} + f_i + f_{i+1}}{\Delta x} \tag{43}$$

where $f_i = v_i \xi_i$

$$f_{i-1} = \begin{cases} v_{i-1} \xi_{i-1} & \text{for } v_{i-1} > 0 \\ 0 & \text{for } v_{i-1} \le 0 \end{cases}$$

$$f_{i+1} = \begin{cases} 0 & \text{for } v_{i+1} \ge 0 \\ v_{i-1} \xi_{i+1} & \text{for } v_{i+1} < 0 \end{cases}$$

Note: Signs in this method are different when Eq. (4) has a different sign for ξ (e.g., [18]).

The main criticism of this method is that these advection terms are only first-order accurate. Nevertheless, one-sided differences with a transportive property have been found to produce results with good absolute accuracy. The second upwind differencing technique, which was called the donar cell method by Gentry et al. [79], is similar to the first method with the added advantage of retaining properties of the second-order accuracy of centered-space derivatives. The second method is also both conservative and transportive.

$$\frac{\Delta \xi}{\Delta t} = -\frac{v_R \xi_R - v_L \xi_L}{\Delta x} + \cdots \tag{44}$$

where $v_R = \dfrac{v_{i+1} + v_i}{2}$

$$v_L = \frac{v_i + v_{i-1}}{2}$$

$$\xi_R = \begin{cases} \xi_i & \text{for } v_R > 0 \\ \xi_{i+1} & \text{for } v_R < 0 \end{cases}$$

$$\xi_L = \begin{cases} \xi_{i-1} & \text{for } v_L > 0 \\ \xi_i & \text{for } v_L < 0 \end{cases}$$

In both these upwind differencing methods, information is advected into a cell only from those cells that are upwind of it. Therefore, information is advected from a cell only into those cells that are downstream of it. This, of course, is physically correct and leads to the following definition of the transportive property. A finite-difference formulation of a flow equation possesses the *transportive property* if the direction of a perturbation in a transport property is advected only in the direction of the velocity. Upwind differencing methods inspired by this physical reasoning possess this property [18]. In any method using space-centered differences for the advective terms, the effect of a perturbation on a transport property is advected upstream against the velocity. The accuracy of these upwind differencing methods in resolving viscous effects usually deteriorates at high Reynolds number because they implicitly introduce an effect that resembles an artificial viscosity effect. Although this numerical viscosity is not unique to upwind differencing [80], it appears to be the predominant source of error in these schemes.

It should be remembered that the accuracy of the result compared to the appropriate physical experiment is the final test of the numerical procedure, and not simply the order of accuracy. Although the upwind differencing methods introduce an artificial viscosity effect that must be carefully examined when assessing the accuracy of results, it has the further advantage that it is not stability limited by a cell Reynolds number as are forward-time space-centered methods. The physical relevance of upwind differencing, which seems particularly appropriate for separated flows with their large backflow regions, has been discussed by many authors (see reference 18 for a list of references). Results using the first and second methods will be discussed in Sec. 3.5.

Dynamic instability is characterized by an oscillatory overshoot with increasing amplitude. Such an instability can be rectified by using a smaller Δt for the vorticity transport equation. A linear stability analysis suggests a critical time step above which dynamic instability will be anticipated.

$$\Delta t_{\text{crit}} = \frac{1}{(2/\text{Re}_R)(1/\Delta z^2 + 1/\Delta r^2) + |v_z|_{\max}/\Delta z + |v_r|_{\max}/\Delta r} \qquad (45)$$

To ensure stability only a fraction of this critical time step is used, usually 85–95% of Δt_{crit}. It has been shown by Underwood [28] that the size of a stable time step does not affect the eventual solution. Both very small (about 20%) and very large (about 95%) fractions of Δt_{crit} were used, and each case converged to the identical solution. Static instability is characterized by an error that grows monotonically. It cannot be removed by decreasing the size of Δt. Static stability is maintained by use of a particular finite-difference method, e.g., an upwind difference method.

Since Eq. (9) is used to determine the values of stream function compatible with the new vorticity values at time level $k + 1$, the equation is rearranged to define stream-function values at neighboring points.

$$\psi_{i,j} = \frac{1}{4}\left[\psi_{i+1,j} + \psi_{i-1,j} + \psi_{i,j+1} + \psi_{i,j-1}\right.$$

$$\left. - \frac{h}{2r}(\psi_{i,j+1} - \psi_{i,j-1}) - \xi_{i,j}rh^2\right] \qquad (46)$$

Equation (46) is solved by using an iterative method of solution called "successive overrelaxation." Successive overrelaxation uses new values of stream function as they become available as the computer sweeps the field in increasing values of i and j. This method speeds the iterative process and also does not require the storage of another steam-function value at different iterate levels.

The new ψ, SNEW, is obtained by subtracting the value of ψ calculated in the previous time step from the right-hand side of Eq. (46). This difference between the old and new ψ is then multiplied by a relaxation factor RF and added to old ψ. In this way, Eq. (46) is manipulated into a form such that successive overrelaxation can be used to hasten convergence.

$$\text{SNEW} = \psi_{i,j} + \text{RF}\left\{\frac{1}{4}\left[\psi_{i+1,j} + \psi_{i-1,j} + \psi_{i,j+1} + \psi_{i,j-1}\right.\right.$$

$$\left.\left. - \frac{h}{2r}(\psi_{i,j+1} - \psi_{i,j-1}) - \xi_{i,j}rh^2\right] - \psi_{i,j}\right\} \qquad (47)$$

For convergence the relaxation factor RF has a value greater than one but less than two. Several computer runs are made to determine an optimum relaxation factor, which requires the fewest number of iterations for convergence.

The determination of the pressure from Eq. (14) requires the normal derivative boundary condition. This Neumann problem, as presented by Fanning [81] and Fanning and Mueller [82], can be expressed by the following two equations:

$$\nabla^2 P = -f \qquad (48)$$

$$\frac{\partial P}{\partial n} = g(s) \qquad (49)$$

Equation (48) is the governing equation in the region Ω, subject to the boundary condition (49) on the boundary σ, but according to Green's first identity the following equation is a necessary auxiliary condition for the solution of Eq. (48):

$$\int_{\Omega} f \, dv = \oint_{\sigma} g(s) \, ds \qquad (50)$$

Adjustments to f and g must be made to ensure that the auxiliary condition (50) is met. The adjustments are made in portions of the flow field where the numerical solution is known to be weak or inaccurate, for example, corners or the outflow. This is part of "the art" of computational fluid dynamics about which Roache speaks [18]. Equation (48) is solved by the Gauss-Seidel iteration technique (equivalent to the SOR method for RF equal to one). The space-centered, second-order accurate finite-difference expressions were used on the boundary as well as in the interior of the field. To avoid using points outside the boundary, which are normally required for centered-difference calculations on the boundary, a separate procedure was used. The terms for those outside points were replaced in the space-centered expression by their equivalent values in terms of the normal derivatives on the boundary and the values of pressure one grid point from the boundary.

An infinite number of solution surfaces are valid solutions for the Neumann problem [81,82]. To tie the numerical result to one solution surface, a constraint was applied. One point in the field was given a constant value. To ensure that the value of pressure at that point satisfied the differential equation, the pressure value was iterated there also. The difference between the constant value assigned and the value calculated there was subtracted from the calculated value of pressure at every other point in the field. This assured two things: (1) the approximate solution was "tied" to one solution surface, and (2) the value of pressure at the point of the "tie" satisfied the differential equation. This technique was used successfully by Fanning and Mueller [82] to calculate the velocity potential for compressible flow. This extension to a pressure solution was also suggested by those authors in their paper.

3.3.2   Boundary and Initial Conditions

As in the solution of any time-dependent partial differential equation, it is necessary to specify the conditions on the boundaries of the system and the initial state of the system.

Upstream boundary   The inflow or upstream boundary is labeled surface 1 in Fig. 12a. The equation for v_z velocity in terms of stream function is integrated to set the inflow boundary condition for stream function.

$$\psi_{1,j} = \frac{Ar^4}{4} + \frac{Br^2}{2} \tag{51}$$

A and B are constants whose values depend on the type of inflow profile desired (parabolic or uniform) and the radius of the tube. The v_z velocity, v_r

velocity, and vorticity were calculated using centered differences in the r direction and forward one-sided differences in the z direction.

$$v_{z_{1,j}} = \frac{\psi_{i,j+1} - \psi_{i,j-1}}{2hr} \tag{52}$$

$$v_{r_{1,j}} = \frac{\psi_{i+1,j} - \psi_{i,j}}{hr} \tag{53}$$

$$\zeta_{1,j} = \frac{v_{r_{i+1,j}} - v_{r_{i,j}}}{h} - \frac{v_{z_{i,j+1}} - v_{z_{i,j-1}}}{2h} \tag{54}$$

The velocity in the radial direction was allowed to adjust at the inflow to give the flow more flexibility in reacting to the obstructions downstream. In one case, to provide maximum flexibility, v_z was also allowed to change at the inflow. For this case stream-function values at the midpoints of the inflow between the centerline and top solid boundary were allowed to float by means of an extrapolation from the points upstream [83]. For a parabolic-inflow initial condition, no significant difference in the results was detected by using this condition, compared to using the condition in which the v_z velocity did not vary. Solutions were obtained using both uniform- and parabolic-inflow velocity profiles with the same mass flow. Although there was only a slight difference in the results downstream of the valve, the uniform inflow produced a significant dip in the streamlines near the inflow [28]. The smooth streamline pattern near the inflow obtained using the parabolic-inflow profile seemed more physically realistic and thus was used for all subsequent calculations.

Centerline   The centerline is labeled surface 2 in Fig. 12a. Axial symmetry was assumed, and a direct consequence of this assumption is zero vorticity and zero radial velocity at the centerline. The centerline stream-function value was arbitrary; so zero was assigned to the stream function at the centerline. The v_z velocity at the centerline was calculated by extrapolating values at grid points near the centerline.

Solid boundary   The solid boundaries are labeled in Fig. 12a. The value for stream function along the top boundary, including sewing ring, was given a constant value found from Eq. (51) for the inflow.

$$\psi_{\text{wall}} = \text{constant}$$

The value of stream function for the disc was that of the centerline.

$$\psi_{\text{disc}} = 0$$

At the solid boundary the velocity in both the radial and axial directions was zero due to the nonslip viscous boundary condition.

$$v_{r_{(solid)}} = 0 \qquad v_{z_{(solid)}} = 0$$

Vorticity on the solid boundary was computed using the definition of vorticity, Eqs. (4) and (9). Since velocity is zero on solid boundaries, and the derivatives along the surface are zero, these equations can be simplified. For example, the top boundary value of vorticity can be expressed by Eq. (55) or (56):

$$\xi_{i,\,JL} = \frac{v_{z_{i,JL-1}}}{h} \tag{55}$$

$$\xi_{i,\,JL} = \frac{2(\psi_{i,\,JL} - \psi_{i,\,JL-1})}{rh^2} \tag{56}$$

Equation (56) is a direct result of expanding $\psi_{i,\,JL-1}$ by a Taylor series. By noting that $\partial\psi/\partial r$ is zero on the boundary, $\partial^2\psi/\partial r^2$ is shown to be $2(\psi_{i,\,JL} - \psi_{i,\,JL-1})/h^2$. Also, $\partial^2\psi/\partial z^2$ and $(1/r)(\partial\psi/\partial r)$ are zero on the boundary. Other solid boundaries have their vorticity expressed in similar ways, differing only in the direction in which the one-sided difference is formed. For all the results presented in this chapter, Eq. (56) was used to obtain the vorticity on the solid boundary.

Corners The stream function at a corner, since it is part of a solid wall, was constant. As in the approach of Roache and Mueller [84], a corner point was assigned two values of vorticity. The vorticity of a point upstream of the corner was computed as if the corner lay on the upstream step surface. However, if the value at the corner was required while computing a point behind the sewing ring and/or disc, for example, the corner vorticity was computed as if it lay on the downstream surface of the boundary wall. Another investigation supporting the use of this double-value treatment of sharp corners has been used by Coder [37] for an orifice problem. Although this treatment of the sharp corners is not strictly conservative, the error introduced has been found to decay rapidly as one moves away from the corner.

Downstream boundary The outflow is labeled surface 4 in Fig. 12a. One of the more challenging problems in computational fluid dynamics is that of the outflow condition. The boundary conditions for stream function and vorticity used by Fanning and Mueller [29] for the outflow were used in this study. A linear extrapolation in the axial direction was made for the value of stream function and vorticity.

$$\psi_{IL,j} = 2\psi_{IL-1,j} - \psi_{IL-2,j} \tag{57}$$

$$\xi_{IL,j} = 2\xi_{IL-1,j} - \xi_{IL-2,j} \tag{58}$$

This condition allows the flow to continue in the general direction dictated by the two interior points adjacent to the outflow. The velocities at the outflow were computed using centered and one-sided differences.

$$v_{z_{IL,j}} = \frac{\psi_{IL,j+1} - \psi_{IL,j-1}}{2hr} \tag{59}$$

$$v_{r_{IL,j}} = \frac{\psi_{IL,j} - \psi_{IL-1,j}}{hr} \tag{60}$$

Pressure boundary conditions For the pressure boundary conditions the normal derivatives of pressure were required. The Navier-Stokes equations containing the derivatives of pressure in the r and z directions were used. For inflow and outflow, the complete equations were required. Since second derivatives necessitate the use of a grid point outside the boundary for central differencing, the assumption was made that the second derivative at the inflow and outflow boundaries did not vary over the first two (and last two) grid points. This condition was not considered too restrictive, and it was used successfully by Campbell for the stream function [83]. The assumption allowed the use of the central-difference technique with the centered point being the grid position one space removed from the boundary. One-sided differencing was used for first derivatives.

Since the velocity on the solid boundaries was zero, the Navier-Stokes equations were simplified for the walls. For the normal pressure derivative in the r direction, the following equation was used:

$$\frac{\partial P}{\partial r} = \frac{1}{Re}\frac{\partial \xi}{\partial z} \tag{61}$$

Similarly, the normal pressure derivative in the z direction at the solid boundary became

$$\frac{\partial P}{\partial z} = -\frac{1}{Re}\frac{\partial \xi}{\partial r} - \frac{\xi}{r\,Re} \tag{62}$$

Initial conditions Two types of initial conditions were used successfully. The first was an extension of the inflow boundary condition throughout the field. The inflow values of stream function were assigned to successive points downstream. In regions of smaller cross-sectional area, a linear variation of stream function between the upper and lower boundary values was determined by interpolation. The vorticity was then found from Eq. (40). The second method was to take the values from a previously converged solution at a different Reynolds number.

Once the new ξ distribution at time $k + 1$ is obtained from Eq. (40), the new ψ distribution is found from Eq. (47) by iteration, using successive overrelaxation. A relaxation factor of from 1.07 to 1.70 was used, and the equation was iterated until the change in the stream-function value at each point in the field was not greater than 10^{-5}. New velocity components were then computed from this new stream-function distribution, and the procedure began again with Eq. (40). This computational procedure continued until the change in total vorticity of the field was of the order of 10^{-3} or less. Both of these convergence criteria produced results that were constant to five significant figures using the IBM 370/158.

3.4 Model Disc Valve in Axisymmetric Aortic-shaped channel

The only difference between the problem discussed in this section and that of the previous section is writing the finite-difference derivatives for a nonuniform mesh. The equations used [i.e., Eqs. (17) and (18)], the type of boundary conditions, initial conditions, and computational procedures were the same as for the uniform-mesh problem just presented. The aortic wall shape was closely approximated using a variable mesh in both Δz and Δr. In addition, it was possible to use finer mesh spacing in areas where large gradients were expected. The nondimensional mesh spacing varied from 0.08 near the inflow to 0.01 in the valve region, for a total of about 20,000 mesh points. The geometry studied is shown in Fig. 12b.

3.4.1 Finite-Difference Derivatives for Nonuniform Mesh

Finite-difference derivatives for a varying mesh are determined from Taylor-series expansions about points separated by different distances (Δx),

$$f_{i+1} = f_i + \Delta x_i\, f_i' + \frac{\Delta x_i{}^2}{2} f_i'' + \frac{\Delta x_i{}^3}{6} f_i''' + \cdots \tag{63}$$

and

$$f_{i-1} = f_i - \Delta x_{i-1}\, f_i' + \frac{\Delta x_{i-1}^2}{2} f_i'' - \frac{\Delta x_{i-1}^3}{6} f_i''' + \cdots \tag{64}$$

The first derivative can be obtained by multiplying Eq. (63) by $(\Delta x_{i-1})^2$ and Eq. (64) by $-(\Delta x_i)^2$, adding the resultant equations, and truncating the term f''' and those of higher order.

$$f'_i = \frac{\Delta x_{i-1} f_{i+1}}{\Delta x_i(\Delta x_{i-1} + \Delta x_i)} - \frac{\Delta x_i f_{i-1}}{\Delta x_{i-1}(\Delta x_{i-1} + \Delta x_i)}$$

$$- \frac{\Delta x_{i-1} - \Delta x_i}{\Delta x_i \, \Delta x_{i-1}} f_i + O(\Delta x^2) \tag{65}$$

Equation (65) has truncation error of the order of Δx^2.

For the second derivative, terms containing f'''' must be retained in Eqs. (63) and (64). Equation (63) is multiplied by $(\Delta x_{i-1})^3$, and Eq. (64) by $(\Delta x_i)^3$. Terms are collected, and Eq. (65) is substituted for the first-derivative expressions. After rearranging and collecting terms, the second derivative is defined as

$$f''_i = \frac{2f_{i+1}}{\Delta x_i(\Delta x_{i-1} + \Delta x_i)} + \frac{2f_{i-1}}{\Delta x_{i-1}(\Delta x_{i-1} + \Delta x_i)} - \frac{2f_i}{\Delta x_i \, \Delta x_{i-1}} \tag{66}$$

To obtain Eq. (66), the following terms must be truncated:

$$- \frac{(\Delta x_{i-1} - \Delta x_i)f'''_i}{3} - \frac{\Delta x_i \, \Delta x_{i-1} \, f''''_i}{24} \cdots$$

It is obvious that Eq. (66) may not be accurate to the order of $(\Delta x)^2$. However, if the grid spacing is designed such that $\Delta x_{i-1} - \Delta x_i$ is of the same order of magnitude as $\Delta x_i \, \Delta x_{i-1}$, truncation at this point gives an accuracy within the order of $(\Delta x)^2$. When $\Delta x_{i-1} = \Delta x_i$, Eqs. (65) and (66) reduce to the equation for a uniform mesh. Equations (65) and (66) were used successfully by Fanning [29]. Roache [18] derived another expression, Eq. (67), for the first derivative:

$$\frac{\partial f}{\partial x_i} = \frac{f_{i+1} - f_{i-1}}{\Delta x_i + \Delta x_{i-1}} + O(\Delta x^2) \tag{67}$$

Both derivatives were tested, but Eq. (65) was found to give a better approximation of the derivative. In tests of polynomial functions, Eq. (65) gave one more order of accuracy than did Eq. (67) in some regions of a varying mesh.

3.4.2 Finite-Difference Equations for a Nonuniform Grid

Equations (17) and (18) are written in finite-difference form for a varying grid in the same manner as Eqs. (40) and (46). Since the varying grid necessitates additional terms and some additional complexity, these equations are broken into like groups and presented in terms of these groups.

The advection terms are written in the first upwind differencing form.

$$\text{ADQR} = \frac{v_{r_{i,j}}\xi_{i,j} - v_{r_{i,j-1}}\xi_{i,j-1}}{\Delta r_{j-1}} \tag{68}$$

$$\text{ADQZ} = \frac{v_{z_{i,j}}\xi_{i,j} - v_{z_{i-1,j}}\xi_{i-1,j}}{\Delta z_{i-1}} \tag{69}$$

$$\text{ADVQ} = \text{ADQR} + \text{ADQZ} \tag{70}$$

ADQR and ADQZ represent these individual terms for the r and z directions. ADVQ is the total advection expression.

The components of the diffusion terms are denoted with a leading letter Q. The second letter represents the coordinate direction, while the last two symbols denote the location of the grid point. For example, QZP1 expresses the value of vorticity at the point $i + 1$ (in the z direction).

$$\text{QZM1} = \frac{\xi_{i-1,j}}{\Delta z_{i-1}(\Delta z_{i-1} + \Delta z_i)} \tag{71}$$

$$\text{QZP1} = \frac{\xi_{i+1,j}}{\Delta z_i(\Delta z_{i-1} + \Delta z_i)} \tag{72}$$

$$\text{QZI} = \frac{\xi_{i,j}}{\Delta z_i \, \Delta z_{i-1}} \tag{73}$$

$$\text{QRM1} = \frac{\xi_{i,j-1}}{\Delta r_{j-1}(\Delta r_{j-1} + \Delta r_j)} \tag{74}$$

$$\text{QRP1} = \frac{\xi_{i,j+1}}{\Delta r_j(\Delta r_{j-1} + \Delta r_j)} \tag{75}$$

$$\text{QRI} = \frac{\xi_{i,j}}{\Delta r_j \, \Delta r_{j-1}} \tag{76}$$

The diffusion terms are represented by the "DUM" series:

$$\text{DUM1} = 2(\text{QZP1} + \text{QZM1} - \text{QZI}) \tag{77}$$

$$\text{DUM2} = 2(\text{QRP1} + \text{QRM1} - \text{QRI}) \tag{78}$$

$$\text{DUM3} = \frac{\xi_{i,j+1} - \xi_{i,j-1}}{r_j(\Delta r_{j-1} + \Delta r_j)} \tag{79}$$

$$\text{DUM4} = \frac{\xi_{i,j}}{r_j^2} \tag{80}$$

The DUM series adds and/or subtracts the Q terms to form the full diffusion expression of the vorticity transport equation. For example, DUM1 represents the second derivative of vorticity with respect to z.
Finally, using this coding, Eq. (17) is written in finite-difference form.

$$\xi_{i,j}^{k+1} = \xi_{i,j}^{k} - \Delta t \, \text{ADVQ}$$

$$+ \frac{\Delta t}{\text{Re}_R}(\text{DUM1} + \text{DUM2} + \text{DUM3} - \text{DUM4}) \tag{81}$$

Since Eq. (18) must be more drastically rearranged for computation than Eq. (17), the terms in finite-difference form are not as obviously discernible. AX and AY, as defined below, are consequences of the grid variation.

$$\text{AX} = \Delta z_{i-1} + \Delta x_i \tag{82}$$

$$\text{AY} = \Delta r_{i-1} + \Delta r_i \tag{83}$$

TERM1 is part of the second derivative of stream function with respect to the z direction, while TERM2 and TERM3 make up the first and second derivatives for the r direction.

$$\text{TERM1} = \frac{\psi_{i+1,j}}{\Delta z_i \, \text{AX}} + \frac{\psi_{i-1,j}}{\Delta z_{i-1} \, \text{AX}} \tag{84}$$

$$\text{TERM2} = \frac{\psi_{i,j+1}(1 - \Delta r_{j-1}/2r_j)}{\Delta r_j \, \text{AY}} \tag{85}$$

$$\text{TERM3} = \frac{\psi_{i,j-1}(1 + \Delta r_j/2r_j)}{\Delta r_{j-1} \, \text{AY}} \tag{86}$$

DN, shown below, is a consequence of separating the stream-function value at an interior point from those values at nearby points. GLOB, as shown below, represents the new value of stream function at a point as a function of the neighboring values of stream function and the new value of vorticity.

$$DN = \frac{1}{\Delta z_i \, \Delta z_{i-1}} + \frac{1}{\Delta r_j \, \Delta r_{j-1}} - \frac{\Delta r_{j-1} - \Delta r_j}{2r_j \, \Delta r_j \, \Delta r_{j-1}} \qquad (87)$$

$$GLOB = \frac{TERM1 + TERM2 + TERM3 + \xi_i r_j/2}{DN} \qquad (88)$$

Finally, the expression for stream function is written so that successive overrelaxation may be used.

$$SNEW = \psi_{i,\,j}(1 - RF) + RF(GLOB) \qquad (89)$$

3.5 Discussion of Results

For the model disc valve in the constant-diameter tube, stream-function, vorticity, and stress solutions were obtained for Reynolds numbers $Re_D = DU_{ave}/\nu$ from 20 to 1300. Preliminary pressure solutions were also obtained using the method of Fanning and Mueller [82]. Some difficulty was experienced in satisfying the auxiliary condition for this problem. Further numerical experimentation is being performed on this difficult problem of pressure extraction. Stream-function plots for both the short and extended outflow cases are shown in Figs. 13 and 14. The extent of the separated-flow regions downstream of the occluder as well as upstream and downstream of

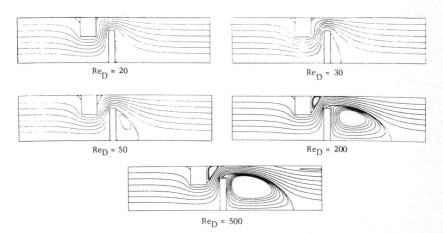

FIG. 13 Stream-function plots for axisymmetric straight-tube geometry with model disc prosthetic heart valve.

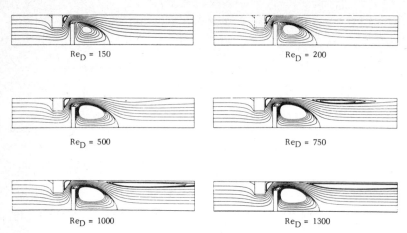

FIG. 14 Stream-function plots for axisymmetric straight-tube geometry with model disc prosthetic heart valve—extended outflow.

the sewing ring are clearly visible in these figures. The location of the separation point on the disc was different for the first and second upwind differencing methods when the Reynolds number was greater than about 100. With the first upwind differencing technique for advection terms, the separation for $Re_D > 100$ remained essentially at the downstream corner of the disc. With the more accurate second upwind method, the separation point moved around this corner to the top surface of the disc, as shown in Figs. 13 and 14 for $Re_D \geq 150$. The experimental location of separation for the higher Reynolds numbers agreed with those obtained with the second upwind differencing method.

At higher Reynolds numbers, the advancing separated-flow region downstream of the disc as calculated with the second upwind method begins to retreat (see Fig. 15). This phenomenon first occurs at Reynolds number 750 and continues at Reynolds numbers 1000 and 1300. The recirculation region along the tube grows with increasing Reynolds number, and this condition causes the separated region along the centerline to shrink accordingly. A case for the first upwind differencing method for Reynolds number 1000 was run for comparison. Unlike the second upwind difference result, the growth of the recirculation region at the higher Reynolds number continued for the first upwind method rather than receding as indicated in Fig. 15. The same phenomenon that caused the flow to detach from the disc at different locations provides this second large variance in results. The stronger upwind effect of the first method sends the closure point of the wake further downstream, and it prevents the separation along the tube from bulging toward the centerline. The appearance of a separated region on the tube wall and subsequent growth with increasing Reynolds number are also evident from the stream-function plots of Figs. 13 and 14. The extent of the separated regions, upstream and downstream of the sewing ring, is shown in Fig. 16.

A large number of flow-visualization experiments using dye injection as well as small nylon spheres have been performed in order to corroborate the numerical results [78,85]. A comparison of the experimental (Fig. 17) and numerical results for the length of the separated region behind the disc is shown in Fig. 15. The overall flow patterns, as well as the occurrence, location, and extent of separated flow regions, obtained numerically agreed very well with the experimental results for the laminar-flow cases. In fact the numerical results were found to be very helpful in positioning the hot-wire anemometer probes in later physical experiments [85].

Examples of composite numerical solutions for the straight-tube geometry for $Re_D = 500$ and 1000 are shown in Figs. 18 and 19, respectively. Forstrom [57] has found that erythrocytes are damaged in flows with no walls present with shear stress in excess of 50,000 dyn/cm². Nevaril et al. [49] reported that shear stresses on the order of 3000 dyn/cm² with walls present are hemolytic. In particular, Nevaril reported that shear stresses between 1500 and 2600 dyn/cm² could damage erythrocytes. The maximum shear stress, in this numerical work, occurred on the upstream corner of the disc, and for Reynolds number 1300 in the straight-tube geometry had a value of $\tau_{zr} = 91.2$ dyn/cm². This value is too low to cause either damage or hemolysis. However, the trend of values from Reynolds number 500

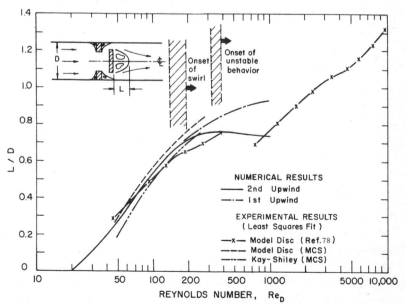

FIG. 15 Comparison of numerical results with least-square fit or experimental data.

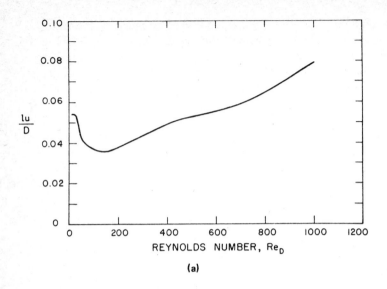

(a)

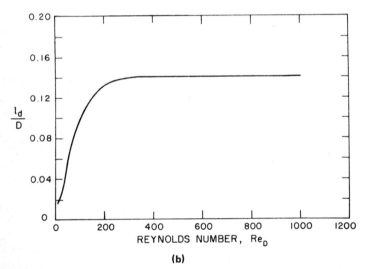

(b)

FIG. 16 Extent of separated flow in vicinity of model disc valve sewing ring in constant-diameter tube geometry. (a) Upstream of sewing ring; (b) downstream of sewing ring.

$(\tau_{zr_{max}} = 33 \text{ dyn/cm}^2)$ to Reynolds number 1300 would indicate that at the maximum Reynolds number found in the aortic region of the body (about $Re_D = 6000$), a shear stress that might damage blood could possibly occur. Furthermore, it is likely that turbulent flow would occur at $Re_D = 6000$ because of the sewing ring and occluder, and turbulent shear stresses are known to be much greater (i.e., approximately 10^2 times greater) than those

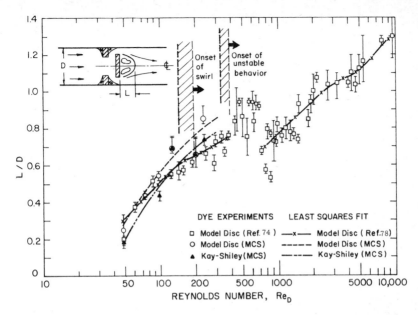

FIG. 17 Experimental data for the length of the disc wake versus Reynolds number for the constant-diameter geometry.

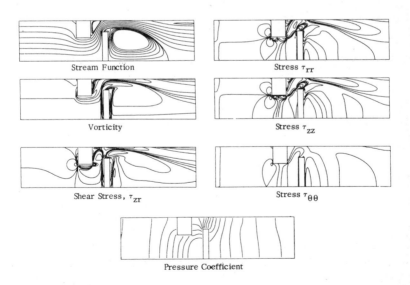

FIG. 18 Stream function, vorticity, stress, and pressure plots for axisymmetric straight-tube geometry with model disc prosthetic heart valve and $Re_D = 500$.

133

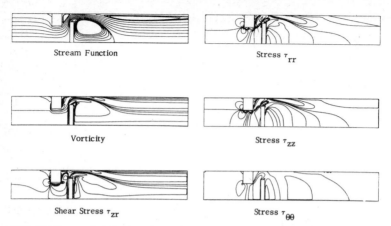

FIG. 19   Stream function, vorticity, and stress plots for the axisymmetric straight-tube geometry with the model disc prosthetic heart valve and $Re_D = 1000$.

in laminar flow. Such turbulent shear stresses could damage and/or hemolyze blood in the presence of walls. A preliminary comparison of numerical and experimental values of the shear stress for $Re_D = 150$ is presented in Fig. 20. The experiments were performed by photographically recording the motion in two perpendicular planes of small neutrally buoyant nylon spheres as a function of time [78]. While the comparison does show several large discrepancies, the results are encouraging. Many of the discrepancies may be traced to the fact that the finite size spheres (1.58 mm) are not able to follow the streamline pattern exactly in some areas. Multisensor hot-wire

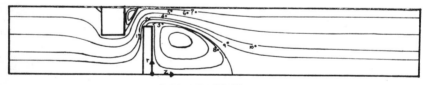

PRELIMINARY DATA

Point	Position (mm)		τ_{rr} (dynes/cm^2)		τ_{zz} (dynes/cm^2)		τ_{zr} (dynes/cm^2)	
	r	z	Exp.	Num.	Exp.	Num.	Exp.	Num.
1	7.0	-3.0	-53.4	9.217	-58.6	-15.883	-62.4	-14.5
2	8.7	-2.3	-11.1	0.0	10.4	-28.3	-12.2	-80.4
3	9.0	+2.0	- 2.0	-4.6	3.85	4.5	1.75	3.3
4	10.3	2.6	- 3.0	-2.5	17.8	3.0	4.0	8.6
5	11.0	3.4	0.82	2.8	- 7.0	- 2.4	- 2.8	- 1.2
6	11.1	6.4	5.8	4.1	-11.7	- 3.7	-10.5	- 5.5
7	11.3	7.7	10.8	3.8	- 4.9	- 3.5	- 7.4	- 7.4
8	4.2	12.3	- 1.3	-1.2	6.0	2.4	2.3	2.2
9	5.9	13.5	———		7.2	1.6	13.2	3.0
10	5.1	18.3	- 1.3	- .30	- 5.5	0.9	2.5	1.8

FIG. 20   Comparison of experimental (Exp.) and numerical (Num.) values of stress for the straight-tube geometry with the model disc valve for $Re_D = 150$.

and hot-film anemometer studies are currently in progress [78]. These experiments will produce more accurate velocity and stress data over a wide range of Reynolds numbers.

Because the shear stress at the wall appears to be related to the atheromatous lesions in humans, the wall-stress distribution for the present case is presented in Figs. 21–23. Although experimental verification of these wall shear stress distributions may be accomplished in a number of ways, electrochemical techniques (e.g., [86]) appear to be an excellent method of obtaining wall shear stress data of this type.

The same type of numerical and experimental results were obtained for the steady flow through the model disc valve in the axisymmetric aortic-shaped channel shown in Fig. 12*b* as for the straight-tube geometry. Numerical solutions for stream function, vorticity, and shear stresses were obtained for Reynolds numbers Re_D from 50 to 300. In this case $Re_D = DU_{ave}/v$, where D is the tissue annulus diameter with no valve present. The stream-function plots indicating the overall flow patterns for $Re_D = 50, 100, 200,$ and 300 are presented in Fig. 24. The extent of the separated flow downstream of the occluder and downstream of the sewing ring is evident in this figure. It is also evident that there is no measurable separated region upstream of the sewing ring, as there is in the straight-geometry case. The separation region on the channel wall just downstream of the axisymmetric sinus of valsalva is also present at $Re_D = 200$ and 300.

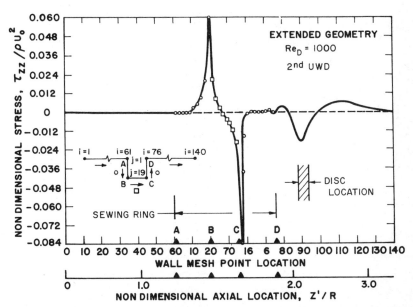

FIG. 21 Numerical wall-stress distribution for the constant-diameter geometry with model disc valve, $\tau_{zz}/\rho U_0^2$.

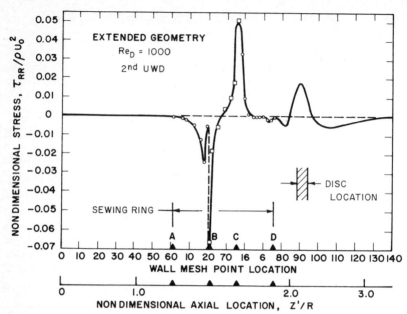

FIG. 22 Numerical wall-stress distribution for the constant-diameter geometry with model disc valve, $\tau_{RR}/\rho U_0^2$.

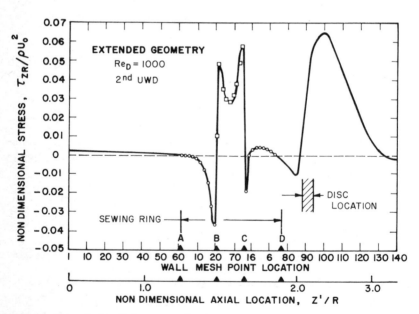

FIG. 23 Numerical wall shear stress for the constant-diameter geometry with model disc valve, $\tau_{ZR}/\rho U_0^2$.

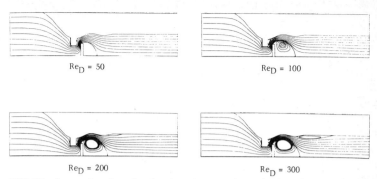

$\mathrm{Re}_D = 50$ $\mathrm{Re}_D = 100$

$\mathrm{Re}_D = 200$ $\mathrm{Re}_D = 300$

FIG. 24   Stream-function plots for the axisymmetric aortic-shaped geometry with model disc heart valve.

A comparison of the numerical and experimental length of the wake behind the disc is shown in Fig. 25a. Very good agreement is obtained for the Reynolds numbers shown for the model disc valve. The numerically determined length of the separated region downstream of the sewing ring is shown in Fig. 25b. As in the case of the straight-tube geometry, the separated region increases in length up to approximately $\mathrm{Re}_D = 200$ and then remains constant as the Reynolds number continues to increase.

For $\mathrm{Re}_D = 200$ and 300, the stream function, vorticity, and shear stress τ_{zr} are presented in Fig. 26. The maximum shear stresses for Reynolds numbers 200 and 300 are 10.5 and 16.5 dyn/cm^2, respectively, for the aortic-shaped geometry and 11.0 and 17.9 dyn/cm^2, respectively, for the straight-tube geometry. It is not surprising that the sinus of the curved geometry allows the fluid to flow through the valve with a smaller shear stress than that obtained with the straight tube.

3.6   Conclusions

The numerical results for the length of separation downstream of the disc in the constant-diameter tube compare very well with experiment through Reynolds number 500. Although the results of the first upwind difference technique diverge from the second upwind difference results, both are within the scatter of experimental data. Since visual experiments show that the flow separates on top of the disc (in agreement with the second method), and since the second method predicts the drop in L/D at high Reynolds number (in agreement with experiments), the second method must be considered to give a better solution.

Limiting cell Reynolds number to 1 or 2 is unnecessary. Excellent results are obtained for cell Reynolds numbers in excess of 7 ($\mathrm{Re}_D = 300$), and as just mentioned *either* differencing technique gives results within the scatter of experimental data for cell Reynolds numbers in excess of 10 ($\mathrm{Re}_D = 500$).

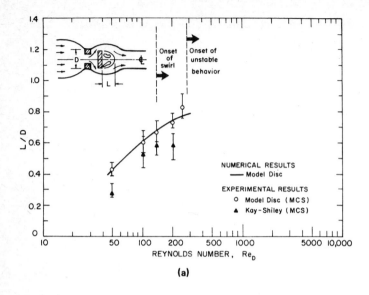

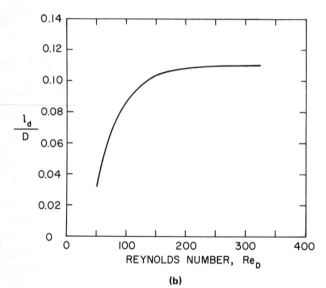

FIG. 25 Extent of separated flow in vicinity of model disc valve in axisymmetric aortic geometry. (a) Behind disc; (b) downstream of sewing ring.

In fact, useful results are obtained for cell Reynolds numbers as high as 20 ($Re_D = 1000$).

The difference in the flow fields for the curved geometry and the straight-tube geometry is slight. Shear stresses were checked throughout both the square and curved fields, and the differences in stress values were within a couple of dynes per square centimeter. This fact is encouraging, since it assures us that simpler approaches often do give useful results. This is particularly valuable since the computer time and storage required for the straight-tube geometry are much smaller than those required for the curved geometry. Most information can be obtained in less time and for a smaller cost using the cruder model.

The shear stresses predicted in the present numerical analysis are not high enough to cause hemolysis. However, the range of Reynolds numbers through an actual prosthetic heart valve is somewhat greater than the range in this study. Damage may occur at the peak Reynolds number *in vivo*. In fact, after the flow becomes turbulent the stresses will increase greatly. The numerical analysis predicts shear stress values near zero on the wall of the artery. Low shear on the surface of blood vessels has been associated with atherosclerosis. As with the pressure solution, numerical study can help choose the valve that does the least damage in the lower Reynolds-number range.

Thrombogenesis is the biggest problem for artificial valves, and this study shows definite potential for thrombogenesis. Separated flows have been linked to thrombus formation, and the flow field about the prosthetic device has as many as four separated flow regions. The numerical analysis predicts recirculation near the prosthesis, where the blood can remain in contact with a foreign surface. For high Reynolds number, the flow develops a separated region along the blood vessel where thromboplastin can be released to enhance clotting. The disc valve's potential for danger is evident. Numerical analysis can be used to expose the various danger points in different designs

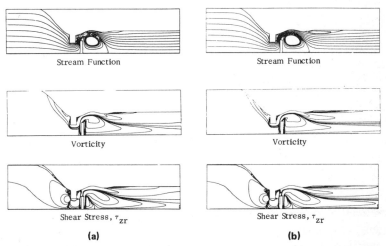

FIG. 26 Stream function, vorticity, and shear-stress plots for the axisymmetric aortic-shaped geometry with the model disc prosthetic heart valve. (a) $Re_D = 200$; (b) $Re_D = 300$.

of prosthetic devices, and an optimum design in terms of fluid mechanics can be chosen. This type of analysis can also be useful in the design of *in vitro* evaluation apparatus.

Interest in the numerical determination of heart-valve flows is very recent, and the results therefore are of a preliminary nature. However, using the numerical techniques developed in this investigation, it is now possible to proceed with a detailed study of the influence of prosthetic heart-valve geometry (i.e., size and shape) on the extent of flow separation and the magnitude of the shear stresses.

4   LOCAL VESSEL CONSTRICTIONS

4.1   Description of the Physiological Problem

Arteries may become locally constricted by the deposition of intravascular plaques, which grow inward from the wall, as shown in Figs. 4*b* and *d*. This plaque begins to form at an early age and continues as the body grows older. Once the local constriction or stenosis becomes large enough to produce a separated-flow region, further and possibly more rapid growth of the stenosis could be induced by this flow pattern. The possibility of thrombus formation in such a slowly recirculating flow region also exists. The weakening and bulging of the artery downstream from the stenosis (i.e., post-stenotic dilatation) is another major complication of this occlusive vascular disease. Local constriction of arteries may also be caused by extravascular compression by ligaments, bone spurs, or measuring instruments. A temporary stenosis is produced by the sphygmomanometer cuff that is commonly used to measure blood pressure.

Experiments with animals (e.g. [87]) indicate that the flow of blood is unaffected until the stenosis has closed off about 80% of the artery. This automatic adjustment of the peripheral vascular conductance so that the blood flow rate remains constant is called "autoregulation." Although the blood flow rate may not immediately be affected, the local velocity, stress, and pressure fields are. Fry [88] has demonstrated that the deterioration and growth of the artery lining (i.e., the endothelial cells) is related to the shear stresses present.

4.2   Description of the Numerical Problem

Analytical studies of mild stenoses have been reported as early as 1968 [89]; however, the first numerical treatment of this problem was published by Lee and Fung [31] in 1970. The method of Thom [90] was used to solve the steady Navier-Stokes equations after the axisymmetric bell-shaped constriction shown in Fig. 27*a* was conformally mapped into a rectangular region.

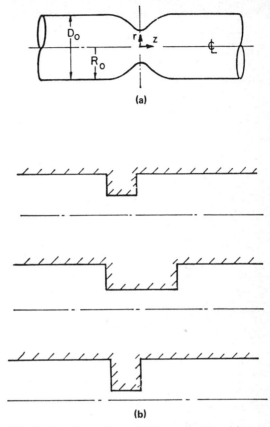

FIG. 27 Local vessel constriction geometries. (a) Bell-shaped axisymmetric constriction; (b) square and rectangular planar constrictions.

Oberkampf and Goh [23] solved the same axisymmetric bell-shaped constriction; however, they use a nonorthogonal transformation to a rectangular region and the semidiscrete method of lines (MOL). In a series of papers, Cheng et al. [34–36] studied the steady, oscillatory, and pulsatile flows past square and rectangular constrictions (Fig. 27*b*), using the explicit method of Fromm [91].

A recent study of the pulsatile flow through constricted rigid and distensible tubes was presented by Daly [92]. The constriction was a circular arc, and the calculation procedure makes use of the arbitrary Lagrangian-Eulerian method for the calculation of transient, multidimensional, viscous flows. Velocity, pressure, and shear-stress plots are presented for the rigid constructed canine femoral artery at four times in the pulsatile cycle. Preliminary velocity-vector and pressure-contour results are also presented for the pulsatile flow through a distensible tube.

4.3   Axisymmetric Bell-shaped Constriction

Lee and Fung [31] expressed the axisymmetric Navier-Stokes equations in terms of the dimensionless stream function ψ and dimensionless vorticity η for steady flow. The physical plane (R, Z) (Fig. 27a) was conformally mapped into a rectangular region, as indicated in Fig. 8e, using the procedure described in detail by Thom [90] for the axisymmetric constriction specified by the equation

$$R = 1 - \tfrac{1}{2} \exp\left(-4Z^2\right) \tag{90}$$

R and Z are dimensionless coordinates, that is, $R = r/(D_0/2)$ and $Z = z/(D_0/2)$. This constriction is bell-shaped and approximately contained in the interval $|Z| \leq 1$. The mesh was uniform and equal to 0.125 for Reynolds numbers from 0 to 25. The iteration process was terminated when the changes in the absolute values of stream function and vorticity were less than 10^{-5} and 5×10^{-4}, respectively. No difficulties were encountered for $Re \leq 15$, but this iterative procedure failed to converge for $Re = 20$. Results for $Re = 25$ were obtained by using underrelaxation. An example of the results obtained for $Re = 25$ is shown in Fig. 28. The separated region downstream of the constriction is evident in the stream-function and velocity-profile plots. Distributions of wall vorticity or shear stress and total head were also obtained for the three cases studied, that is, $Re = 0$, 10, and 25. Although the numerical procedure used is not suitable for higher Reynolds numbers or unsteady flows, this work has served as a stimulus for subsequent numerical and experimental work.

A recent experimental study of the problem was presented by Bentz and Evans [94]. Velocity profiles were obtained using a laser Doppler velocimeter for exactly the same geometry constriction used by Lee and Fung. Steady-flow results were obtained for the Reynolds-number range $2 < Re < 170$. A comparison between the experimental velocity profiles and those obtained by Lee and Fung is shown in Figs. 29 and 30. Considering the inherent difficulties in exactly duplicating the numerical conditions in the experiments, the velocity profiles for $Re \sim 2$ and 26 show good agreement. Experience with this type of problem would indicate that more time, energy, and money were expended in the experiments than in the numerical study.

Oberkampf and Goh [23] also studied the axisymmetric bell-shaped constriction described by Eq. (90). They used the explicit method of lines, which is capable of extension to include time-dependent boundaries. The physical plane (Fig. 27a) was mapped into a rectangular region using the mapping functions $\bar{R} = R/R(Z)$, $\bar{Z} = \tanh(\alpha Z)$, and $\bar{t} = t$, where $R(Z)$ is the local radius of the tube, α is the stretching factor, and t is the dimensionless time. These mapping functions transform the curved boundary $R(Z)$ into a straight line $\bar{R} = 1.0$, and the inflow/outflow boundaries at $\pm\infty$ are mapped

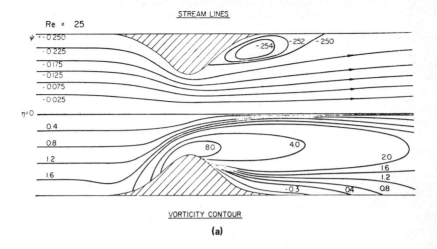

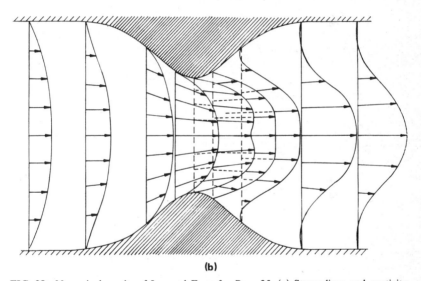

FIG. 28   Numerical results of Lee and Fung for Re = 25. (a) Streamlines and vorticity contours; (b) velocity profiles. Taken from [31].

to $\bar{Z} = \pm 1$. The conservative vorticity transport equation (21) and stream-function equation (22) in weak conservation form were transformed using the generalized transformation equations given by Oberkampf [20]. The steady-state convergence criterion used was

$$\left| \zeta_{i,j}^{n+1} - \zeta_{i,j}^{n} \right| < \frac{10^{-4}}{IJ} \sum_{i,j=1}^{I,J} \left| \zeta_{i,j}^{n+1} \right|$$

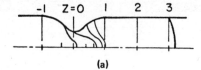

(a)

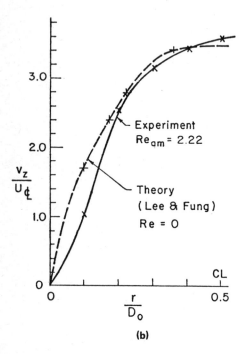

(b)

FIG. 29 Experimental and numerical velocity profiles for Reynolds number approximately equal to 2. (a) Nondimensional velocity profiles v_z/U_C; (b) comparison of experimental and numerical results for $Z = z/R_0 = 0$. Replotted from [94].

where IJ was the total number of points and n the time-step index.

Numerical results were obtained for both a single and a double constriction using MOL, with a fourth-order accurate Runge-Kutta integrator and a 21×21 uniform mesh. Streamlines and vorticity contours for the single constriction for Re = 25, and the double constriction for Re = 10, are shown in Fig. 31. While the streamline results of Fig. 31 compare favorably with the results of Lee and Fung, the authors found significant differences in the vorticity contours. Their explanation for these differences revolve around the different manner in which the downstream boundary condition was formulated. An outflow type boundary condition was used by Lee and Fung, while Oberkampf and Goh used an infinity condition, which in the transformed plane is applied at a finite location. The finite-different method (MOL) used in this case is stable at high Reynolds numbers.

4.4 Planar Rectangular Constriction

In a continuing study, Cheng et al. [34–36] used the explicit method of Fromm and a nonconservative form of the planar vorticity transport equation and stream-function equation to obtain results for steady, oscillatory, and pulsatile flow past rectangular constrictions. The pressure gradient for the oscillatory and arterial pulsatile (see Fig. 9b) flow was introduced as a forcing function by means of Eq. (39). The mesh was uniform an one-fortieth of the channel width D, and the dimensionless time increment was 0.0001. The convergence criterion used was

$$\left| \psi_{i,j}^{n \Delta T + 0.2} - \psi^{n \Delta T} \right| < 5 \times 10^{-4}$$

where n is the time index and 0.2 is the period of one cycle. According to the authors, less than half a period was required for the whole flow field to reach

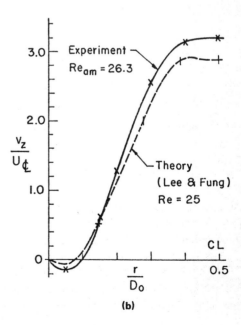

FIG. 30 Experimental and numerical velocity profiles for Reynolds number approximately equal to 26. (a) Nondimensional velocity profiles v_z/U_C; (b) comparison of experimental and numerical results for $Z = z/R_0 = 0.40$. Replotted from [94].

Vorticity Contours

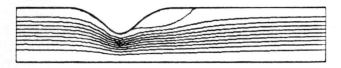

Streamlines

(a)

Vorticity Contours

Streamlines

(b)

FIG. 31 Numerical results of Oberkampf and Goh. (a) Single constriction, Re = 25; (b) double constriction, Re = 10. Taken from [23].

steady state. For each time step, the iteration process was limited by the criterion

$$|\psi_{i,j}^{k+1} - \psi_{i,j}^{k}| < \frac{|\psi_{\text{centerline}}|}{1000}$$

where k is the iteration number. If $|\psi_{\text{centerline}}|/1000 < 5 \times 10^{-4}$, then $|\psi_{i,j}^{k+1} - \psi_{i,j}^{k}| < 5 \times 10^{-4}$ was used. Examples of the type of results obtained for the oscillating flow near the square constriction [34] are shown in Figs. 32 and 33 for the Karman number $K = 1000$ and the Stokes number $S = \beta = 10\pi$. (Other β's were also studied in reference [35].) Figure 32a shows the path of the center of the principal vortex as it moves on the right side of the square constriction, while Fig. 32b illustrates the longitudinal pressure variations along the centerline at various times; the shear-stress and vorticity profile at three times are shown in Fig. 33.

A summary and comparison of numerical results for steady, oscillatory, and pulsatile flow past a square constriction are presented in reference [36].

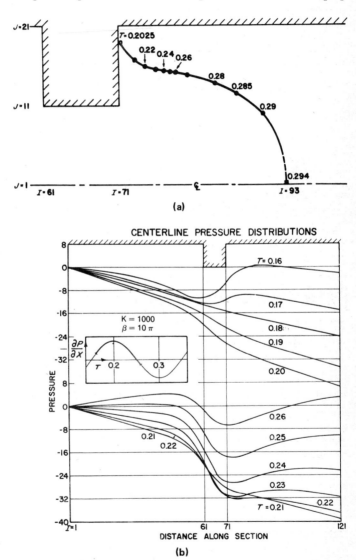

(a)

CENTERLINE PRESSURE DISTRIBUTIONS

(b)

FIG. 32 Numerical results for the oscillatory flow through a planar square constriction. (a) Vortex center at various times; (b) pressure distributions along centerline at various times. Taken from [34].

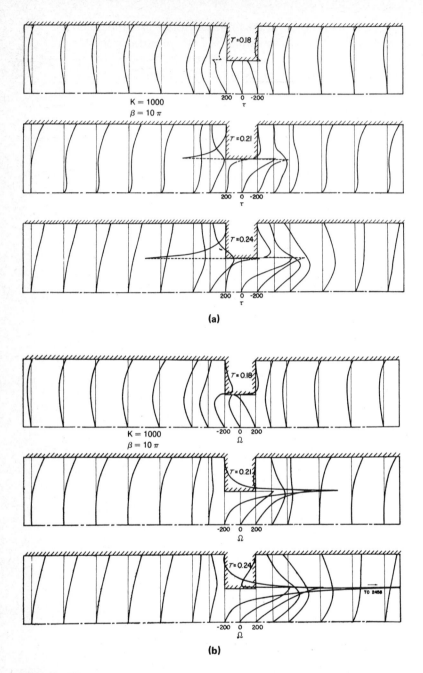

FIG. 33 Shear-stress and vorticity profiles at various times for oscillatory flow through a planar square constriction. (a) Shear-stress profiles; (b) vorticity profiles. Taken from [34].

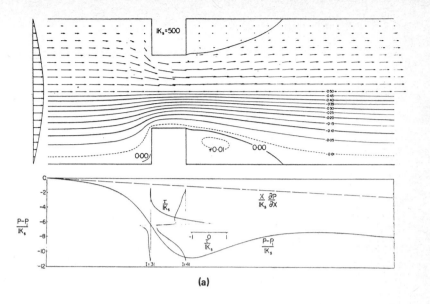

(a)

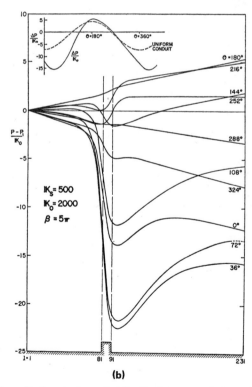

(b)

FIG. 34 Numerical results for steady and pulsatile flow through a planar square constriction. (a) Flow pattern and centerline pressure for steady flow; (b) centerline pressure at various times. Taken from [36].

Examples of these results are illustrated in Fig. 34. Detailed information of the type shown in Figs. 32–34 clearly indicates the extent of flow separation, the location and magnitude of shear-stress maxima, and other characteristics of this type of flow.

4.5 Conclusions

These sample results demonstrate the potential in studying complex steady and unsteady separated flows numerically. Obtaining this type of detailed information solely from physical experiments would be a far more time-consuming and costly task. The favorable comparison of the steady-flow numerical results of Lee and Fung with recent experiments adds a great deal of confidence concerning the absolute accuracy of the numerical methods used.

5 CONCLUDING REMARKS

From this study of complex physiological flows using standard numerical techniques, it is apparent that the numerical approach has important advantages that are just beginning to be exploited. The inherent flexibility in studying planar and axisymmetric, and steady and unsteady, flows numerically has added a great deal to our understanding of several physiological flow problems. The favorable comparison of numerical and experimental results indicates that the absolute accuracy of these techniques is at least as good as that of experimental techniques. Furthermore, they allow the study of complex unsteady separated-flow phenomena, which are more difficult to produce and study in physical experiments. The application of numerical methods to physiological flows is still in the development stage, and improvements can be expected. However, it is also evident that both numerical experiments of this type and physical experiments are needed if we are to arrive at a satisfactory level of understanding in the shortest possible time.

REFERENCES

1 Cokelet, G. R.: The Rheology of Human Blood, in Y. C. Fung, N. Perrone, and M. Anliker (eds.), " Biomechanics," pp. 63–103, Prentice-Hall, Englewood Cliffs, N.J., 1972.
2 Chaffee, E. E., and E. M. Greisheimer: "Basic Physiology and Anatomy," Lippincott, Philadelphia, 1964.
3 Lightfoot, E. N.: "Transport Phenomena and Living Systems," Wiley, New York, 1974.
4 Charm, S. E., and G. S. Kurland: " Blood Flow and Microcirculation," Wiley, New York, 1974.
5 Talbot, L., and S. A. Berger: Fluid-Mechanical Aspects of the Human Circulation, *Am. Sci.*, **62**(6): 671–682, 1974.
6 Ganong, W. F.: " Review of Medical Physiology," 5th ed., Lange Medical Publications, Los Altos, Calif., 1971.

7   Walburn, F. J., and D. J. Schneck: A Constitutive Equation for Whole Human Blood. ASME Paper No. 75-WA/BIO-12, 1975.

8   Surgenor, D. MacN.: "The Red Blood Cell," 2d ed., parts I and II, Academic, New York, 1975.

9   Glassman, E., M.D., and G. S. Roberts, M.D.: Hemolysis after Heterograft and Prosthetic Valve Replacement. *Am. Heart J.* **79**(2): 281, 1970.

10   Andersen, M. N., M.D., E. Gabrieli, M.D. (by invitation), and J. A. Zizzi, M.D. (by invitation): Chronic Hemolysis in Patients with Ball-Valve Prostheses, *J. Thorac. Cardiovasc. Surg.*, **50**(4): 501, 1965.

11   Johnson, S. A., and M. M. Guest (eds.): "Dynamics of Formation and Dissolution," Lippincott, Philadelphia, 1969.

12   Caro, C. G., and R. M. Nerem: Transport of ^{14}C-4-Cholesterol Between Serum and Wall in the Perfused Dog Common Carotid Artery, *Circ. Res.*, **32**: 187–205, 1973.

13   Lutz, R. J., J. N. Cannon, J. E. Fletcher, and D. L. Fry: The Measurement of Wall Shear Stress in Model Arteries by an Electrochemical Technique, *Proc. 27th ACEMB*, Philadelphia, Pa., October 6–10, 1974, p. 279.

14   Lighthill, M. J.: "Mathematical Biofluiddynamics," Regional Conference Series in Applied Mathematics, SIAM, Philadelphia, 1975.

15   Piacsek, S. A.: Some Problems in the Numerical Solution of Three-dimensional Incompressible Fluid Flows, *Proc. 7th Symp. Naval Hydrodyn.*, Rome Italy, August 25–30, 1968. ONR, Departments of the Navy, DR-148, pp. 1616–1618.

16   Roache, P. J.: Recent Developments and Problem Areas in Computational Fluid Dynamics, Lecture Notes in Mathematics 461, in " Computational Mechanics," pp. 195–256, Springer, Berlin, 1975.

17   Staff of Langley Research Center: Numerical Studies of Incompressible Viscous Flow in a Driven Cavity, NASA SP-378, 1975.

18   Roache, P. J.: "Computational Fluid Dynamics," Hermosa, Albuquerque, N.M., 1972.

19   Richtmyer, R. D., and K. W. Morton: "Difference Methods for Initial-Value Problems," 2d ed., Interscience, New York, 1957.

20   Oberkampf, W. L.: Domain Mappings for the Numerical Solution of Partial Differential Equations, *Int. J. Numer. Methods Eng.* **10**: 211–223, 1976.

21   Oberkampf, W. L.: Private communication, 1976.

22   Vinokur, M.: Conservation Equations of Gasdynamics in Curvilinear Coordinate Systems. *J. Comput. Phys.*, **14**: 105–125, 1974.

23   Oberkampf, W. L., and S. C. Goh: Numerical Solutions of Incompressible Viscous Flow in Irregular Tubes, *Proc. Int. Conf. Comput. Methods Nonlinear Mech.*, University of Texas, Austin, 1974, pp. 569–579.

24   Cheng, S. I.: A Critical Review of Numerical Solution of Navier-Stokes Equations, Lecture Notes in Physics 41, in " Progress in Numerical Fluid Dynamics," pp. 79–225, Springer, Berlin, 1975.

25   Piacsek, S. A., and G. Williams: Conservation Properties of Convection Difference Schemes, *J. Comput. Phys.*, **6**: 393–405, 1970.

26   Roache, P. J.: Private communication, 1976.

27   Mueller, T. J.: On the Fluid Dynamics of Prosthetic Heart Valve Flow—A Preliminary Numerical and Experimental Study, von Karman Institute for Fluid Dynamics, technical note 101, June, 1974.

28   Underwood, F. N., Jr.: "A Numerical Study of the Steady, Axisymmetric Flow through a Disc-Type Prosthetic Heart Valve," Ph.D. dissertation, Department of Aerospace and Mechanical Engineering, University of Notre Dame, Notre Dame, Ind., 1975.

29   Fanning, A. E., and T. J. Mueller: Numerical and Experimental Investigation of the Oscillating Wake of a Blunt-Based Body. *AIAA J.*, **11**(11): 1486–1491, November, 1973.

30   Gillani, N. V.: "Time-Dependent Laminar Incompressible Flow through Spherical Cavity," Ph.D. dissertation, Washington University, St. Louis, Mo., 1974.

31   Lee, J. S., and Y-C. Fung: Flow in Locally Constricted Tubes at Low Reynolds Numbers, *ASME J. Appl. Mech.*, **E37**(1): 9–16, 1970.

32   Thompson, J. F., F. C. Thames, and C. W. Mastin: Automatic Numerical Generation of Body-Fitted Curvilinear Coordinate System for Fields Containing Any Number of Arbitrary Two-Dimensional Bodies, *J. Comput. Phys.*, **15**(3): 299–319, 1974.

33   Thompson, J. F., et al.: Use of Numerically Generated Body-Fitted Coordinate Systems for Solution of the Navier-Stokes Equations, *Proc. AIAA 2d Comput. Fluid Dyn. Conf.*, Hartford, Conn, June 19–20, 1975, pp. 68–80.

34   Cheng, L. N., M. E. Clark, and J. M. Robertson: Numerical Calculations of Oscillating Flow in the Vicinity of Square Wall Obstacles in Plane Conduits, *J. Biomech.*, **5**: 467–484, 1972.

35   Cheng, L. C., J. M. Robertson, and M. E. Clark: Numerical Calculations of Plane Oscillatory Non-Uniform Flow: II, Parametric Study of Pressure Gradient and Frequency with Square Wall Obstacles, *J. Biomech.* **6**: 521–538, 1973.

36   Cheng, L. C., J. M. Robertson, and M. E. Clark: Calculation of Plane Pulsatile Flow past Wall Obstacles, *Comput. Fluids*, **2**: 363–380, 1974.

37   Coder, D. W., and F. T. Buckley, Jr.: Implicit Solutions of the Unsteady Navier-Stokes Equation for Laminar Flow through an Orifice within a Pipe, *Comput. Fluids*, **2**: 295–315, 1974.

38   Roache, P. J.: Finite Difference Methods for the Steady-State Navier-Stokes Equations, "Lecture Notes in Physics," vol. 18, pp. 138–145, Springer, New York, 1973.

39   Roache, P. J.: A Review of Numerical Techniques, *Proc. First Conf. Numer. Ship Hydrodyn.*, NSRDC, Carter Rock, Md., October 20–22, 1975.

40   Morihara, H., and R. T. Cheng: Numerical Solution of the Viscous Flow in the Entrance Region of Parallel Plates, *J. Comput. Phys.*, **11**: 550–572, 1973.

41   Lomax, H., and E. D. Martin: Fast Direct Numerical Solution of the Nonhomogeneous Cauchy-Riemann Equations, *J. Comput. Phys.*, **15**(1): 55–80, 1974.

42   Sator, F. G.: Application of the Finite Element Method to the Calculation of Plane Pulsatile Flow through Artificial Heart Valves, *GAMM Conf. Numer. Methods Fluid Mech.*, October 8–10, 1975, pp. 173–180.

43   Robertson, J. M., and J. F. Herrick: Turbulence in Blood Flow?, T. & A.M. report No. 401, Department of Theoretical and Applied Mechanics, University of Illinois, Urbana, Illinois, April, 1975.

44   McGoon, D. C.: On Evaluating Valves, *Mayo Clin. Proc.* **49**: 233–235, 1974.

45   Harken, D. E.: Heart Valves, Values and Vacuums, *Med. Instrum.*, **7**: 276, November–December, 1973.

46   Yacoub, M. H., and D. H. Keeling: Chronic Haemolysis following Insertion of Ball Valve Prostheses, *Br. Heart J.*, **30**: 676–678, 1968.

47   Silver, M. D.: Cardiac Pathology—A Look at the Last Five Years: Part II, The Pathology of Cardiovascular Prostheses, *Hum. Pathol.* **5**: 127–128, March, 1974.

48   Wallace, H. W., D. L. Kenepp, and W. S. Blakemore: Quantitation of Red Blood Cell Destruction Associated with Valvular Disease and Prosthetic Valves, *J. Thorac. Cardiovasc. Surg.*, **60**: 843–852, December, 1970.

49   Nevaril, C. G., E. C. Lynch, C. P. Alfrey, Jr., and J. D. Hellums: Erythrocyte Damage and Destruction Induced by Shearing Stress. *J. Lab. Clin. Med.*, **71**: 784–790, May, 1968.

50   Nuter-Hauge, K., K. V. Hall, T. Froysaker, and L. Efskind: Aortic Valve Replacement One-Year Results with Lilleher-Kaster and Bjork-Shiley Disc Prosthesis, *Am. Heart J.*, **88**, July, 1974.

51   Fernandez, J., D. Morse, V. Maranhao, and A. S. Gooch: Results of Use of the Pyrolytic Carbon Tilting Disc Bjork-Shiley Aortic Prosthesis, *Chest.* **65**: 640–645, June, 1974.

52   Ben-Zvi, J., F. J. Hildner, P. A. Chandravatna, and P. Samet: Thrombosis on Bjork-Shiley Aortic Valve Prosthesis—Clinical, Arteriographic, Echocardiographic, and Therapeutic Observations in Seven Cases, *Am. J. Cardiol.*, **34**, October 3, 1974.

53 Brais, M. P., and Braunwald, N. S.: Tissue Acceptance of Materials Implanted within the Circulatory System, *Arch. Surg.*, **109**: 351–358, September, 1974.

54 Monsler, M., W. Morton, and R. Weiss: The Fluid Mechanics of Thrombus Formation, AIAA Third Fluid and Plasma Dynamics Conference, Los Angeles, June 29–July 1, 1970, AIAA paper No. 70-787.

55. Lee, S. J. K., G. Lees, J. C. Callaghan, C. M. Couves, L. P. Sterns, and R. R. Rossall: Early and Late Complications of Single Mitral Valve Replacement Comparison of Eight Different Prostheses, *J. Thorac. Cardiovasc. Surg.*, **67**: 920–925, June, 1974.

56 Sullivan, J. M., D. E. Harken, and R. Gorlin: Effect of Dipyridamole on the Incidence of Arterial Emboli after Cardiac Valve Replacement, *Circulation*, **40** (Suppl. I): I-149–I-153, May, 1969.

57 Forstrom, R. J.: "A New Measure of Erythrocyte Membrane Strength—The Jet Fragility Test," Ph.D. thesis, University of Minnesota, Minneapolis, 1969.

58 Keshaviah, P. K., P. L. Blackshear, and R. Forstrom: Red Cell Destruction in Regions of Accelerated Flow, ASME 1975 Advances in Bioengineering, pp. 49–53, Winter Annual Meeting, Houston, Texas.

59 Knapp, C F.: "An Experimental Investigation of the Mechanics of Hemolysis in Couette Flow," Ph.D. thesis, University of Notre Dame, 1968.

60 Shapiro, E. I., and M. C. Williams: Hemolysis in Simple Shear Flows, *AIChE J.*, **16**: 575, 1970.

61 Blackshear, P. L., F. D. Dorman, J. H. Steinback, E. J. Maybach, and R. E. Collingham: Shear, Wall Interaction and Hemolysis, *Trans. Am. Soc. Artif. Int. Organs*, **12**: 113, 1966.

62 Bernstein, E. F., P. L. Blackshear, and K. H. Keller: Factors Influencing Erythrocyte Destruction in Artificial Organs, *Am. J. Surg.*, **114**: 126, 1967.

63 Lefemine, A. E., M. Miller, and G. C. Pinder: Chronic Hemolysis Produced by Cloth-Covered Valves: The Effect of Design and Valve Position. *J. Thorac. Cardiovasc. Surg.*, **67**: June, 1974.

64 Greenfield, H., and W. Kolff: The Prosthetic Heart Valve and Computer Graphics, *J. Am. Med. Assoc.*, **219**(1): 69–74, Jan. 3, 1972.

65 Underwood, F. N.: "Numerical and Experimental Studies of the Steady, Axisymmetric Flow through a Disc-Type Prosthetic Heart Valve," M.S. thesis, Department of Aerospace and Mechanical Engineering, University of Notre Dame, August, 1972.

66 Underwood, F. N., and T. J. Mueller: Numerical Studies of the Steady, Axisymmetric Flow through a Disc-Type Prosthetic Heart Valve, *Proc. 25th ACEMB*, 1972, p. 273.

67 Schuessler, G. C., and T. K. Hung: Hemodynamics during the Opening of a Heart Valve: A Computational Analysis, *Proc. 25th ACEMB*, 1972, p. 271.

68 Skaland, E., and G. Moskowitz: Plane Flow through a Prosthetic Aortic Valve, *Proc. 25th ACEMB*, 1972, p. 272.

69 Skaland, E.: "Analysis of Plane Flow through a Prosthetic Aortic Valve," Ph.D. dissertation, Drexel Institute of Technology, Philadelphia, June, 1971.

70 Peskin, C. G.: Flow Patterns around Heart Valves: A Numerical Method, *J. Comput. Phys.*, **10**: 252–271, 1972. Also "Flow Patterns and Heart Valves: A Digital Computer Method for Solving the Equations of Motion," Ph.D. thesis, Yeshiva University, New York, 1972.

71 Krause, E., and U. Giese: Theoretische Untersuchung der Strömung dürch kuntstliche Herzklapper, *Dtsch. Ges. Biomed. Tech.*, 20 (Erganzungsband): 9–10, 1975.

72 Viecelli, J. A.: A Method for Including Arbitrary External Boundaries in the MAC Incompressible Fluid Computing Technique, *J. Comput. Phys.*, **4**: 543–551, 1969.

73 Viecelli, J. A.: A Computing Method for Incompressible Flows Bounded by Moving Walls, *J. Comput. Phys.*, **8**: 119–143, 1971.

74 Mueller, T. J., J. R. Lloyd, J. L. Lower, W. T. Struble, and F. N. Underwood: On the Separated Flow Produced by a Fully Open Disc-Type Prosthetic Heart Valve, *Biomech. Symp.*, AMD, **2**: 97–98, 1973.

75   Lambert, J.: A New Experimental Set-Up for the Direct Microscopic Observation of the Flow Behavior of Blood in Models of Natural and Artificial Vessel Systems, *Biomed. Tech.*, **20**(4): 139–144, 1975.

76   Richardson, E.: Deformation and Haemolysis of Red Cells in Shear Flow, *Proc. Roy. Soc., London*, A338: 129–153, 1974.

77   Richardson, E.: Applications of a Theoretical Model for Haemolysis in Shear Flow, *J. Biorheol.*, **12**: 27–37, 1975. Also presented at Euromech 67, Aachen, West Germany, September 15–18, 1975.

78   Figliola, R. S.: Unpublished data, Department of Aerospace and Mechanical Engineering, University of Notre Dame, 1975.

79   Gentry, R. A., R. E. Martin, and B. J. Daly: An Eulerian Differencing Method for Unsteady Compressible Flow Problems, *J. Comput. Phys.*, **1**: 87–118, 1966.

80   Roache, P. J.: On Artificial Viscosity, *J. Comput. Phys.*, **10**: 169–184, 1972.

81   Fanning, A. E.: "An Analytic Method for the Calculation of Flow Fields in the Transonic Region of Nozzles," Ph.D. dissertation, Department of Aerospace and Mechanical Engineering, University of Notre Dame, May, 1973.

82   Fanning, A. E., and T. J. Mueller: On the Solution of a Neumann Problem for an Inhomogeneous Laplace Equation, *J. Comput. Phys.* **13**(3), 450–454, 1973.

83   Campbell, D. R., and T. J. Mueller: Effects of Mass Bleed on an Internal Separated Flow, *Proc. Symp. Appl. Comput. Fluid Dyn. Anal. Des.*, Polytechnic Institute of Brooklyn, New York, January 3–4, 1973.

84   Roache, P. J., and T. J. Mueller: Numerical Solutions of Laminar Separated Flows, *AIAA J.*, **8**(3): 530–538, 1970.

85   Struble, W. T.: "An Experimental Investigation of the Flow through Disc-Type Prosthetic Heart Valves," M.S. thesis, University of Notre Dame, January, 1976.

86   Lutz, R. J., J. N. Cannon, J. E. Fletcher, and D. L. Fry: The Measurement of Wall Shear Stress in Model Arteries by an Electrochemical Technique, *Proc. 27th ACEMB*, Philadelphia, October 6–10, 1974, p. 279.

87   May, A. G., J. A. De Weese, and C. G. Rob: Hemodynamic Effects of Arterial Stenosis, *Surgery*, **53**: 513–524, 1963.

88   Fry, D. L.: Acute Vascular Endothelial Changes Associated with Increased Blood Velocity Gradients, *Circ. Res.* **22**: 165–197, 1968.

89   Young, D. F.: Effect of a Time-Dependent Stenosis on Flow through a Tube, *ASME J. Eng. Ind.*, **90**: 248–254, 1968.

90   Thom, A.: The Flow past Circular Cylinder at Low Speeds, *Proc. Roy. Soc., London*, A126: 651–669, 1933.

91   Fromm, J.: The Time Dependent Flow of an Incompressible Viscous Fluid, in B. Alder, S. Fernbach, and M. Rotenberg (eds.), "Methods in Computational Physics," pp. 345–382, Academic, New York, 1964.

92   Daly, B. J.: A Numerical Study of Pulsatile Flow through Constricted Arteries, "Lecture Notes in Physics," pp. 117–124, Proceedings of the Fourth International Conference on Numerical Methods in Fluid Dynamics, Springer, New York, 1975.

93   Thom, A., and C. J. Apelt: "Field Computations in Engineering and Physics," Van Nostrand, London, 1966.

94   Bentz, J. C., and N. A. Evans: Hemodynamic Flow in the Region of a Simulated Stenosis, ASME Paper No. 75-WA/B10-10, presented at ASME, Winter Annual Meeting, Houston, November 30–December 4, 1975.

CHAPTER 3

Some Recent Progress in Transonic Flow Computations

William F. Ballhaus

1 INTRODUCTION

1.1 Description of the Flow Field

The most distinguishing feature of transonic flows is their mixed subsonic-supersonic flow character. This is illustrated schematically in Fig. 1. The acceleration of the initially subsonic flow over the front portion of the airfoil is sufficient to produce an embedded region of supersonic flow adjacent to the airfoil surface. This supersonic region is terminated by a shock wave that recompresses the flow. The strength and extent of the shock wave increase with free-stream Mach number M_∞ and angle of attack α.

Flow in the boundary (or viscous) layer on the airfoil surface is retarded both by interaction with the shock wave and by the adverse pressure gradient over the aft portion of the airfoil. The boundary layer can separate at any point between the shock and trailing edge, depending on the strength of these retarding influences and the Reynolds number. The presence of any separation can significantly affect the entire inviscid flow field near the airfoil.

Shock waves and viscous effects are of interest to the aerodynamicist because they influence the forces acting on the airfoil. The qualitative behavior of lift and drag coefficients (C_l and C_d) as functions of M_∞ is shown in Fig. 2 [1]. The critical Mach number M_{cr} is the value of M_∞ for which an embedded supersonic region first appears. As the Mach number increases beyond M_{cr}, the supersonic region grows, increasing the strength and extent of the terminating shock; C_l also increases, and C_d remains essentially constant. As M_∞ increases beyond M_d, the drag-rise Mach number, shock and viscous influences cause a rapid increase in drag and, eventually, a decrease in lift.

155

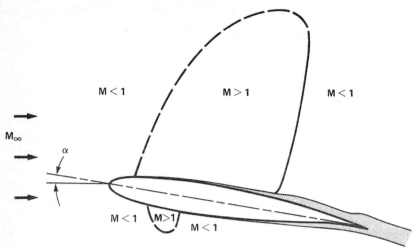

FIG. 1 Mixed flow over an airfoil.

Transonic flow fields are characterized by two complicating features in addition to mixed-flow and viscous effects; that is, they tend to be more *unsteady* and *three-dimensional* than their subsonic and supersonic counterparts. Any complete analysis of transonic flows should be capable of treating the combined unsteady, three-dimensional, viscous, mixed-flow character of such flows.

1.2   Aerodynamic Performance and Design

The Mach-number range just above M_d is known to be one of the most efficient regimes of flight, and it is this feature that makes the analysis of

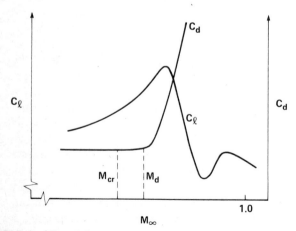

FIG. 2 Lift and drag at transonic speeds.

transonic flow fields one of the most studied problems in fluid dynamics. For example, optimum aircraft cruise performance, which is achieved at the value of M_∞ that maximizes $M_\infty C_l/C_d$, is encountered in this range. For maximum maneuverability, high C_l is important, and this is also achieved in the same range.

As M_∞ increases beyond the optimum performance range, adverse transonic effects in the form of increased drag, shock-induced separation, and buffet are encountered. Performance deteriorates drastically. The onset of these adverse effects can impose limitations on the operating Mach-number range of airplanes, helicopter rotors, propellers, inlets, and compressors.

The value of M_∞ at which these performance-limiting conditions are encountered depends strongly on configuration. One of the principal objectives of transonic configuration design is to maximize this value of M_∞ and minimize the rate of performance deterioration with M_∞ beyond this point.

Until recently, the transonic design process consisted almost entirely of cut-and-try wind-tunnel simulations. However, the high cost of transonic wind-tunnel test time and of manufacturing fully instrumented models severely limits the number of configurations that can be considered in the search for an optimum design.

1.3   The Development of a Computational Approach

Considerable attention has been directed in recent years to developing theoretical methods for the analysis and design of aerodynamic bodies to supplement wind-tunnel simulations. Such methods would reduce the amount of wind-tunnel test time required for aircraft development, thereby reducing costs while allowing the designer to consider a larger class of candidate configurations. Theoretical methods would also provide an additional data base to check wind-tunnel measurements, which can be influenced by errors due to wind-tunnel wall interference, incorrect Reynolds-number scaling, and model and sting deflections under load.

In spite of an extensive effort, no general analytical solution method has been developed for the type of flow field illustrated in Fig. 1 because the governing partial differential equations are nonlinear. Fortunately, attempts to obtain solutions numerically have met with more success. The first major breakthrough was made by Magnus and Yoshihara [2], who demonstrated that solutions for two-dimensional, inviscid, transonic flow fields could be computed using finite-difference algorithms programmed to run on digital computers. Steady-state solutions were obtained as large time-asymptotic solutions to a time-accurate formulation of the problem.

A method requiring about an order of magnitude less computer time per solution was subsequently introduced by Murman and Cole [3], who developed a finite-difference relaxation procedure to solve the steady form of the equations of motion. Type-dependent finite-difference operators, which account for the fundamentally different behavior of subsonic and supersonic flow, were applied to solve a simplified version of the equations solved by Magnus and Yoshihara.

These initial successes gave birth to a new field of study, *computational transonic aerodynamics*, and a new faith that a supplement to wind-tunnel simulations could indeed be developed. At the same time, advances in computer technology were substantially decreasing the required processor time and the cost involved in obtaining solutions, allowing more ambitious computations to be attempted.

To date, no complete treatment has been developed for transonic flows that includes three-dimensional, unsteady, and viscous effects. However, considerable progress has been made in each of these areas, and some of the highlights are reviewed here. To begin with, a small-disturbance procedure for steady, inviscid, *three-dimensional* transonic flows is outlined in Sec. 2. The use of small-disturbance theory simplifies the enforcement of boundary conditions but imposes some restrictions on the class of aerodynamic configurations and flow situations that can be properly treated. Computed and experimental results are compared for several wing and wing-fuselage configurations. Next, the treatment of inviscid, primarily two-dimensional, *unsteady* transonic flows is reviewed in Sec. 3. The emphasis is on deriving simplified equations of motion and on solving these equations by (1) time integration and (2) separation of the steady and unsteady components of the solution. Computed results are presented for various types of airfoil motions, including the simulated motion of a helicopter rotor blade in forward flight. Section 4 reviews recent progress in an effort to solve two-dimensional, steady, *viscous* flows about airfoils. Computed results are compared with experimental measurements for flows that include trailing-edge and shock-induced separation of a turbulent boundary layer.

Finally, two techniques are discussed that should prove useful in computational transonic aerodynamics applications. The first, the finite-volume method, is reviewed in Sec. 5. This procedure simplifies the application of boundary conditions without introducing the restrictions associated with small-disturbance theory. The finite-volume method was used to obtain the viscous-flow solutions reported in Sec. 4 and should eventually prove useful in providing accurate treatment of unsteady and three-dimensional flows. The second technique, configuration design by numerical optimization, is reviewed in Sec. 6. This method can be used in aircraft design to develop configurations that satisfy specified geometric and performance constraints. Two examples of airfoil design by numerical optimization are presented.

This review is intended to provide the reader with an indication of the state of the art in computational transonic aerodynamics. It is by no means all-inclusive, the most notable omission being the work of Prof. Antony Jameson and his associates at Courant Institute. This work, along with related efforts by other researchers, is presented in Chap. 1 by Prof. Jameson.

2   STEADY THREE-DIMENSIONAL FLOWS ABOUT WINGS AND WING-FUSELAGE CONFIGURATIONS

The solution method outlined here, which is an extension of the Murman-Cole finite-difference relaxation procedure to three dimensions, is based primarily on the work of Bailey and Ballhaus [4–7]. Computed and experimental results are compared for an isolated wing, followed by a discussion of techniques for improving the predictive capabilities of the small-disturbance theory. Finally, several comparisons of computed and experimental surface pressures for wing-fuselage configurations are presented.

2.1   The Small-Disturbance Approach for Wings

The principal advantage of a small-disturbance approach is the flexibility that results from the simplified treatment of boundary conditions. This flexibility has motivated the development of two-dimensional small-disturbance relaxation algorithms and their application to a variety of practical problems. In such applications, coordinate transformations are not usually required because the airfoil boundary condition is applied, in terms of airfoil slopes, on some mean-surface slit instead of on the airfoil surface. Consequently, problems with multiple boundaries (e.g., a configuration with multiple lifting surfaces and mounted in a wind tunnel) can be handled rather easily. Since the primary difficulty encountered in the treatment of three-dimensional flows for wing and wing-fuselage configurations is the enforcement of boundary conditions, it is not surprising that all the transonic relaxation methods developed to date for such configurations [4–11] are based on the small-disturbance approach, with the exception of the method of reference 12, which is restricted to yawed wings.

The classical three-dimensional transonic, small-disturbance equation is

$$[(1 - M_\infty^2)\phi_x - \tfrac{1}{2}(\gamma + 1)M_\infty^2\phi_x^2]_x + (\phi_y)_y + (\phi_z)_z = 0 \qquad (1)$$

where M_∞, γ, and ϕ are the free-stream Mach number, specific heat ratio, and disturbance velocity potential. The equation is derived [13] from the Eulerian gas dynamic equations by satisfying conservation of mass, momentum, and energy to lowest order under the assumptions of small-disturbance transonic flow about high-aspect-ratio (AR) wings; that is,

$$\tau^{2/3} \sim 1 - M_\infty^2 \sim \frac{1}{(AR)^2} \ll 1 \qquad (2)$$

where τ is a representative thickness-to-chord ratio for the wing. The shock conditions implied by Eq. (1) are transonic approximations to the Rankine-Hugoniot shock relations.

Computed results are usually reported in terms of a pressure coefficient

$$C_p = \frac{p - p_\infty}{\frac{1}{2}\rho_\infty U_\infty{}^2} \simeq -2\phi_x \qquad (3)$$

where p, ρ, and U are pressure, density, and velocity, and the subscript ∞ denotes free-stream conditions.

Here we consider a semispan wing as shown in Fig. 3. The planform may be swept and tapered and may have curved leading and trailing edges. The airfoil section and incidence may vary with distance along the span. If the wing surface is given by $f(x, y, z) = 0$, then the (linearized) wing boundary condition consistent with the assumptions of Eq. (2) is

$$f_x + (\phi_z + \alpha)f_z = 0 \qquad (4)$$

This condition is enforced on a flat mean-surface approximation to the wing in the $z = 0$ plane. The angle of attack α is the incidence angle of the free-stream flow relative to this plane.

It is well known that the small-disturbance assumptions that led to the derivation of Eqs. (1) and (4) are violated at the wing leading edge. Experience in two-dimensional computations indicates a resulting mesh dependency of the solution (especially for lifting, blunt-nosed airfoils); that is, the computed airfoil surface pressures depend on (1) the grid spacing near the airfoil leading edge, and (2) the location of the leading edge relative to the mesh. This mesh dependency affects the solution over most of the airfoil surface for coarse meshes. However, for fine meshes near the leading edge, the solutions generally exhibit mesh dependency in only a very small region.

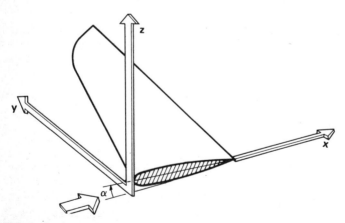

FIG. 3 Semispan wing.

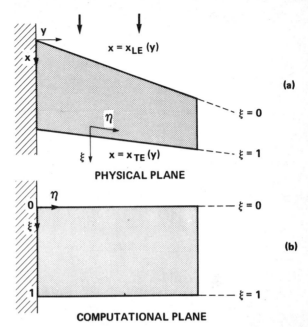

FIG. 4 Wing planform transformation. (a) Physical plane; (b) computational plane.

The dependence of the solution on leading-edge mesh spacing poses a problem in the application of Eq. (1) to swept wings. To maintain a sufficiently fine mesh spacing along the leading edge of a swept wing in a Cartesian coordinate system is impractical because of the large number of points required. Thus, three-dimensional computations must be performed on what would be considered a coarse leading-edge grid in two dimensions [8,10]. An alternative approach was developed by Ballhaus and Bailey [4]. A transformation was used to map a trapezoidal planform wing in the physical plane into a rectangle in the computational plane, as shown in Fig. 4, where

$$\xi(x, y) = \frac{x - x_{LE}(y)}{x_{TE}(y) - x_{LE}(y)}$$

$$\eta(y) = y \tag{5}$$

The leading and trailing edges lie on the coordinate lines $\xi = 0, 1$, respectively. Use of the transformation permits a more efficient distribution of mesh points, and each span station ($\eta = $ constant) has the same number of chordwise mesh points on the wing surface. The complicating effect of geometry has been transferred from the boundary conditions to the

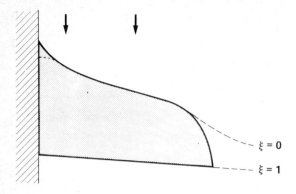

FIG. 5 Planform transformation for wing with regions of large sweep.

governing equation, which can be written in conservation form as follows

$$\left[(1 - M_\infty{}^2)\xi_x\phi_\xi - \frac{\gamma + 1}{2}M_\infty{}^2\xi_x{}^2\phi_\xi{}^2 + \frac{\xi_y}{\xi_x}(\xi_y\phi_\xi + \phi_\eta)\right]_\xi$$

$$+ \left[\frac{1}{\xi_x}(\xi_y\phi_\xi + \phi_\eta)\right]_\eta + \left(\frac{1}{\xi_x}\phi_z\right)_z = 0 \qquad (6)$$

The terms within the brace come from the ϕ_{xx} terms in Eq. (1). Note that, for large sweeps, the angle between the ξ and η coordinate directions becomes more acute (Fig. 5), and, at some point, the coordinate lines must deviate from the planform boundaries [11].

The Kutta condition requires that the pressure ϕ_x be continuous at the trailing edge. This fixes the section circulation $\Gamma(y)$, which is equal to the difference in potential across the wing mean surface at the trailing edge. The potential jumps are uniform in x (or ξ) on the trailing vortex sheet (the plane $z = 0$ downstream of the wing), across which both pressure and downwash ϕ_z are continuous.

Conditions must also be satisfied on the boundaries of the computational domain. At the wing symmetry plane $y = \eta = 0$,

$$\phi_y = \phi_\eta + \xi_y\phi_\xi = 0 \qquad (7)$$

must be enforced. At the far-field boundaries, located at some large distance from the wing, an asymptotic far-field approximation [4] is applied; this approximation depends on the wing loading $\Gamma(y)$.

2.2   Differencing of the Small-Disturbance Equation

Following the mixed-difference procedure of Murman and Cole, central or backward x differences are used in regions where the flow is elliptic or hyperbolic (corresponding to subsonic or supersonic flow), respectively. Central differences are used throughout the flow field to approximate the y and z derivatives.

A schematic of the hyperbolic (supersonic) computational molecule in a $z = $ constant plane is illustrated in Fig. 6. To maintain stability, the domain of dependence of this molecule must include the domain of dependence of the partial differential equation, given by the characteristic traces in a $z = $ constant plane as

$$\frac{dx}{dy} = \pm\{-[1 - M_\infty{}^2 - (\gamma + 1)M_\infty{}^2\phi_x]\}^{1/2} \tag{8}$$

The characteristics are real, and hence Eq. (1) is hyperbolic, with x the timelike (marching) direction, when the term in the brackets is positive; the characteristic traces are symmetrical about the free-stream (x) direction. Note that the use of central differences for the y and z derivatives results in an implicit computational molecule (i.e., the solution at each point in an $x = $ constant plane depends on the solution at every other point in that plane). Hence, the domain of dependence of the computational molecule always includes that of the equation, even in sonic regions, where the charac-

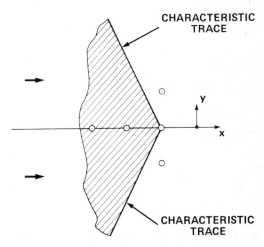

FIG. 6   Difference molecule and characteristic traces for a $z = $ constant plane in a supersonic region.

teristic traces are nearly normal to the x direction; and stability is assured regardless of mesh distribution.

Differencing of the governing equation is complicated by the shearing transformation introduced in the previous section. In this case, a "rotated differencing" procedure, an extension of the Murman-Cole mixed-difference procedure to arbitrary coordinate systems (developed by Jameson [14] and Albone [15]) is used. Mixed differences are used only to approximate derivatives in the marching direction; all other derivatives are approximated by central differences. Hence, all the terms that come from x derivatives in Eq. (1) are mixed-differenced in the ξ-η plane [the braced terms in Eq. (6)]. The remaining terms are central-differenced.

It is now well known that the original Murman-Cole mixed-difference scheme does not maintain proper conservative form at the shock point (the first mesh point downstream of a supersonic-to-subsonic shock wave), where the x difference switches from backward to central. Consequently, computed shock jumps to not satisfy the shock conditions associated with Eq. (1), and in some instances nonunique flow-field solutions have been obtained. Murman remedied the situation by applying a "shock-point operator," which is the sum of the elliptic and hyperbolic x operators at the first mesh point downstream of the shock. In Murman's terminology, the original Murman-Cole scheme and the new conservative scheme are referred to as "nonconservative relaxation" (NCR) and "fully conservative relaxation" (FCR) schemes, respectively. The FCR scheme is consistent (in the limit of vanishing mesh spacing) with both the governing equation and its corresponding shock conditions.

A majority of the two- and three-dimensional calculations reported to date have been performed using the NCR procedure. In the absence of viscosity corrections, the NCR solutions generally agree better with experimental pressure measurements than the FCR solutions. The reason is that spurious sources result from the failure of the NCR procedure to maintain conservative form at the shock point. These sources reduce the computed shock strengths to values that, coincidentally, are nearly equal to those obtained experimentally, which are, of course, weakened by the interaction of the shock and boundary layer. This remarkable (but fortuitous) agreement is illustrated in Fig. 7 [16]. Airfoil surface-pressure jumps across the shock are plotted from a variety of computed and experimental work. Note first of all that none of the *experimental* pressure jumps agree with the Rankine-Hugoniot pressure jump; the only *computed* results that agree were obtained from fully conservative schemes or from schemes that "fit" the shocks as internal boundaries. Note also that for shock mach numbers greater than about 1.3, the experimental pressure ratios level off, an indication that the shock has separated the boundary layer. Further discussion of NCR and FCR schemes is deferred to the next section, where results from both are compared with experiment.

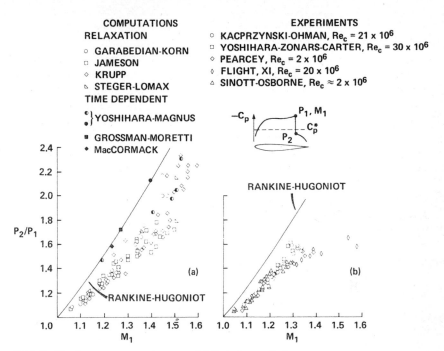

FIG. 7   Shock-wave pressure jump on airfoils in transonic flows.

2.3   Comparisons of Computed and Experimental Results for Wings

The agreement of computed and experimental results can be adversely affected by inherent differences in the two simulation procedures. The computational procedure is designed to simulate the inviscid flow field (infinite Reynolds number) about a wing immersed in a uniform free stream of infinite extent. On the other hand, the experimental results are obtained from wind-tunnel simulations at finite Reynolds numbers that are usually considerably lower than those encountered in flight. Hence, comparisons of experimental (viscous) and computed (inviscid) results can be expected to suffer when viscous effects strongly influence experimental surface pressures, that is, for low Reynolds numbers, high lift, and large shock Mach numbers. Furthermore, wind-tunnel conditions are usually influenced by interference from the tunnel walls. Although attempts are made to correct for such influences, the corrected free-stream Mach number and angle of attack do not always correspond to the equivalent free-air values. Also, as the free-stream Mach number approaches 1, the wall effects on the embedded supersonic region and shock locations become even more significant and difficult to assess.

An additional source of disagreement in comparisons of computed and experimental results is violation of the assumptions, in Eq. (2), under which the transonic small-disturbance theory is derived. Such violations can result in significant departures from correct inviscid free-air solutions. For example, the assumption of small disturbances is seriously violated near blunt leading edges and at high angles of attack. Thus, erroneous results can be obtained in the leading-edge region. Also, the pressure differences across shock waves predicted by the small-disturbance equation become significantly greater than those predicted by the Rankine-Hugoniot relations as the shock Mach number increases past about 1.3. Finally, the small-disturbance shock conditions do not adequately approximate the Rankine-Hugoniot relations for oblique shocks at angles in excess of about 20°; this deficiency is discussed at length in the next section.

Keeping in mind the possible sources of disagreement alluded to above, we now consider the comparison of computational and experimental surface pressures shown in Fig. 8 [7] for the ONERA M6 wing. The experimental data [17] were obtained in the ONERA S2 transonic tunnel at $Re_{\bar{c}} \sim 2.5 \times 10^6$. The computations were performed on a sequence of three x, y, z grids ($35 \times 20 \times 24$; $45 \times 23 \times 30$; and $66 \times 37 \times 32$). That is, a course-grid solution was obtained first and then used, along with an interpolation, to provide an initial guess to start the computation on the next (finer) grid, and so on. Solutions were obtained using both the NCR and FCR procedures, and each solution required approximately 5 min on a CDC 7600 computer.

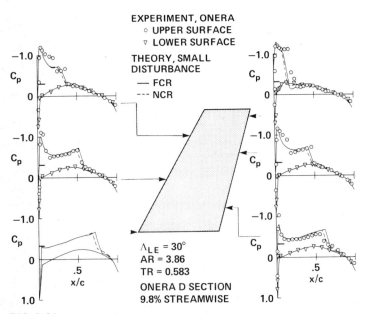

FIG. 8 Comparison of computed and experimental pressure coefficients for the ONERA M6 wing at $M_\infty = 0.84$ and $\alpha = 3°$.

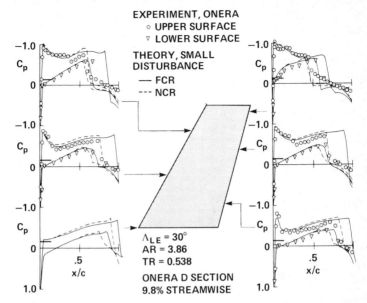

FIG. 9 Comparison of computed and experimental pressure coefficients for the ONERA M6 wing at $M_\infty = 0.92$ and $\alpha = 3°$.

The FCR method predicts a slight downstream shift in shock location relative to the NCR results for the rather weak shock waves in the $M_\infty = 0.84$, $\alpha = 3°$ case. This observation is consistent with the results obtained by Murman [18] in two-dimensional numerical experiments. Upper and lower surface pressures, as well as the location and strength of the shock terminating the supersonic region, are generally adequately predicted by the NCR and FCR methods. The primary deficiency is the failure of the computed results to capture properly the weak, highly oblique (36°) shock waves in the supersonic region.

A comparison at the same angle of attack, but for a higher free-stream Mach number, is shown in Fig. 9. In this case, the FCR shock locations are significantly downstream from those of the NCR method. The NCR results, with the weaker shocks, generally agree better with the experimental data but are still in serious disagreement. The increasingly large discrepancy observed here between FCR and NCR shock strengths and locations with increasing shock strength is consistent with Murman's observations in reference 18. Again, neither method captures the forward oblique shock that is clearly evident in the experimental data at the three outboard span stations. The disagreement in upper and lower surface (supersonic to subsonic) shock locations can probably be attributed primarily to a decrease in wing lift caused by trailing-edge viscous effects. This decrease in lift moves the upper surface shock upstream, and the lower surface shock downstream.

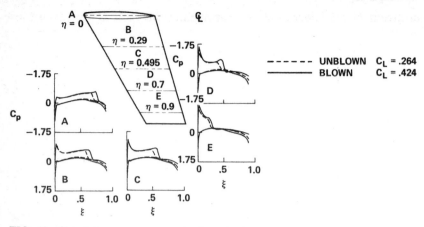

FIG. 10 Chordwise pressure on a jet-flapped tapered wing with $\Lambda = 30°$, $\lambda = 0.538$, $AR = 3.86$, $M_\infty = 0.84$, $\alpha = 3°$, $C_j = 0.03$, ONERA M6 section, and full span blowing.

Computed pressures for the ONERA M6 wing with a trailing-edge jet flap have recently been reported by Murphy and Malmuth [19]. In this work, the Bailey-Ballhaus transonic wing code [4] has been modified to permit trailing-edge blowing. Blown and unblown results for the $M_\infty = 0.84$, $\alpha = 3°$ case are compared in Fig. 10. Considerable lift enhancement due to blowing is evident in the resulting surface pressures and lift coefficients. Unfortunately, no experimental jet-flap results are available for comparison.

2.4   Improvements to Classical Small-Disturbance Theory

One of the major deficiencies in classical small-disturbance theory is its treatment of shock waves. However, this treatment can be substantially improved by modifying the theory. Here, modifications are discussed first for normal shocks and then for oblique (or swept) shocks. The shock waves to be considered are assumed to be sufficiently weak so that the surrounding flow is isentropic and irrotational.

2.4.1   Normal Shocks

The small-disturbance theory is derived under the assumption that M_∞ is nearly unity. In flows for which this is not the case (e.g., $M_\infty = 0.7$), good agreement between the small-disturbance shock conditions and the (exact) Rankine-Hugoniot shock relations cannot be expected. Krupp [20] was able to extend the Mach-number range over which good agreement can be obtained by "tuning" the small-disturbance theory, that is, by scaling variables in the problem by powers of M_∞. A more direct but similar approach is taken here.

First note that, for a shock wave to exist in a steady flow, the flow upstream of the shock must be supersonic; that is, the flow must have expanded beyond $C_p = C_p^*$, the "critical" or "sonic" pressure coefficient. For the exact equations,

$$C_{P_E}^* = \frac{2}{\gamma M_\infty^2} \left\{ \left[\frac{2}{\gamma + 1} \left(1 + \frac{\gamma - 1}{2} M_\infty^2 \right) \right]^{\gamma/(\gamma - 1)} - 1 \right\} \tag{9}$$

This expression can be expanded for small $s = 1 - M_\infty^2$ to give

$$C_{P_E}^* = \frac{-2s}{\gamma + 1} \left[1 + \frac{(2\gamma + 1)s}{2(\gamma + 1)} \right] + O(s^3) \tag{10}$$

Now let us determine the conditions that a normal shock must satisfy. For simplicity, we assume that the upstream flow is uniform with a Mach number equal to M_∞. Then, from the Rankine-Hugoniot relations,

$$\frac{p_2}{p_1} = 1 + \frac{2\gamma}{\gamma + 1} (M_\infty^2 - 1) \tag{11}$$

and so

$$\overline{C_{P_E}} \equiv \frac{1}{2} (C_{p_1} + C_{p_2}) = \frac{p_2 - p_\infty}{\rho_\infty U_\infty^2} = \frac{p_2/p_\infty - 1}{\gamma M_\infty^2} \qquad C_{p_1} = 0 \tag{12}$$

where subscripts 1 and 2 indicate values just upstream and downstream of the shock. Expanding for small $s = 1 - M_\infty^2$ gives

$$\overline{C_{P_E}} = \frac{-2s}{\gamma + 1} (1 + s) + O(s^3) \tag{13}$$

Note that, from Eqs. (10) and (13),

$$C_{P_E}^* = \overline{C_{P_E}} + O(s^2) \tag{14}$$

that is, for weak shocks, the average C_p across the shock is nearly equal to C_p^*.

Now let us determine the values of C_p^* and $\overline{C_p}$ corresponding to the small-disturbance equation

$$\left(1 - M_\infty^2 - \frac{\gamma + 1}{2} M_\infty^m \phi_x^2 \right)_x + (\phi_y)_y + (\phi_z)_z = 0 \tag{15}$$

where m remains undefined for the time being. However, we do require that

$\lim_{M_\infty \to 0}(M_\infty{}^m) = 0$ and $\lim_{M_\infty \to 1}(M_\infty{}^m) = 1$. Equation (15) changes type when $C_p = C_p^* = -2\phi_x^*$, which is given by

$$C_{P_{SD}}^* = \frac{-2(1 - M_\infty{}^2)}{(\gamma + 1)M_\infty{}^m} = \frac{-2s}{\gamma + 1}\left(1 + \frac{m}{2}s\right) + O(s^3) \tag{16}$$

The normal shock relations corresponding to Eq. (15) are given by

$$(1 - M_\infty{}^2)[[\phi_x]] - \frac{\gamma + 1}{2} M_\infty{}^m[[\phi_x{}^2]] = 0 \tag{17}$$

where $[[\phi_x{}^p]]$ indicates a jump in $\phi_x{}^p$ across the shock. Now

$$[[\phi_x{}^2]] = [[\phi_x]](\phi_{x_1} + \phi_{x_2}) = [[\phi_x]]2\bar{\phi}_x \tag{18}$$

This relation, along with $C_p = -2\phi_x$, can be substituted into Eq. (17). The result, expanded for small $s = 1 - M_\infty{}^2$, is

$$\bar{C}_{P_{SD}} = \frac{1}{2}(C_{p_1} + C_{p_2}) = \frac{-2(1 - M_\infty{}^2)}{(\gamma + 1)M_\infty{}^m} = \frac{-2s}{\gamma + 1}\left(1 + \frac{m}{2}s\right) + O(s^3) \tag{19}$$

Several interesting observations can be made by comparing Eqs. (10), (13), (16), and (19). First,

$$C_{P_{SD}}^* = \bar{C}_{P_{SD}} \tag{20}$$

exactly, whereas this is only true to lowest order for the exact equations, as indicated by Eq. (14). Second, $C_{P_{SD}}^*$ agrees with $C_{P_E}^*$ to first order for any m and agrees to second order for $m = (2\gamma + 1)/(\gamma + 1)$ ($= 1.583$ for $\gamma = 1.4$). Third, $\bar{C}_{P_{SD}}$ agrees with $\bar{C}_{P_E}$ to first order for any m and agrees to second order for $m = 2$. Hence, m can be selected to match either $C_{P_E}^*$ or $\bar{C}_{P_E}$ to second order, but not both.

The use of $m = 2$ can cause significant differences between the values of $C_{P_{SD}}^*$ and $C_{P_E}^*$ when M_∞ is not close to unity. Consequently, small-disturbance theory may predict a flow field that is entirely subsonic when the exact equations indicate a substantial embedded supersonic region. On the other hand, if m is chosen so that $C_{P_{SD}}^*$ matches $C_{P_E}^*$, then the resulting small-disturbance shock jump can be considerably larger than the exact one (assuming the same C_{p_1} in both cases) unless the shocks are weak.

These effects are illustrated in Fig. 11, which compares small-disturbance and exact surface pressures for an NACA 64A410 airfoil [21] at $\alpha = 0°$ and $2°$, and $M_\infty = 0.72$. For the results shown in Fig. 11a, $m = 1.69$ so that $C_{P_{SD}}^* = C_{P_E}^*$ exactly (not just to second order). For $\alpha = 0°$, the results generally compare favorably. The disagreement at the shock is caused by "smearing" due to the coarseness of the small-disturbance grid near the shock. (The disagreement near the trailing edge is due to the coarseness of the grid in that region for the exact procedure. The violation of $C^*_{P_{SD}} = \bar{C}_{P_{SD}}$ in the computations is also caused by the coarseness of the grid.) In this case, the shock Mach number is nearly 1.3, the point at which rotational and nonisentropic effects become important. At $\alpha = 2°$, the small-disturbance shock is considerably stronger (and hence located further downstream) than the Rankine-Hugoniot (exact) shock. Here, the shock Mach number is about 1.45. In Fig. 11b, a similar comparison is made using $m = 2$. In this case, the shock agreement is improved for $\alpha = 2°$ at the expense of the agreement in the $\alpha = 0°$ case. For $m = 2$, $C^*_{P_E} = -0.700$, while $C^*_{P_{SD}} = -0.774$.

The value of m to be selected in a particular application depends, in a sense, on the solution. If only weak shocks are expected, then m should be

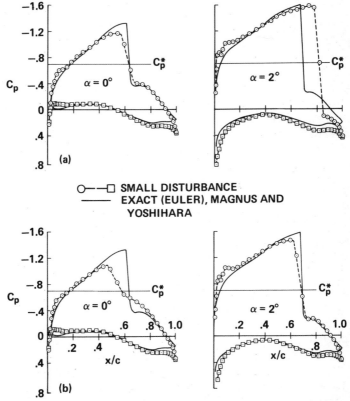

FIG. 11 Surface pressures for an NACA 64A410 airfoil. (a) $m = 1.69$; (b) $m = 2$.

chosen to approximate C_p^*. In cases with stronger shocks, $m = 2$ should be used. Krupp [20] developed a scaling that, in effect, used a value of $m = 1.75$. This would seem to be a reasonable compromise between the values of m required to match the $\overline{C}_{p_E}$ ($m = 2.0$) and $C_{p_E}^*$ ($m = 1.583$) to second order. Krupp's scaling also divides the airfoil surface slope boundary conditions by $M_\infty^{1/4}$. Hence, for $M_\infty < 1$, Krupp "thickens" the airfoil. This seems to improve the agreement in comparisons of small-disturbance results with exact results by compensating for errors introduced by the mean-surface boundary-condition approximation. This procedure was followed in generating the small-disturbance surface pressures shown in Fig. 11. Finally, note that, as M_∞ approaches unity, the effects of tuning (i.e., the $M_\infty{}^m$ in the governing equation and the $M_\infty{}^{1/4}$ in the boundary conditions) diminish.

2.4.2 Oblique Shock Waves

The computed results shown in Figs. 8 and 9 for the ONERA M6 wing failed to resolve the oblique-leading-edge shock wave that was found to exist experimentally. It was previously mentioned that this failure was caused by a deficiency in the three-dimensional small-disturbance theory. The source of this deficiency is identified here. Also, it is shown that this situation can be corrected by retaining certain higher-order terms in the governing equation.

The classical small-disturbance equation (1) was derived for flows that are fully three-dimensional. However, in some cases, there may exist regions of nearly two-dimensional flow in which the spanwise component of velocity is nearly uniform and gradients of flow quantities in the span direction are very small. In these regions, two-dimensional sweep theory can be applied; that is, the flow field, which is nearly uniform in the spanwise direction, can be determined by solving a two-dimensional problem in a plane normal to the span direction.

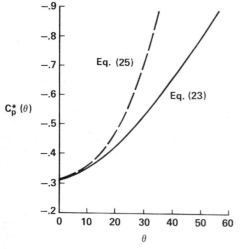

FIG. 12 $C_p^*(\theta)$ versus θ for $M_\infty = 0.85$.

Let us investigate the shock conditions for such a two-dimensional flow region with leading edge, trailing edge, and shock sweep all equal to some angle θ. We assume, for convenience, that the shock is vertical (it should be nearly so). The two-dimensional small-disturbance shock condition (in a plane normal to the span direction) is given by [see Eqs. (17) and (18)]

$$(\overline{C_p})_n = \frac{1}{2}[(C_{p_1})_n + (C_{p_2})_n] = \frac{-2(1 - M_n^{2})}{(\gamma + 1)M_n^{2}} \tag{21}$$

where M_n is the component of free-stream Mach number normal to the leading edge, and m has been set equal to 2. The values of pressure coefficient and Mach number in the normal direction can be related to values in the free-stream direction by

$$C_p \equiv \frac{p - p_\infty}{\frac{1}{2}\rho_\infty U_\infty^{2}} = \frac{(p - p_\infty)\cos^2 \theta}{\frac{1}{2}\rho_\infty U_n^{2}} = (C_p)_n \cos^2 \theta$$

$$U_\infty = \frac{U_n}{\cos \theta} \tag{22}$$

$$M_\infty = \frac{M_n}{\cos \theta}$$

Hence, from Eqs. (21) and (22), we can define

$$\overline{C_p}(\theta) = (\overline{C_p})_n \cos^2 \theta = \frac{-2(1 - M_\infty^{2} \cos^2 \theta)}{(\gamma + 1)M_\infty^{2}} \tag{23}$$

From this equation, we can determine to what degree the flow must expand for a shock with sweep θ to exist. Recall that for the small-disturbance theory, $C_p^* = \overline{C_p}$, that is, the critical pressure coefficient is equal to the average pressure coefficient across the shock. Hence,

$$C_p^*(\theta) = \overline{C_p}(\theta) \tag{24}$$

and, for a shock with sweep θ to exist, the flow must expand beyond $C_p^*(\theta)$.

Now let us investigate the shock relations corresponding to the classical three-dimensional small-disturbance equation. With the directional derivatives in the span direction set equal to zero, the shock conditions reduce to

$$\overline{C_p}(\theta) = \frac{-2(1 - M_\infty^{2} \cos^2 \theta)}{(\gamma + 1)M_\infty^{2} \cos^2 \theta} \tag{25}$$

The difference between Eqs. (23) and (25), illustrated in Fig. 12, is the source of the deficiency in the classical three-dimensional small-disturbance theory for oblique shocks. The disagreement becomes appreciable for θ greater than about 20°. Clearly, the three-dimensional small-disturbance theory can fail to predict the existence of moderately swept shocks (20–45°) for cases in which, according to sweep theory, such shocks do exist.

This deficiency was first pointed out in reference 16, and a new equation was derived to be consistent to the first order with the full potential equation (the Euler equations for isentropic flow), irrespective of θ, in terms of $C_p^*(\theta)$, $C_p(\theta)$, and characteristic traces in the x-y plane.

$$\left[(1 - \mathbf{M}_\infty^2)\phi_x - \underbrace{\frac{3 - \gamma}{2}\mathbf{M}_\infty^2\phi_y^2} - \frac{\gamma + 1}{2}\mathbf{M}_\infty^2\phi_x^2\right]_x$$

$$+ \left[\phi_y - \underbrace{(\gamma - 1)\mathbf{M}_\infty^2\phi_x\phi_y}\right]_y + (\phi_z)_z = 0 \qquad (26)$$

Equation (26) retains higher-order terms (braced) that were previously neglected. (Note that this equation can be "tuned" by adjusting the exponents of $\mathbf{M}_\infty$ in the nonlinear terms.) The $\overline{C_p}(\theta)$ and $C_p^*(\theta)$ for this equation agree exactly with sweep theory. Other researchers have since derived equivalent small-disturbance equations for improving the treatment of oblique shocks by adding higher-order terms to the classical three-dimensional small-disturbance equation [11,22–24].

Proper consideration of oblique shocks in the governing equation is not sufficient to guarantee that these shocks will be properly resolved computationally. Difference operators used to solve the governing equations must be capable of enforcing the correct shock conditions and must resolve the discontinuity with minimum distortion due to dissipative and dispersive errors. For example, the Murman upwind difference operator, used in supersonic regions, is highly dissipative and tends to "smear out" weak oblique supersonic-to-supersonic shocks. Resolution of these shocks can be improved by refining the mesh.

Recently, van der Vooren et al. [23] have attempted to capture this type of shock, using an improved version of Eq. (26). Computed results for the 44% semispan station on the ONERA M6 wing are shown in Fig. 13. The experimental oblique supersonic-to-supersonic shock is stronger for this $\alpha = 6°$ case than for the $\alpha = 3°$ case previously shown (Fig. 8). Note that there is a hint of the existence of this shock in the computed results. To establish that this is indeed a shock, computations should be performed on successively refined grids to determine if this compression localizes and steepens.

The evidence presented here indicates that the addition of higher-order terms to the governing equations can improve the treatment of oblique shocks in the three-dimensional small-disturbance theory. However, the addition of these terms could substantially complicate the solution process in

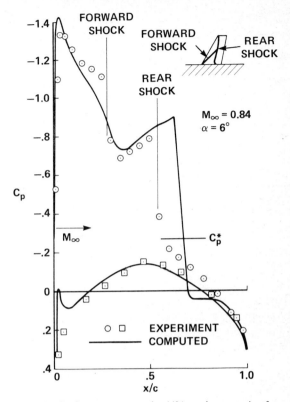

FIG. 13   Surface pressures at the 44% semispan station for the ONERA M6 wing.

some cases. Recall that the characteristic traces for the classical small-disturbance theory are symmetrical about the x direction (Fig. 6), and domains of dependence are easily satisfied in the difference scheme. For Eq. (26), however, the characteristic traces are symmetrical about the local flow direction, which varies throughout the flow field. Consequently, domains of dependence are more difficult to account for in the difference algorithm, and failure to do so can cause instabilities.

2.5   The Treatment of Wing-Fuselage Configurations

In transonic flight, the flow field near the wing can be strongly influenced by the geometry of the fuselage. Hence, proper wing-fuselage junction design is important for achieving suitable aircraft performance. An illustration (Fig. 14) of the strong effect of fuselage shape on the resulting flow field is provided by the first transonic relaxation solution reported for wing-cylinder configurations [5]. The calculations were performed for a nonlifting

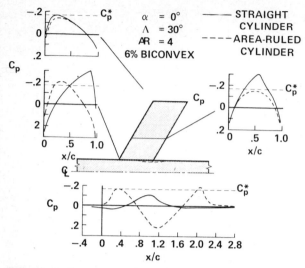

FIG. 14 Surface pressures for a wing-cylinder configuration.

swept wing that was midmounted on an infinitely long cylinder. Surface pressures are compared for two fuselage geometries with circular cross sections: (1) straight cylinder and (2) symmetrically indented cylinder based on Mach 1 area ruling. The area ruling completely eliminates the embedded shock waves on the wing.

More recently, small-disturbance relaxation computations for wing-fuselage configurations have been reported in references 7, 22, and 24–26. Only those from references 7 and 22 are reproduced here, and we begin by

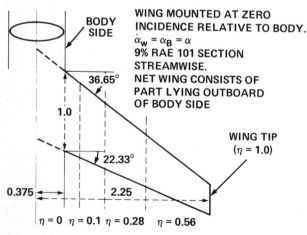

FIG. 15 Wing-body configuration with wing midmounted on cylindrical body of circular cross section.

discussing the results of Albone et al. of the Royal Aircraft Establishment, England. The configuration is shown in Fig. 15. Computed surface pressures shown in Fig. 16 for the wing-body and net-wing (without body) configurations illustrate the effect of the fuselage on the solution. The fuselage does not strongly affect the flow field on the wing in this subcritical case, and its effect diminishes with distance along the span such that there is no plottable difference at $\eta = 0.52$. For the supercritical case shown in Fig. 17, the fuselage influence is much stronger. The agreement between computed and measured surface pressures is less satisfactory than in the subcritical case, but is considered acceptable in view of uncertainty over wind-tunnel corrections.

A comparison of lift coefficient C_L as a function of M_∞ for the wing-body and net wing is shown in Fig. 18. It is interesting to note that the wing-body C_L peaks at a higher M_∞ than does the net wing. The loss in lift as M_∞ approaches 1 is explained in reference 22 as follows: Lift increases with M_∞ until the upper surface shock reaches the vicinity of the trailing edge. If M_∞ is further increased, the lower surface shock moves toward the trailing

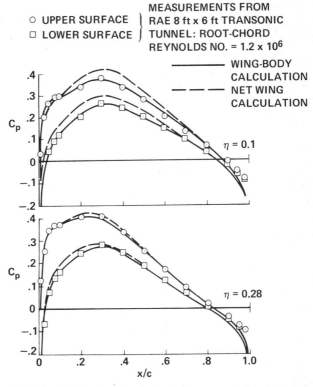

FIG. 16 Calculated and measured pressures at two spanwise stations for $M_\infty = 0.8$ and $\alpha = 1°$.

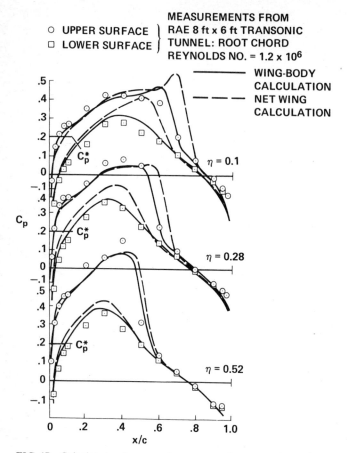

FIG. 17 Calculated and measured pressures at three spanwise stations for $M_\infty = 0.9$ and $\alpha = 1°$.

edge, thus increasing the total suction on the lower surface while the suction on the upper surface remains essentially unchanged. Hence, the lift decreases. Since the shock waves are initially further downstream on the net wing than on the wing-body configuration, the reduction in lift occurs at a lower M_∞ for the net wing.

In reference 22, the fuselage boundary condition is applied on the sides of a rectangular prism defined by $y = y_0$, $z = z_0$ in the alternative forms

$$(\bar{\phi}_y)_{y=y_0} = M_n{}^{r-4/3} \frac{\partial y_B}{\partial x} + \left[M_n{}^{r-4/3} \alpha_B + (\bar{\phi}_z)_{y=y_0} \right] \frac{\partial y_B}{\partial z}$$

or   (27)

$$(\bar{\phi}_z)_{\bar{z}=\bar{z}_0} = M_n{}^{r-4/3} \left(\frac{\partial z_B}{\partial x} - \alpha_B \right) + (\bar{\phi}_y)_{\bar{z}=\bar{z}_0} \frac{\partial z_B}{\partial y}$$

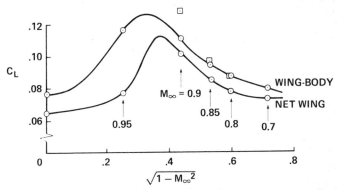

FIG. 18 Variation with M_∞ of lift at wing-body junction and at centerline of net wing at $\alpha = 1°$.

where $z = z_B(x, y)$ and $y = y_B(x, z)$ define the body surface when the angle of incidence of the body α_B is zero, and $\bar{\phi} = M_n{}^{r-4/3}(\phi - x)$, $\bar{z} = M_n{}^{2/3-r/2}z$, and $M_n = M_\infty \cos \Lambda$. Here Λ is an angle that is zero at the body centerline and tends to the local sweep angle of the midchord line of the wing away from the body; r is an empirically determined parameter.

Similar wing-fuselage computations have been reported in reference 7. The guiding philosophy in this effort has been to treat the fuselage with the simplest computational method possible to determine its effect on the flow field over the wing. Consequently, the fuselage boundary condition is satisfied in a Cartesian grid system. The mesh is constructed so that points lie reasonably close to the fuselage surface, as shown in Fig. 19. Points may

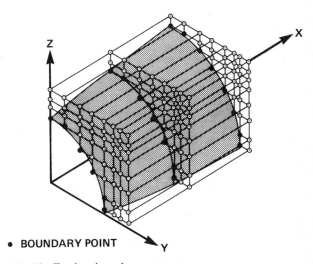

FIG. 19 Fuselage boundary treatment.

lie either inside or outside the fuselage. For a body described by
$f(x, y, z) = 0$, the small-disturbance boundary condition (with $\alpha = 0$) is

$$f_x + \phi_y f_y + \phi_z f_z = 0 \qquad (28)$$

The value of ϕ at the boundary is found by using three point-extrapolated
differences (at the boundary point) in Eq. (28). Results computed using this
approach are compared (Fig. 20) with experiment and with an axisymmetric
computation from reference 27 for a parabolic-arc body with sting. Pressures at the body surface and in the flow field at two body diameters from
the centerline are plotted for $M_\infty = 0.99$. Good agreement is indicated between the present Cartesian grid (NCR) method, the axisymmetric (NCR)
method, and measured surface pressures. The x-mesh spacing used in the
Cartesian grid calculation is 2.5 times coarser than that used in the axisymmetric one and accounts for the difference in shock resolution between the
two methods. The results plotted in Fig. 20 indicate that the simplified
approach for satisfying the fuselage boundary condition is acceptable.

Figure 21 compares NCR, FCR, linear theory, and experimental
pressures [28] for a rectangular-wing–Sears-Haack-fuselage combination at
$M_\infty = 0.90$ and $\alpha = 0°$. The data were obtained in the AEDC 16-ft wind
tunnel at $Re_{\bar{c}} = 3.0 \times 10^6$. As one would expect, the plotted results show the
FCR-predicted shock locations downstream of the NCR prediction, and the
linear-theory result shows no shock.

Finally, NCR results are compared with experimentally measured pressures (Fig. 22) for a swept-wing–fuselage configuration at $M_\infty = 0.93$ and
$\alpha = 0°$. The uncorrected experimental data [29] were obtained in the Langley 8-ft tunnel (solid wall) at $Re_{\bar{c}} = 2.0 \times 10^6$. The agreement with experi-

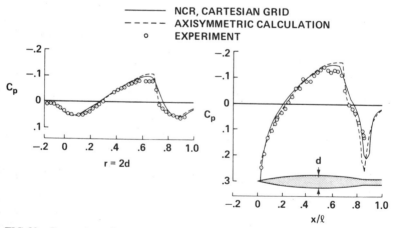

FIG. 20 Comparison of computed and experimental pressure coefficients for a parabolic
arc of revolution, a fineness ratio of 10, and $M_\infty = 0.99$.

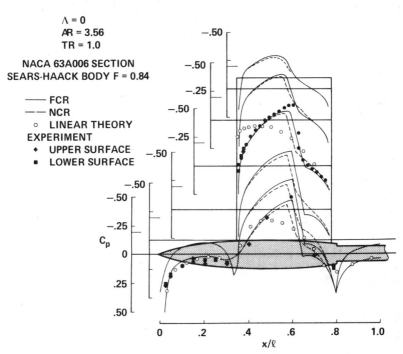

FIG. 21 Comparison of computed and experimental pressure coefficients for a rectangular-wing–fuselage configuration at $M_\infty = 0.90$ and $\alpha = 0°$.

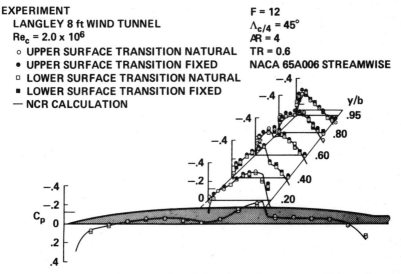

FIG. 22 Comparison of computed and experimental pressure coefficients for a swept-wing–fuselage configuration at $M_\infty = 0.93$ and $\alpha = 0°$.

ment on the fuselage centerline and the two inboard panels is good. In the computed results, the wing-root shock extends laterally to about 60% semi-span, but the experimental shock dissipates before reaching this point. The source of disagreement is not clear but could well be an effect of viscosity.

The computation discussed in the previous paragraph was performed on a Cartesian grid for the entire flow field. For tapered configurations, the number of chordwise mesh points on the wing surface decreases with distance along the span (in the present case, only 12 points near the tip, as opposed to 23 points near the wing root). This could in part be responsible for the insufficient expansion near the tip in the computed results shown in Fig. 22. Recently, the code has been generalized to treat wings using the planform transformation, Eq. (5). The fuselage is still treated using a Cartesian grid, and this simple treatment is considered adequate for many applications. The code has also been extended to treat lifting configurations.

3   ANALYSIS OF INVISCID UNSTEADY TRANSONIC FLOWS

This section begins with a discussion of the nature of unsteady transonic flows, followed by some examples of practical applications. It is shown that, for some of these applications, the flow-field solution can be separated into steady-state and unsteady components, thereby simplifying the solution process. Next, different forms of governing partial differential equations and boundary conditions, with varying degrees of simplification for certain special cases, are investigated. This discussion is followed by a review of computed solutions for flows resulting from unsteady airfoil motion and from simulated helicopter rotor-blade motion. The section closes with a review of finite-difference methods designed to solve efficiently the physically interesting case of low-frequency motion, for which shock excursion amplitudes (and hence, aerodynamic force amplitudes) are large.

3.1   Nature of Unsteady Transonic Flows

In transonic flight, the unsteady motion of a body strongly affects the resultant aerodynamic forces acting on that body. The reason is that surface pressures, and hence aerodynamic forces, are extremely sensitive to perturbations in boundary conditions. For example, the well-known expression relating the lift coefficient C_l to the angle of attack α for a flat plate, $C_l \sim \alpha/\sqrt{1 - M_\infty^2}$, indicates that changes in C_l due to changes in α become more pronounced as the free-stream Mach number M_∞ approaches unity. Similarly, changes in airfoil velocity or shape (e.g., flap deflection) can be expected to produce large changes in aerodynamic forces, especially for motions that induce large excursions of embedded shock waves. Such large shock excursions might be expected to occur, for example, in the case of low-frequency oscillations of supercritical type airfoils with flat upper-surface pressure distributions. Furthermore, multiple embedded supersonic regions with shock waves could perhaps appear and subsequently disappear during the course of the motion cycle.

In the analysis of unsteady flows, surface pressures and aerodynamic forces must be considered, not only in terms of magnitude, but also in terms of phase relative to the motion of the body. In the transonic case, both the changes in magnitude and the phase lags are usually large compared to those in subsonic or supersonic flows; that is, transonic flow fields are relatively slow to adjust to unsteady perturbations. This behavior is to be expected, since disturbances generated on or near the body must travel upstream at the local speed of sound against a nearly sonic flow, and, should they encounter a supersonic region, they must propagate around it. The slow upstream propagation rate, then, allows a disturbance to affect the flow field near the airfoil for a period of time that is large compared to that in either a purely subsonic or supersonic flow.

3.2 Examples of Unsteady Transonic Flows of Engineering Interest

The unsteady transonic-flow applications considered here can be divided into two classes that are based on the amplitude of the motion: (1) finite (but small) amplitude, for which the amplitude-to-chord ratio is of the same order as the thickness-to-chord ratio τ of the body (τ is assumed to be much less than 1) and (2) infinitesimal amplitude, for which, practically speaking, the amplitude-to-chord ratio is much less than τ.

3.2.1 Helicopter Rotors

An example of a *finite-amplitude* motion of practical interest is that of a helicopter rotor in forward flight. In this case, a rotor-blade element encounters a sinusoidally varying free-stream velocity that is the vector sum of the rotational (ΩR) and forward-flight (V) velocities. The maximum and minimum velocities are attained on the advancing and retreating sides of the rotor disk, respectively. To avoid roll moments, a nearly uniform lift must be maintained, and hence the blade incidence is varied nearly sinusoidally, 180° out of phase with the velocity. For large advance ratios ($\mu = V/\Omega R$), the retreating blade operates at angles of attack that are beyond the steady-state stall point, while the tip region on the advancing side operates beyond the drag-rise Mach number. Attempts to increase the forward flight speed beyond this operating range can trigger dynamic stall and buffet on the retreating and advancing sides, respectively, resulting in increased power requirement, lift reduction, and severe vibration and noise.

If, on the advancing side, the onset of adverse transonic effects could be delayed to higher tip Mach numbers, then an increase in forward flight speed could be compensated by an increase in rotational speed to maintain the same relative velocity on the retreating side. This would require a redesign of the tip region including planform variation (e.g., tip sweep) and proper spanwise variation of airfoil section. (Of course, the resulting retreating-blade stall properties of any design modification would also have to be considered.) Detailed knowledge of the complex (nonlinear, three-dimensional, unsteady, viscous) transonic flow field would be required. The difficulty and expense of obtaining such detailed information experimentally is sufficient motivation for the development of a computational approach to the problem.

3.2.2   Flutter

Motions of *infinitesimal amplitude* are considered in flutter analysis. The object is to predict the conditions that permit (unstable) growth of these motions in the interaction between aerodynamic forces and elastic deformations of the structures on which the forces act. Oscillatory aerodynamic coefficients for generalized forces are usually computed and tabulated as functions of free-stream Mach number and frequency and then used in a flutter analysis to identify regions of destabilizing aerodynamic forces. Generally, such tabulations can be applied because the equations governing the aerodynamics are assumed to be linear, and thus the superposition principle can be used.

3.3   The Linear-Perturbation Assumption and Superposition

In what follows, it is shown that the classification of a particular motion according to amplitude determines, to a great extent, how the resulting flow field should be treated computationally. To begin with, let us assume that an unsteady flow field can be described by a linear initial-value problem. The solution can then be expressed as the sum of two components, one representing the steady state, and the other, the unsteadiness. The effects of thickness, camber, and mean angle of attack can all be considered independently and then superposed to form the steady-state component of the solution. The unsteady component, which is independent of the other, is the solution for the unsteady motion of a flat plate. This approach is usefully applied in the case of subsonic and supersonic flows, for which the governing equations are linear. However, the equations governing transonic flows are nonlinear, and, hence, the superposition principle cannot be so generally applied. For example, the effects of airfoil thickness, camber, and mean angle of attack cannot be considered separately and then added to form a steady-state solution. Solutions can, however, be separated into steady-state and unsteady components when the unsteadiness in the flow field can be treated as a linear perturbation about some steady-state condition.

The linearity of the unsteadiness of a flow field can be estimated experimentally. For the sinusoidal motion of an airfoil, if the unsteady effects are linear, then the resulting surface pressures and forces should also vary sinusoidally (though possibly with some shift in phase), and their amplitudes should depend linearly on the amplitude of the motion. Violation of either of these conditions, for example, resulting forces with higher harmonic content or ones that depend nonlinearly on the amplitude of the motion, would be indications of nonlinearity.

Recent detailed experimental observations by Tijdeman [30,31] give an indication of the effect of free-stream Mach number on linearity (Fig. 23). Results are given in terms of quasisteady wing-moment derivative m_c and hinge-moment derivative n_c. The model has a trailing-edge quarter-chord flap, and here three values of mean flap deflection were considered. The results shown have been obtained by considering vanishingly small flap deflections about δ_0. The results indicate that wing-moment derivatives for various mean angles δ_0 are identical up to about the critical Mach number $M_\infty = 0.85$, which means that the moment varies linearly with amplitude δ_0

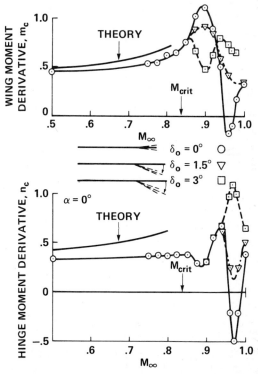

FIG. 23 Wing-moment and hinge-moment derivatives as functions of Mach number for three mean flap deflections.

up to $M_\infty = 0.85$ for amplitudes up to at least 3°. At transonic Mach numbers, the derivatives depend strongly on the mean flap position, and linearity can be assumed only for vanishingly small amplitudes around the given mean flap position. This sensitivity of the wing-moment derivative to mean flap position is to a large extent due to the relatively large changes in mean shock position with mean flap angle, as illustrated in Fig. 24. The hinge-moment derivatives for various δ_0 diverge at a higher Mach number than the curves for the wing-moment derivatives. This is to be expected because changes in hinge moment due to changes in shock positions only become important when the shocks approach the hinge axis, which occurs at about $M_\infty = 0.94$.

Tijdeman's results indicate that, for subcritical flows, forces vary linearly with amplitude, and, hence, one would expect that unsteady forces could be computed by superposing the unsteady and steady-state solutions. However, for supercritical flows, forces do not vary linearly with flap-deflection amplitude, and thus one would expect superposition to be applicable only for very small amplitudes of unsteady motion.

Tijdeman's observations further imply that, for an oscillatory motion, the

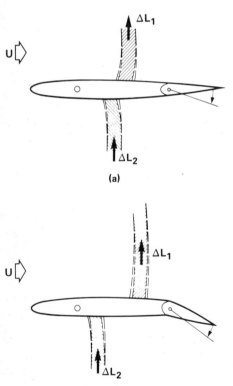

FIG. 24 Effect of flap angle on shock position. (a) Flap deflection about zero mean flap angle; (b) flap deflection about nonzero mean flap angle.

shock excursion amplitude is an important factor in assessing the validity of the linear unsteady perturbation assumption; that is, the motion amplitude range over which the linearity assumption can be considered valid depends on the shock excursion amplitude produced by the motion. Certainly, surface pressures in regions traversed by a shock wave will not be linear functions of the motion. The larger these regions become, the greater effect they have on forces, which are integrated surface pressures. The amplitude restriction can then be stated as follows: For the unsteadiness in a flow field to be considered a linear perturbation about some mean steady state, the amplitude of the motion must be sufficiently small that no appreciable shock motion results.

Of course, factors other than motion amplitude affect the shock excursion amplitude. For example, Tijdeman found that the shock excursion amplitude increases with a decrease in either motion frequency or chordwise Mach-number gradient upstream of the shock. For example, one would expect the amplitude restriction to be most severe for low-frequency oscillations of supercritical airfoils, which have relatively flat upper-surface chordwise Mach-number distributions.

Conclusions to be drawn from Tijdeman's experiments provide some guidelines for the development of computational algorithms for transonic flows:

1   For oscillation amplitudes that are of the same order as the thickness ratio of the airfoil (or the mean angle of attack, whichever is larger), the solution cannot be separated into steady and unsteady components unless some additional restriction, for example, high frequency, is imposed such that the resulting shock excursion amplitude is very small. Time-accurate integration of the equations governing unsteady transonic flows is required.

2   For infinitesimal amplitude disturbances, the unsteady flow can be treated as a linear perturbation about a mean (nonlinear) steady-state flow. The computational procedure can be greatly simplified in this case because superposition is applicable. However, the unsteady component of the solution depends strongly on the steady component. This, unfortunately, requires that a large number of steady-state conditions be considered to define flutter boundaries adequately, since steady-state solutions cannot be superposed in the transonic case.

3.4   Two Procedures for Separating the Solution into Steady and Unsteady Components

3.4.1   Harmonic Approach

For subcritical flows the most common procedure used to separate the solution into steady and unsteady components is the *harmonic* approach [32,33]. It is assumed that the solution for some sinusoidal motion of an airfoil with frequency ω can be expressed in the form

$$\phi(x, y, t) = \phi_0(x, y) + \phi_1(x, y)e^{i\omega t} \qquad (29)$$

where ϕ is some dependent variable, for example, the disturbance velocity potential. Here ϕ_0 is the steady-state solution, which is a function only of the spatial coordinates x and y. The unsteady component ϕ_1 is complex, thus allowing for phase differences between the motion and the resulting flow field. If the governing equations are linear, as in the case of purely subsonic or supersonic flow, ϕ_1 represents the unsteady motion of a flat plate and is independent of ϕ_0. Solutions for arbitrary airfoil motion can be constructed by superposing solutions obtained for each ω in the frequency spectrum of the motion.

In the transonic case, a similar, but less general, approach can be used [34–37]. The disturbance velocity potential for the sinusoidal motion of an airfoil is expressed as

$$\phi(x, y, t) = \phi_0(x, y) + \varepsilon\phi_1(x, y)e^{i\omega t} + \varepsilon^2\phi_2(x, y)e^{2i\omega t} + \cdots \qquad (30)$$

where ε is related to the amplitude of the motion. Usually terms of order ε^2 and higher are neglected, thus invoking the linear-perturbation assumption with the accompanying amplitude restrictions previously discussed. In this case, the unsteady component ϕ_1 depends on ϕ_0. This approach has the advantage that ϕ_1 can be computed using essentially the same well-known finite-difference relaxation algorithms used to compute the mean steady-state condition ϕ_0. It has the disadvantage that ϕ_1 depends on ω, so that a solution ϕ_1 for the complete flow field must be computed for each frequency considered.

3.4.2   Indicial Approach

An alternative method for separating the solution into steady and unsteady components is the *indicial approach* [38–39]. If the indicial response for an instantaneous change in some motion (pitch, plunge, etc.) is known, then the solution for any arbitrary schedule of that motion can be found with the aid of Duhamel's integral. For example, consider some arbitrary variation of angle of attack α as a function of time, as shown in Fig. 25a; and suppose that the lift-coefficient response to a change in angle of attack is given as shown in Fig. 26. The motion can be divided into equal $\Delta\alpha$ increments. The value of C_l (Fig. 25b) at some time t is then given by the sum of the increments ΔC_l due to the $\Delta\alpha$ increments (the governing equations are assumed linear). The increments in C_l for the various $\Delta\alpha$ steps at time t are equivalent to increments in the first step (indicated by the arrows) at $t - \tilde{t}$ where $\tilde{t}$ is the

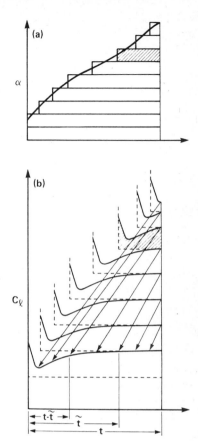

FIG. 25 C_l response for arbitrary angle-of-attack schedule by super-position. (a) α versus time; (b) C_l versus time.

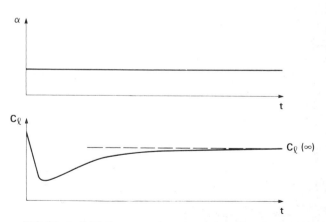

FIG. 26 Indicial C_l response to a step change in angle of attack.

time at which the corresponding $\Delta\alpha$ is initiated. The total lift at time t then is given by the summation [39]

$$C_l(t) = C_{l_\alpha}(t)\alpha(0) + \sum_0^t C_{l_\alpha}(t - \tilde{t})\frac{\Delta\alpha}{\Delta t}(\tilde{t})\,\Delta\tilde{t} \tag{31}$$

where $C_{l_\alpha}(t)$ is the indicial response to a unit change in α. Introducing a change of variable and letting the time increment approach zero leaves

$$C_l(t) = C_{l_\alpha}(t)\alpha(0) + \int_0^t C_{l_\alpha}(\tau)\frac{d}{dt}\alpha(t - \tau)\,d\tau \tag{32}$$

Once the indicial response $C_{l_\alpha}(t)$ is known, the lift coefficient for any $\alpha(t)$ schedule can be found from Eq. (32). For example, this approach can be used to determine the surface pressures and forces resulting from the sinusoidal oscillation of an aerodynamic configuration. Let the motion be given by

$$\alpha(t) \equiv \alpha_0 + \alpha_1 e^{i\omega t} \tag{33a}$$

and define

$$\Delta C_{l_\alpha}(\tau) \equiv C_{l_\alpha}(\infty) - C_{l_\alpha}(\tau) \tag{33b}$$

where $C_{l_\alpha}(\infty)$ is the steady-state change in lift due to a unit change in α. Then, substituting into Eq. (32) and taking the limit as $t \to \infty$ (since we are only interested in the periodic solution) gives

$$C_l(t) = \alpha_1 e^{i\omega t}\left[C_{l_\alpha}(\infty) - i\omega \int_0^\infty \Delta C_{l_\alpha}(\tau)e^{-i\omega t}\,d\tau\right] + C_{l_\alpha}(\infty)\alpha_0 \tag{34}$$

The last term represents the mean steady-state lift, and hence the expression for $C_l(t)$ has been separated into steady and unsteady components. The solution can be expressed in terms of real and imaginary components

$$\bar{C}_l \equiv \frac{C_l - C_{l_\alpha}(\infty)\alpha_0}{\alpha_1 e^{i\omega t}}$$

$$\mathrm{Re}[\bar{C}_l] = C_{l_\alpha}(\infty) - \omega \int_0^\infty \Delta C_{l_\alpha}(\tau)\sin\omega\tau\,d\tau \tag{35}$$

$$\mathrm{Im}[\bar{C}_l] = -\omega \int_0^\infty \Delta C_{l_\alpha}(\tau)\cos\omega\tau\,d\tau$$

In practice, the integrals are evaluated over a finite time interval that is at least as great as the period of the motion considered [40]. Other aerodynamic forces and the surface pressures, as well as other types of motion, can be handled in a similar manner using Duhamel's principle, providing the governing equations are linear as in the purely subsonic or purely supersonic cases.

This approach can also be used to treat transonic flows if the unsteadiness can be considered a linear perturbation about a nonlinear steady state; and it is interesting to compare the harmonic and indicial approaches in terms of computer efficiency for transonic-flow applications. Recall that the harmonic approach requires that a relaxation solution of the flow field for the unsteady component be computed for each frequency of interest. In the indicial approach, a flow-field solution (i.e., the indicial response) must also be computed. However, with this single flow-field computation, the solution for any frequency can be computed by evaluating the simple integrals in Eq. (35), and the amount of computational effort involved is negligible compared to that required for a relaxation solution. Thus, a large number of frequencies can be considered for a cost slightly greater than considering only one. The question of efficiency, then, depends on (1) the number of frequencies to be considered, and (2) the relative cost of computing an indicial response as opposed to relaxation solutions.

3.5 The Case of Finite-Amplitude Shock Excursions

It has already been established that time-accurate integrations of the governing gas dynamic equations are required for unsteady flows with shock excursion amplitudes that are other than very small. In the construction of numerical schemes to treat such cases, some insight can be gained from experimental measurements. Here again, we rely on Tijdeman's observations of the flow field about an airfoil with an oscillating trailing-edge flap.

Tijdeman has identified three types of periodic shock motion resulting from sinusoidal flap motion (Fig. 27).

Type A, sinusoidal shock-wave motion The shock moves nearly sinusoidally (only the lowest harmonic was measured) but with a phase shift relative to the sinusoidal flap motion. There also exists a phase shift between the shock motion and its strength; that is, the maximum shock strength is not encountered when the shock reaches its maximum downstream position, as in the steady case, but at a later time, during its upstream motion. Shock excursion amplitudes decrease for higher frequencies and for larger chordwide Mach-number gradients upstream of the shock.

Type B, interrupted shock-wave motion The shock moves as in type A, but now the oscillatory shock strength is of the same magnitude as the mean steady-shock strength. Hence, the shock weakens such that it disappears during the downstream-moving portion of its cycle.

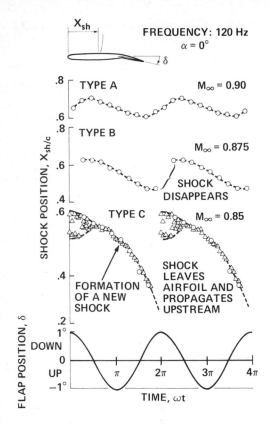

FIG. 27 Experimentally observed shock-wave motions.

Type C, upstream propagating shock waves At slightly supercritical conditions, a shock wave that moves upstream is periodically formed. Eventually, it propagates off the front of the airfoil and continues to travel upstream into the incoming flow.

(Tijdeman points out that, in principle, these three types of shock motion could also be produced for a fixed Mach number by variation of the amplitude and frequency, instead of by a variation in Mach number as in Fig. 27.) For computational algorithms that "capture" shocks, as opposed to those that "fit" them as internal boundaries, spatial grid points should be concentrated in the region traversed by a shock during its motion cycle. The extent of this region must be anticipated a priori, and it can be quite large, especially for low-frequency type A and B motions. Type C motions pose a greater difficulty because the shock actually travels off the surface of the airfoil.

3.6 Simplifications of the Governing Equations and Boundary Conditions

In practical transonic-flow applications, three-dimensional and viscous effects are often as important as nonlinear (mixed-flow) effects. For example, in Tijdeman's experiment, boundary-layer influences were found to be of the same order as thickness (nonlinear) effects. Presently, the coupling of unsteady and nonlinear effects is just beginning to be studied in detail computationally, and just recently, some results have even been reported for a simple three-dimensional configuration. Continued progress can be expected in these areas during the next few years. However, the computational treatment of viscous effects, especially separation, for either three-dimensional or unsteady flows is much more difficult, and empirical treatment will probably be required, at least in the near future. Here we consider only inviscid, two-dimensional, unsteady transonic flows, which are assumed to be governed by the Eulerian gas dynamic equations.

The Euler equations expressing conservation of mass, x and y momentum, and energy are written in the familiar conservation form as

$$\frac{\partial}{\partial t} \begin{Bmatrix} \rho \\ \rho u \\ \rho v \\ e \end{Bmatrix} + \frac{\partial}{\partial x} \begin{Bmatrix} \rho u \\ \rho u^2 + p \\ \rho u v \\ u(e + p) \end{Bmatrix} + \frac{\partial}{\partial y} \begin{Bmatrix} \rho v \\ \rho u v \\ \rho v^2 + p \\ v(e + p) \end{Bmatrix} = 0 \tag{36}$$

with the equation of state

$$p = (\gamma - 1)[e - \tfrac{1}{2}(\rho u^2 + \rho v^2)] \tag{37}$$

where ρ, p, and e are the density, pressure, and total energy per unit volume, and γ is the ratio of specific heats, equal to 1.4 for air.

The boundary conditions associated with Eq. (36) are that there be no flow through the surface of the body and that the pressure and the flow direction be continuous at the trailing edge. In practice, conditions are also specified at the outer boundaries of the computational domain, which are far removed from the body. The Rankine-Hugoniot relations must be satisfied across shock waves to ensure conservation of mass, momentum, and energy. In cases in which only a steady state is required, the energy equation can be replaced [41] by the expression for constant total enthalpy H,

$$H = \frac{\gamma}{\gamma - 1} \frac{p}{\rho} + \frac{u^2 + v^2}{2} = \text{constant} \tag{38}$$

thereby reducing the number of dependent variables from four (ρ, u, v, e) to three (ρ, u, v).

Various degrees of approximation can be made, to simplify both Eq. (36) and the accompanying boundary conditions. These simplifications usually impose some limitations on the applicability of the resulting equation. For example, for small-disturbance transonic flows, $\tau^{2/3} \sim 1 - M_\infty^2 \ll 1$, where τ is the body thickness-to-chord ratio and M_∞ is the free-stream Mach number, and the approximation takes the form [42]

$$k^2 M_\infty^2 \phi_{tt} + 2k M_\infty^2 \phi_{xt} = V_c \phi_{xx} + .\phi_{yy} \qquad (39)$$

where V_c is $1 - M_\infty^2 - (\gamma + 1) M_\infty{}^m \phi_x$; t has been scaled by $(\omega)^{-1}$; and k is the reduced frequency. (For an airfoil of chord length c, traveling with speed U_∞, and executing some unsteady oscillatory motion of frequency ω, $k \equiv \omega c / U_\infty$ and thus is a measure of the degree of unsteadiness of the flow field, since it is a ratio of the time scale of the airfoil flight speed c/U_∞ to that of the motion $1/\omega$. The units of k are radians of oscillatory motion per chord length of airfoil travel.) Equation (39) contains shock conditions that are transonic small-disturbance approximations to the Rankine-Hugoniot equations. The number of dependent variables has now been reduced to one, the disturbance velocity potential ϕ; the velocity can be expressed as the vector sum of the free-stream velocity U_∞ and the gradient of ϕ. The right-hand side of Eq. (39) is the familiar two-dimensional transonic small-disturbance equation for steady flows.

The small-disturbance approximation results in a significant simplification in the application of the condition of no flow through the body surface. The condition is enforced by specifying ϕ_y, which depends only on the local airfoil slope, on a straight-line mean-surface approximation to the airfoil, and this mean surface is independent of any unsteady motion of the body. For lifting bodies, a constant jump in potential across the vortex sheet $y = 0$ must be enforced along free-stream particle paths downstream of the body. Values of ϕ or derivatives of ϕ (with ϕ defined for at least one point) are also specified on the boundaries of the computational domain. The boundary conditions are illustrated in Fig. 28.

A further simplification can be achieved for the physically interesting low-frequency case $k \ll 1$ by dropping the first term in Eq. (39). To understand the implications of such a simplification, it is helpful to investigate its effect on the characteristic surfaces, the wave fronts along with information is propagated throughout the flow field. Let us suppose that, at $t = 0$, an instantaneous disturbance of infinitesimal strength occurs at the point $x = 0$, $y = 0$ in some uniform stream with velocity u and speed of sound a; we can then rewrite Eq. (39) in the form

$$A\phi_{tt} + 2B\phi_{xt} = C\phi_{xx} + \phi_{yy} \qquad (40)$$

where now, if we drop the scaling $1/\omega$ for t and replace M_∞ by u/a, A

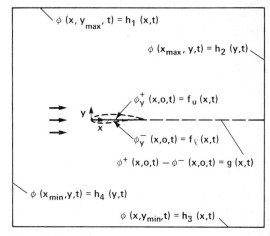

FIG. 28 Boundary conditions for the small-disturbance equation.

becomes $1/a^2$, B is u/a^2, and C is $1 - (u/a)^2$. The characteristic equation for Eq. (40) is

$$Ct^2 - Ax^2 - (AC + B^2)y^2 + 2Bxt = 0 \qquad (41)$$

At some time $t > 0$, the disturbance front is given by

$$\left(x - \frac{Bt}{A}\right)^2 + \frac{AC + B^2}{A}y^2 = \frac{AC + B^2}{A^2}t^2 \qquad (42a)$$

or, equivalently,

$$(x - ut)^2 + y^2 = a^2t^2 \qquad (42b)$$

The disturbance front propagates at the speed of sound relative to the fluid; it is a circle with radius at and center at $x = ut$, $y = 0$ as shown in Fig. 29. The disturbance center corresponds to the location of the fluid particle that was at the point of the disturbance at $t = 0$, and its moves with velocity u. In the plane $y = 0$, the effect of the disturbance propagates upstream (for $u < a$) with velocity $u - a$, and downstream with velocity $u + a$. The same results would be obtained from the complete Euler equations (36), and can be summarized as follows:

1 Disturbance-front radius is $\sqrt{AC + B^2}\, t/A = at$.
2 Center of disturbance front is $y = 0$, $x = (B/A)t = ut$.
3 Speed of sound is $\sqrt{AC + B^2}/A = a$.
4 Particle speed is $B/A = u$. $\qquad\qquad (43)$
5 Upstream propagation rate for $y = 0$ is $(B - \sqrt{B^2 + AC})/A = u - a$.
6 Downstream propagation rate for $y = 0$ is $(B + \sqrt{B^2 + AC})/A = u + a$.

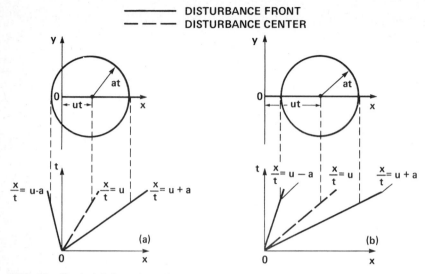

FIG. 29 Characteristic surfaces for the small-disturbance equation. (a) Subsonic, $u < a$; (b) supersonic, $u > a$.

Now consider the effect of neglecting the ϕ_{tt} term in Eq. (39) by taking $A = 0$ in Eq. (40). Equations (43) become, respectively,

$$\lim_{A \to 0} \frac{\sqrt{AC + B^2}\, t}{A} = \infty$$

$$y = 0, \qquad x = \lim_{A \to 0} \frac{Bt}{A} = \infty$$

$$\lim_{A \to 0} \frac{\sqrt{AC + B^2}}{A} = \infty$$

$$\lim_{A \to 0} \frac{B}{A} = \infty \qquad\qquad (44)$$

$$\lim_{A \to 0} \frac{B - \sqrt{B^2 + AC}}{A} = -\frac{C}{2B^2} = \frac{a + u}{2u}(u - a)$$

$$\lim_{A \to 0} \frac{B + \sqrt{B^2 + AC}}{A} = \infty$$

The sound and particle speeds become infinite, and hence the downstream propagation rate, which is the sum of the two, is infinite. However, the upstream propagation rate, the difference between the two, is finite and is a good approximation to the upstream propagation rate $u - a$ in Eq. (43), since $u \sim a$ for transonic flows. The disturbance front at any time t is a parabola as shown in Fig. 30.

The low-frequency approximation assumes that the time scale of the motion is of the same order as the time scale associated with upstream disturbance propagation and that these scales are much larger than the time scales associated with downstream propagation and convection; that is,

$$\frac{1}{\omega} \sim \frac{c}{u - a} \gg \frac{c}{u + a} \sim \frac{c}{u}$$

where c is the chord length of the body. Hence, the low-frequency condition is $k \equiv \omega c / u \ll 1$.

The infinite downstream convection speed resulting from the low-frequency approximation allows an additional simplification in the case of lifting bodies. The jump in potential can be assumed uniform in x from the trailing edge to downstream infinity as in the steady case; that is, the downstream flow adjusts instantaneously to changes in lift on the body. The reason is that the rate of change in lift due to the motion of the body is much smaller than the rate at which the effect of these changes is propagated downstream.

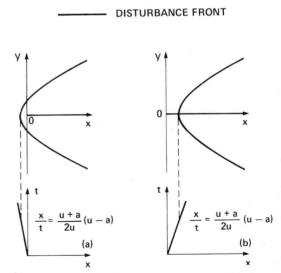

FIG. 30 Characteristic surfaces for the low-frequency approximation. (a) Subsonic, $u < a$; (b) supersonic, $u > a$.

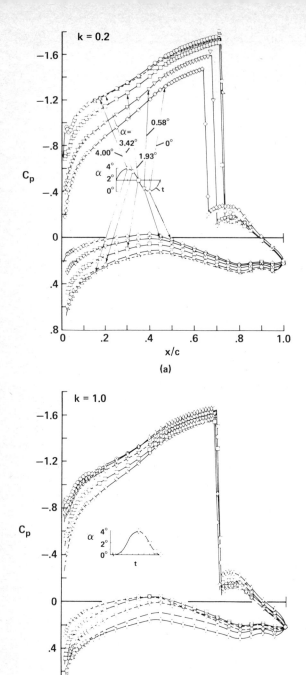

FIG. 31 Unsteady pressure distributions. (a) $k = 0.2$; (b) $k = 1.0$.

3.7 Computed Results for Unsteady Airfoil Motions

3.7.1 Euler Equations

Computed results presented in this and the following section give an indication of the present state of the art in the numerical simulation of unsteady transonic flows. Time-accurate solutions to the Euler equations (36) are considered first, followed by solutions that are obtained using some of the approximations discussed previously.

The most ambitious computations undertaken to date solved the Euler equations (36) for sinusoidal pitching oscillations (from 0 to 4°) of the NACA 64A410 airfoil [43]. A two-step Lax-Wendroff difference scheme was used, and the airfoil boundary condition was enforced on the surface of the airfoil at its mean location. Resulting surface pressures for reduced frequencies of 0.2 and 1.0 are shown in Fig. 31. The pitching motion in this case produced a shock motion that corresponds to Tijdeman's type A; the higher reduced frequency in the Magnus-Yoshihara computations resulted in a smaller shock excursion amplitude (as Tijdeman noted in the oscillating-flap experiment). Magnus and Yoshihara report that normal forces and surface pressures (except in the regions traversed by shock waves) varied nearly sinusoidally with varying degrees of phase shift. However, the axial forces, pitching moments, shock locations, and surface pressures near shocks all showed significant deviation from pure sinusoidal behavior, especially for the lower frequencies. This is evident in Fig. 32, which shows the unsteady variation of pitching moment, shock location, and pressure coefficient (on the upper surface of the airfoil at 80% chord), each compared with a corresponding least-square fit to a sine curve. As noted earlier, the presence of

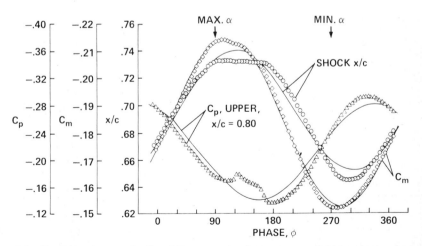

FIG. 32 Selected unsteady $(k = 0.2)$ traces showing presence of higher harmonics. Least-squares-fitted fundamentals are shown as light solid lines.

higher harmonics in these quantities is an indication of significant nonlinear behavior over the amplitude range considered.

The Magnus-Yoshihara procedure has also been used to compute the indicial flow-field response to a step change in angle of attack of an NACA 64A410 airfoil. In reference 21, particular attention is paid to the various stages of unsteady flow-field development, and in reference 43, a preliminary effort was made to compute solutions for oscillatory motion using the indicial method. Results for the normal force compared more favorably with time-accurate solutions than did the results for pitching moment. This is to be expected since the pitching-moment response, as indicated earlier, showed more evidence of nonlinearity. In the application of this or any other method to practical problems, two questions naturally arise: (1) How accurately do the computed results simulate this flow field? (2) How much does it cost in terms of workhours and computer hours to compute a solution? First of all, solutions to inviscid governing equations cannot be expected to predict real flow fields that are dominated by viscous effects. The flow fields computed by Magnus and Yoshihara [43] all have embedded shock waves with Mach numbers greater than about 1.3 throughout the entire motion cycle. The authors point out that, for such cases, the boundary layer would almost certainly be separated at the shock, and shock strengths and locations would be quite different from the inviscid case. For flows with weaker shocks, their method would be expected to provide good quantitative results. However, the cost of obtaining solutions precludes its use in most engineering applications. For example, the most expensive case, $k = 0.2$, required 7 h of CDC 7600 computer time (changing compilers and reprogramming have since reduced the requirement to 2 h). The Magnus-Yoshihara code was actually never intended to be used for engineering applications but was designed to compute a limited number of standard solutions to evaluate faster approximate methods.

Two approaches for solving the Euler equations more efficiently in certain limited applications have been investigated by Beam and Warming. To date, their work has been limited to parabolic-arc airfoils, for which they apply the boundary condition in terms of airfoil slopes on the airfoil mean surface. In their first investigation, Beam and Warming used the indicial method with an explicit finite-difference scheme providing the steady-state and indicial solutions [40]. They were able to predict negative aerodynamic damping in cases where it was not predicted by linear theory but was found to exist experimentally. More recently, they have developed an implicit finite-difference algorithm designed to solve the Euler equations efficiently for low-frequency motions. Computed results (the algorithm is outlined in Sec. 3.9) compared with linear theory for plunging airfoil oscillations illustrate the strong influence of shock waves on the resulting aerodynamic forces—especially on the out-of-phase components, which represent aerodynamic damping (cf. [44,45]).

3.7.2 Small-Disturbance Equations

The only time-accurate method developed to date to solve the unsteady small-disturbance equation (39) was reported in Ballhaus and Lomax [46]. Computed surface pressures were compared with results from linear theory and from a numerical solution to the low-frequency equation for the thickening motion of a parabolic-arc airfoil.

Ehlers [34], and more recently Weatherill et al. [35], have computed solutions to the small-disturbance equation using the harmonic approach with a finite-difference relaxation method. (A very similar approach has been applied by Traci et al. [36,47] to solve the low-frequency equation.) Results from reference 31 of Ehler's computations for Tijdeman's airfoil with an oscillating flap are shown in Fig. 33 in terms of surface-pressure differences across the airfoil. Also plotted are experimental and linear (flat-plate) theoretical results. The comparisons indicate the influence of nonlinearity (note the effect of the shock near midchord), and Ehler's results show good qualitative agreement with experiment. Surface-pressure differences at the plane of symmetry of an aspect-ratio-5 wing with an NACA-64A006 airfoil section are compared with two-dimensional results in Fig. 34. The motion is pitching oscillations [35] at a reduced frequency of 0.12. A significant reduction in shock influences caused by three-dimensional " relief " is evident.

The methods of Ehlers and Traci et al. should be easily extendable to the treatment of more practical three-dimensional configurations and should provide considerable insight into the nature of the resulting flow fields. However, there are two notable deficiencies in their approach. The first,

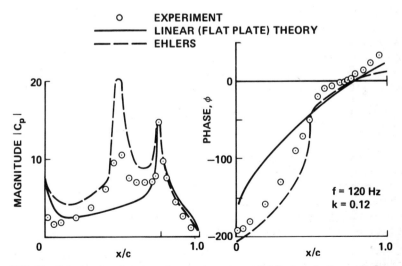

FIG. 33 Calculated and measured unsteady pressure distributions in transonic flow (NACA 64A006 airfoil with flap).

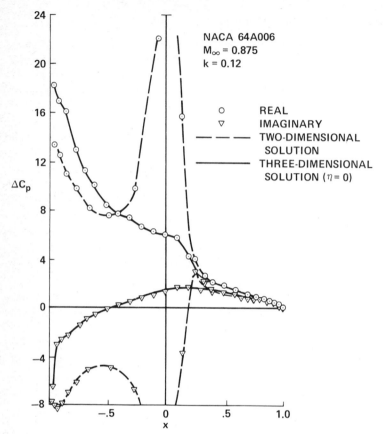

FIG. 34 Pressure-coefficient distributions for an aspect-ratio-5 rectangular wing in pitch.

discussed previously, is an inherent limitation in the harmonic assumption itself; that is, shock locations must remain essentially unchanged during the motion cycle. This increasingly limits the permissible motion amplitude as lower frequencies are considered. The second deficiency is in the finite-difference (relaxation) algorithm used to obtain the steady and unsteady components of the solution. Proper conservation form of the difference equations is not maintained at shocks, and, hence, erroneous shock jumps and speeds are likely to be predicted (e.g., shock speeds that depend on mesh spacing). In some steady-state results, this type of nonconservative differencing has provided what might be considered empirical (but fortuitous) corrections for the effect of the boundary layer on the shock strength and location (cf. Fig. 7). However, there is nothing at this point to suggest that the use of nonconservative differencing would also fortuitously provide a good correction for boundary-layer influences on shock speeds.

3.7.3 The Low-Frequency Approximation

Time-accurate schemes for solving the low-frequency equation are reported in Ballhaus and Lomax [46] and Ballhaus and Steger [47]. Computed (linear and nonlinear) and exact (linear) surface pressures were compared [46] for the case of an impulsively started parabolic-arc airfoil and that of a thickening parabolic-arc airfoil. The algorithm was then extended to the lifting case and the computed results compared with solutions to the Euler equations [44] for plunging oscillations of a parabolic-arc airfoil. The comparison indicated that the low-frequency approximation is valid for reduced frequencies (based on the chord) up to about 0.2. Of course, the low-frequency approximation is also limited by the usual transonic small-disturbance restrictions. The most noticeable breakdown of the approximation resulting from violation of these restrictions was for flows with shock Mach numbers greater than about 1.3, in which case the small-disturbance shock conditions are a poor approximation to the Rankine-Hugoniot conditions, the shock conditions associated with the Euler equations. This also accounted for the poor agreement [21] obtained in comparisons of the low-frequency approximation results with Magnus and Yoshihara's solutions of the Euler equations for pitching oscillations of the NACA 64A410 airfoil. Steady-state comparisons at $\alpha = 0$, 2, and 4° showed similar discrepancies. Poor agreement in such extreme cases is no reason to abandon the small-disturbance approaches, however, since without viscous corrections that include the effect of shock-induced separation, neither the Euler nor the small-disturbance approach can be expected to provide results that compare favorably with experiment.

3.8 Computed Results for Helicopter Rotor Blades

A small-perturbation analysis of the complex flow field near the advancing blade tip of a helicopter rotor in forward flight was carried out [48], and a governing small-perturbation equation was derived. It was concluded that for transonic tip Mach numbers, the flow field would likely be higher nonlinear, unsteady, and three-dimensional. Several results reported for rotors in hover (in which case the governing equations are steady in a coordinate system fixed with respect to the blade) confirm the importance of nonlinear and three-dimensional effects [49–51]. More recently, computations have been performed to assess the combined effects of nonlinearity and unsteadiness.

The unsteady motion of a rotor blade in forward flight was simulated computationally [44] by an NACA 0010 airfoil undergoing simultaneous sinusoidal Mach-number and incidence variations with a phase difference of 180°, as shown in Fig. 35. The governing equation, a two-dimensional version of the small-disturbance equation derived in reference 48, has the same

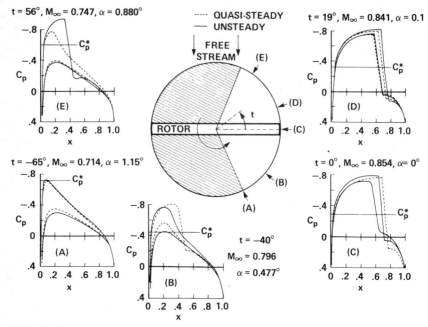

FIG. 35 Simultated helicopter motion (NACA 0010 airfoil).

form as the low-frequency equation, except that the term $M_T(1 + \mu \cos t)$, which represents the instantaneous free-stream Mach number, appears in the boundary conditions and in the coefficients of ϕ_{xt} and ϕ_{xx}. The quantity M_T is the tip Mach number due to rotation, μ is the advance ratio and is equal to the forward-flight Mach number divided by M_T, and the time t is represented by azimuth angle on the rotor disk. Quasisteady results were computed for comparison by solving the same equation with $\phi_{xt} = 0$ so that time appeared only as a parameter. Computed results are shown in Figs. 35 and 36. The shaded region indicates the portion of the cycle for which the flow is subcritical; there the unsteady values of C_p and x_p (center of pressure) are nearly quasisteady. In the supercritical portion, however, the shock-wave locations, as well as C_p and x_p, lag the quasisteady results.

The direction and speed of propagation of the shock can be determined by the differences between C_p^* and the average C_p across the shock. If it is assumed that the shock is nearly normal, then from the shock relations for the governing equations it can be shown that the shock is moving upstream, stationary, or moving downstream, as the average C_p across the shock is greater than, equal to, or less than C_p^*, respectively. Hence, in Fig. 35, the shock on the upper surface is traveling downstream at points B and C, is nearly stationary at point D (corresponding to the maximum rearward drift in center of pressure), and is traveling upstream at point E. In fact, the shock subsequently continues to propagate upstream off the front of the airfoil and into the oncoming flow. This corresponds to Tijdeman's experimentally observed type C shock motion, and it produces the overshoot in x_p and the irregularity in C_l evident in Fig. 36. A similar observation was reported [52] for the three-dimensional flow field about a nonlifting helicopter rotor with a circular-arc section (Fig. 37). In this (nonlifting) case, x_p indicates the center of absolute rather than differential pressure. (Some recent in-flight measurements [53] of the noise generated by a helicopter rotor seem to substantiate the existence of type C motion at high advance ratios.)

A better picture of this type C shock motion is given in reference 47. Here, the motion is simplified to that of a constant-chord nonlifting parabolic-arc airfoil that thickens during the first 15 chord lengths of airfoil travel and thins during the subsequent 15 chord lengths of travel (Fig. 38). This motion produces a flow field similar to that generated by the accelerating and subsequently decelerating motion of a helicopter rotor in forward flight. During the thickening portion of the motion (Fig. 39*a*), a shock wave forms and

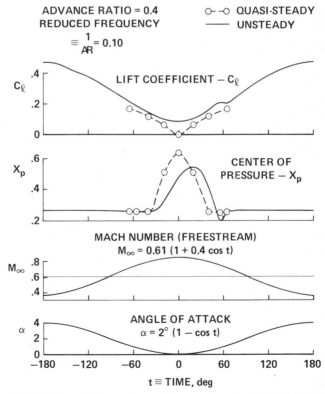

FIG. 36   Lift coefficient and center of pressure for a simulated helicopter-blade element (NACA 0010 airfoil).

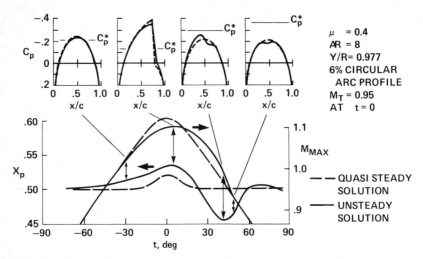

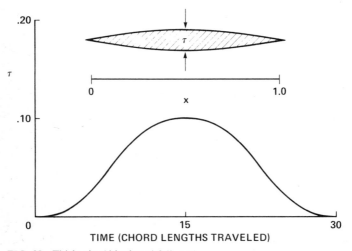

FIG. 37 Center of pressure and maximum Mach number from a helicopter rotor simulation.

propagates downstream, reaching its maximum downstream location at $t = 18.25$ chord lengths of airfoil travel. As the airfoil thins, the shock reverses direction and eventually propagates off the front of the airfoil $(x = 0)$, chasing the expansion wave that precedes it. Again, the difference between C_p^* and the average C_p across the shock is an indication of the speed and direction of shock travel. Note that after about 27 chord lengths of airfoil travel, the pressures are all greater than C_p^*, indicating that the flow is entirely subsonic relative to the airfoil.

FIG. 38 Thickening/thinning airfoil motion.

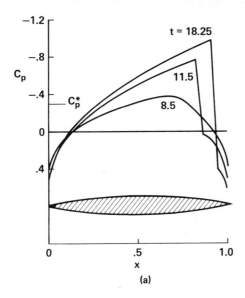

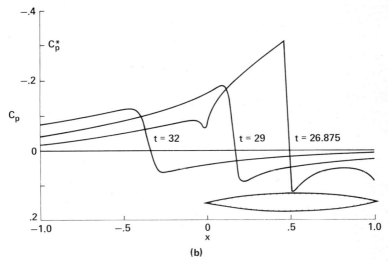

FIG. 39 Pressure coefficient for thickening/thinning airfoil motion at $M = 0.85$. (a) Downstream propagation of the shock wave; (b) upstream propagation of the shock wave.

3.9 Efficient Algorithms for Low-Frequency Motions

Explicit finite-difference methods applied to low-frequency transonic flow are notoriously inefficient because they have an integration time-step restriction for stability that is much more severe than the one required to resolve the unsteady flow field adequately. This situation is known as "parasitic stiffness." The stability restriction is based on the smallest time scale in the problem, that associated with downstream disturbance propagation (along "advancing" waves). On the other hand, the requirement for adequate flow-field resolution is based on the time scale of the motion, which (for low-frequency motion) is of the same order as the time scale corresponding to upstream disturbance propagation (along "receding" waves). A standard approach used to increase computational efficiency in parasitic stiff problems is to remove the severe time-step restriction by the use of implicit difference operators. Several such methods have been developed for the efficient solution of low-frequency transonic flows.

"Semi-implicit" difference operators were developed for both the small-disturbance equation (39) and its low-frequency approximation by Ballhaus and Lomax [46]. The schemes were designed to (1) have a time-step restriction for stability based on the receding rather than the advancing wave propagation time scale, (2) be as consistent as possible with the characteristics of the corresponding partial differential equation, and (3) reduce to the Murman scheme [18] in the steady-state case. Both schemes are implicit in the y direction, thus avoiding a severe time-step restriction due to the shallowness of the characteristic traces in the y-t plane. The differencing in the x-t plane is implicit in the upstream direction and explicit in the downstream direction; that is, to update the solution at some point in the flow field, upstream information at the new time level is required as well as old time-level information. The difference equations must be solved by marching both in x (the free-stream direction) and t. For a given time level, tridiagonal equations for ϕ along y lines (x = constant) are solved successively, marching in the x direction.

The use of semi-implicit operators for low-frequency flows results in a considerable increase in efficiency (usually by more than an order of magnitude) over explicit methods applied to the Euler equations. For example, the $k = 0.2$ case for pitching oscillations of the NACA 64A410 airfoil required 6 min of CDC 7600 run time per cycle for the Ballhaus-Lomax procedure as opposed to 210 min for the Magnus-Yoshihara procedure. Yet the low-frequency approach can be made even more efficient.

An unnecessarily severe time-step limitation is imposed with the semi-implicit schemes near singular points, such as the airfoil leading and trailing edges, where the small-disturbance assumptions break down. To avoid complications in coding, this restrictive time step is used in updating every point in the flow field. To overcome this deficiency, Ballhaus and Steger [47] developed fully implicit approximate-factorization schemes that have no time-step limitation based on a linear stability analysis. One of these schemes was used to compute the solution shown in Fig. 39. In principle, relatively large time steps can be taken with the factored schemes if the flow-field response to some motion has only low-frequency content. In practice, however, the time step for accuracy (and, in some cases, even stability) is limited by high-frequency content in the solution due to the motion of shock waves, although this restriction is not nearly so severe as the one associated with the semi-implicit difference schemes. The restriction could be removed entirely by "fitting" embedded shock waves as internal boundaries rather than "capturing" them automatically with the difference scheme.

The implicit approximate-factorization methods of Ballhaus and Steger for the low-frequency equation are sufficiently similar to the implicit scheme constructed for the Euler equations by Beam and Warming [45] that only the latter will be reproduced here. (Note that the discussion in the preceding paragraph concerning time-step restrictions based on shock motion also applies to the Beam and Warming algorithm.) The Euler equations can be expressed in the conservation-law form

$$\frac{\partial u}{\partial t} + \frac{\partial F(u)}{\partial x} + \frac{\partial G(u)}{\partial y} = 0 \tag{45}$$

where u is an unknown p-component vector, and F and G are given vector functions of the components of u. Let $u(t) = u(n\,\Delta t) = u^n$, where Δt is the integration time step. The time differencing is the trapezoidal formula

$$u^{n+1} = u^n + \frac{\Delta t}{2}\left[\left(\frac{\partial u}{\partial t}\right)^n + \left(\frac{\partial u}{\partial t}\right)^{n+1}\right] + O(\Delta t^3) \tag{46}$$

Inserting Eq. (45) in Eq. (46) gives

$$u^{n+1} = u^n - \frac{\Delta t}{2}\left[\left(\frac{\partial F}{\partial x} + \frac{\partial G}{\partial y}\right)^n + \left(\frac{\partial F}{\partial x} + \frac{\partial G}{\partial y}\right)^{n+1}\right] + O(\Delta t^3) \tag{47}$$

The nonlinearity of the problem presents an obvious difficulty since we want to solve for u^{n+1}, but $F^{n+1} = F(u^{n+1})$ and $G^{n+1} = G(u^{n+1})$ are nonlinear functions of u^{n+1}. The nonlinearity can be removed while maintaining the order of accuracy by the use of the Taylor-series expansions

$$F^{n+1} = F^n + A^n(u^{n+1} - u^n) + O(\Delta t^2)$$
$$G^{n+1} = G^n + B^n(u^{n+1} - u^n) + O(\Delta t^2) \tag{48}$$

where A and B are Jacobian matrices $A = \partial F/\partial u$ and $B = \partial G/\partial u$. It can be shown [45] that for the Euler equations

$$F = Au$$
$$G = Bu \tag{49a}$$

and, hence, Eqs. (48) simplify to

$$F^{n+1} = A^n u^{n+1} + O(\Delta t^2)$$
$$G^{n+1} = B^n u^{n+1} + O(\Delta t^2)$$

$$(49b)$$

Equation (47) can now be written

$$\left[I + \frac{\Delta t}{2} \left(\frac{\partial}{\partial x} A^n + \frac{\partial}{\partial y} B^n \right) \right] u^{n+1} = \left[I - \frac{\Delta t}{2} \left(\frac{\partial}{\partial x} A^n + \frac{\partial}{\partial y} B^n \right) \right] u^n + O(\Delta t^3)$$

$$(50)$$

The system of algebraic equations generated by Eq. (50), after $\partial/\partial x$ and $\partial/\partial y$ are replaced by difference approximations, is formidably large and cannot be solved efficiently. However, the equation can be expressed in the factored form

$$\left(I + \frac{\Delta t}{2} \frac{\partial}{\partial x} A^n \right) \left(I + \frac{\Delta t}{2} \frac{\partial}{\partial y} B^n \right) u^{n+1}$$
$$= \left(I - \frac{\Delta t}{2} \frac{\partial}{\partial x} A^n \right) \left(I - \frac{\Delta t}{2} \frac{\partial}{dx} B^n \right) u^n + O(\Delta t^3)$$

$$(51)$$

where the additional third-order term

$$\frac{\Delta t^3}{4} \frac{\partial}{\partial x} \dot{A}^n \frac{\partial}{\partial y} B^n \frac{u^{n+1} - u^n}{\Delta t} = \frac{\Delta t^3}{4} \frac{\partial}{\partial x} A^n \frac{\partial}{\partial y} B^n \frac{\partial u^n}{\partial t} + O(\Delta t^4)$$

has been added. The factored form, Eq. (51), reduces the large-bandwidth matrix-inversion problem to 2 one-dimensional problems. The algorithm can be rewritten in the form

$$\overline{u^{n+1}} = \left(I - \frac{\Delta t}{2} \frac{\partial}{\partial y} B^n \right) u^n$$
$$\left(I + \frac{\Delta t}{2} \frac{\partial}{\partial x} A^n \right) \overline{\overline{u^{n+1}}} = \left(I - \frac{\Delta t}{2} \frac{\partial}{\partial x} A^n \right) \overline{u^{n+1}} + O(\Delta t^3)$$

$$(52)$$
$$\left(I + \frac{\Delta t}{2} \frac{\partial}{\partial y} B^n \right) u^{n+1} = \overline{\overline{u^{n+1}}}$$

If three-point central differences are used to approximate the spatial differences, then the second and third steps in Eqs. (52) require the solution of a block tridiagonal matrix, first for the x direction and then for the y direction. To improve the resolution of shock waves, upwind (backward) and central spatial differences are used in supersonic and subsonic regions of the flow field, respectively. Transition operators are required to maintain proper conservation form of the difference scheme when switching between upwind and central spatial differences.

4   COMPUTATION OF SEPARATED TRANSONIC TURBULENT FLOWS ABOUT AIRFOILS

There are several computer codes in existence that treat viscous transonic flow over airfoils; they have been developed primarily to provide inexpensive flow-field predictions for engineering use. Most of these codes have been constructed by linking a boundary-layer code with an inviscid transonic full-potential or small-disturbance code. Computed results are generally satisfactory for cases with boundary-layer separation confined to small regions. These separated regions, as well as the shock–boundary-layer interaction region, trailing-edge region, and wake region, are treated using various approximate procedures.

A more complete treatment of viscous transonic flows about airfoils has been attempted by Deiwert, who solves the time-dependent Reynolds-averaged Navier-Stokes equations with turbulent closure achieved by means of model equations for the Reynolds stresses. The purpose of this effort is to determine if a computational approach using an eddy viscosity model is adequate for simulating separated transonic turbulent flows.

The viscous transonic flow field about an 18% circular-arc airfoil, illustrated schematically in Fig. 40, has been investigated computationally and experimentally, and results are reported in references 54–57.

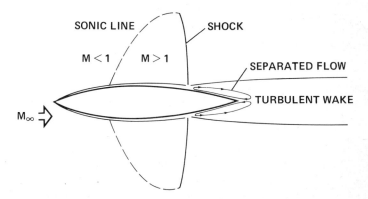

FIG. 40   Schematic of transonic flow over an 18% thick circular-arc airfoil.

4.1 Governing Equations and Method of Solution

The equations governing the two-dimensional, turbulent, compressible flow are

$$\frac{\partial}{\partial t} \iiint\limits_{vol} U \, d \text{ vol} + \oiint\limits_{s} \bar{G} \cdot \bar{\mathbf{n}} \, ds = 0$$

where
$$U = \begin{bmatrix} \rho \\ \rho u \\ \rho v \\ e \end{bmatrix} \qquad \bar{G} = \begin{bmatrix} \rho \bar{q} \\ \rho u \bar{q} + \bar{\bar{\tau}} \cdot \hat{\mathbf{i}}_x \\ \rho u \bar{q} + \bar{\bar{\tau}} \cdot \hat{\mathbf{i}}_y \\ e \bar{q} + \bar{\bar{\tau}} \cdot \bar{q} - k \, \nabla T \end{bmatrix} \tag{53}$$

and
$$\bar{q} = u \hat{\mathbf{i}}_x + v \hat{\mathbf{i}}_y$$

$$\bar{\bar{\tau}} = \sigma_x \hat{\mathbf{i}}_x \hat{\mathbf{i}}_x + \tau_{xy} \hat{\mathbf{i}}_x \hat{\mathbf{i}}_y + \tau_{yx} \hat{\mathbf{i}}_y \hat{\mathbf{i}}_x + \sigma_y \hat{\mathbf{i}}_y \hat{\mathbf{i}}_y$$

k is the thermal-conductivity coefficient, ∇T is the temperature gradient, $\hat{\mathbf{i}}_x$ and $\hat{\mathbf{i}}_y$ are unit vectors, $\bar{\mathbf{n}}$ is a unit normal vector, and σ and τ are the normal and shear stresses. These equations are solved using the finite-volume method discussed in Sec. 5. Some special treatment is required in differencing the viscous derivatives; this is covered in detail in reference 54.

To simulate boundary-layer separation reliably for turbulent flow requires resolution of the boundary layer to the sublayer scale. This sublayer scale is proportional to $1/\sqrt{Re_c}$ (Re_c is the chord Reynolds number) so that, for the high-Reynolds-number flows of interest, the mesh distribution near the surface must be extremely fine. The turbulence model uses a van Driest formulation for the wall region and a Clauser formulation with an intermittancy factor in the outer region (other models have been considered with no notable improvement over the one outlined here). The turbulence modeling is incorporated in the shear-stress term τ_{xy} and τ_{yx} in the form of an eddy viscosity coefficient ε as

$$\tau = \tau_{xy} = \tau_{yx} = (\mu + \varepsilon)\left(\frac{\partial u}{\partial x} + \frac{\partial v}{\partial y}\right) \tag{54}$$

Different values of ε are used for different flow regions:

1   For the wall region,

$$\varepsilon = \rho l^2 \left[\left(\frac{\partial u}{\partial y}\right)^2 + \left(\frac{\partial v}{\partial x}\right)^2\right]^{1/2}$$

where
$$l = 0.41 y[1 - \exp{(-y/A)}] \tag{55a}$$

and
$$A = 26 \frac{\mu_w}{\rho_w} \sqrt{\frac{\rho_w}{|\tau_w|}}$$

2   For the outer region,

$$\varepsilon = \frac{\rho 0.0168 u_\delta \, \delta_i^*}{1 + [(y - y_0)/\delta]^6}$$

where
$$\delta_i^* = \int_{y_0}^{\delta} \left(1 - \frac{u}{u_\delta}\right) dy \tag{55b}$$

3   For the separation bubble, the same as the wall region except

$$\varepsilon = l^2 \max \left\{ \left[\left(\frac{\partial u}{\partial y}\right)^2 + \left(\frac{\partial v}{\partial x}\right)^2 \right]^{1/2}, \frac{u_\delta}{\delta} \right\} \tag{55c}$$

4   For the wake region, the same as the outer region

The molecular and turbulent Prandtl numbers are 0.72 and 0.90. The subscript w refers to wall-surface values, y_0 is the furthest point across the boundary layer at which the velocity is zero (for attached boundary layers, $y_0 = 0$), u_δ is the velocity at the edge of the wake or boundary layer, and δ is the boundary-layer thickness.

4.2   Computed and Experimental Results

An experimental program is in progress in the NASA-Ames high-Reynolds-number channel to provide rigorous verification of viscous transonic flow-simulation codes. The initial results were obtained for an 18% thick circular-arc airfoil, selected for its simplicity and the anticipated presence of both shock-induced and trailing-edge separation in the Mach-number range of interest. In this effort, special precautions were taken to minimize wind-tunnel wall interference effects. For example, the upper and lower tunnel walls were contoured to fit inviscid free-air streamlines (with a wall boundary-layer correction added) generated by Deiwert's computer code for $M_\infty = 0.775$ flow over the subject airfoil. Measured results have been obtained for cases that include both trailing-edge and shock-induced separation.

Computed viscous and inviscid surface pressures from reference 57 are compared with measured pressures for $M_\infty = 0.775$ and $Re_c = 2 \times 10^6$ in Fig. 41. Here, the shock–boundary-layer interaction is weak; that is, the shock is not sufficiently strong to separate the boundary layer. Note, however, that the effect of viscosity is to weaken the shock, displacing it

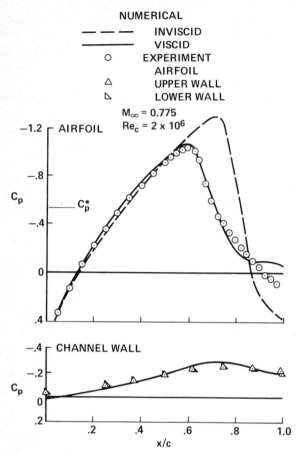

FIG. 41 Computed and experimental pressures for trailing-edge-separation example.

upstream relative to the inviscid results. The computed viscous shock location and strength agree well with experiment. Trailing-edge separation is evident in both the computed viscous and experimental results; the separation effects are more severe in the computed pressures. A detailed investigation of the experimental flow field indicated that there was asymmetric, periodic, unsteady flow in the separated trailing-edge region for the Mach-number range between $M_\infty = 0.76$ and $M_\infty = 0.78$, and this could have contributed to the disagreement. [For $M_\infty < 0.76$, the flow was found to be steady with trailing-edge separation. For $M_\infty > 0.78$, the flow was steady (for $Re_c > 1 \times 10^6$) with the separation point fixed at the base of the shock.] Computed (free-air) and measured tunnel-wall surface pressures agree well (Fig. 41), indicating that the attempt to minimize wind-tunnel wall interference by contouring the tunnel walls was probably successful.

A comparison of surface pressures for $M_\infty = 0.786$ and $Re_c = 10 \times 10^6$ is shown in Fig. 42. In this case, the flow is separated over the entire aft portion of the airfoil, starting at the foot of the shock. The computed viscous solution does not correctly predict the shock location and strength, and it over-predicts the pressure recovery over the aft portion of the airfoil. This is probably due primarily to inadequate turbulence modeling throughout the separated region. The chordwise growth of the separation region in the computations was not large enough to turn the shock near the airfoil surface as in the experiment. The (oblique) experimental shock is thus weaker and located upstream of the (normal) computed shock. (Recently computed results for shock-induced laminar separation do show an oblique shock.) Some wall-interference effects due to the slightly off-design Mach number are probably present, as indicated by the wall surface-pressure comparison.

A qualitative comparison [56] for two similar cases is shown in Fig. 43. For the lower Mach-number case, the shadowgraph of the aft airfoil region indicates the existence of a nearly normal shock and of trailing-edge separation (illustrated by the sudden change in curvature of the boundary-layer

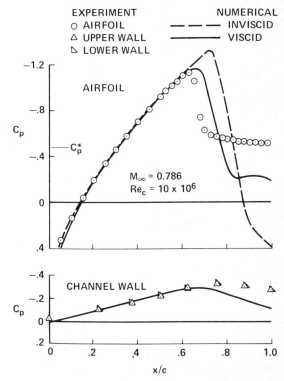

FIG. 42 Computed and experimental pressures for shock-induced-separation example.

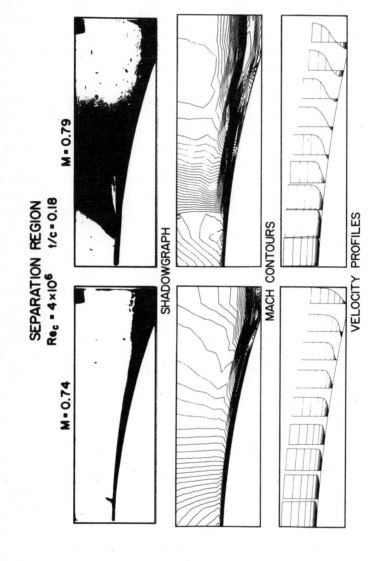

FIG. 43 Flow-field details for the aft portion of an 18% thick circular-arc airfoil.

edge). Computed lines of constant Mach number, in increments of 0.02, show a nearly normal shock and a separated region near the trailing edge. For the higher-Mach-number case, the shadowgraph indicates an oblique shock in the interaction region. The flow behind the shock separates and reattaches downstream in the wake. The computed Mach-number contours show the same type of separation phenomena. However, the shock appears to be normal and, consequently, would be expected to be stronger and located downstream of the experimentally measured shock.

It is clear from the results presented here that it is possible to simulate the basic features of separated, transonic, turbulent flows using the simple turbulence model given by Eq. (55). Computed results are quantitatively correct when the shock–boundary-layer interaction is weak and when there is no appreciable separation. The results are only qualitatively correct for cases with severe separation. Efforts to develop more accurate turbulence models for use in Deiwert's code are continuing, and the code is being modified to treat lifting airfoils.

5 FINITE-VOLUME METHOD

In the computational solution of transonic problems of engineering interest, proper treatment of boundary conditions is often more difficult than solution of the governing equations. Boundary conditions are usually treated by either coordinate transformations, approximations (as in the small-disturbance approaches), or both. An alternative approach is provided by the finite-volume method, introduced by MacCormack and Paullay [58], modified [59], and applied to blunt-body problems [60]. The discussion and results presented here are based on the work of Rizzi [61]. The principal advantage of the method is that the governing equations are solved in a Cartesian coordinate system using a body-oriented and shock-oriented mesh network. One need only know the volume and surface normal directions of the volume elements.

5.1 Governing Equations

The generalized Eulerian formulation expressing conservation of mass, momentum, and energy can be written

$$\frac{\partial}{\partial t} \iiint_{\text{vol}} \begin{bmatrix} \rho \\ \rho\bar{q} \\ e \end{bmatrix} d \text{ vol} + \oiint_{S(t)} \begin{bmatrix} \rho \\ \rho\bar{q} \\ e \end{bmatrix} (\bar{q} - \bar{\lambda}) \cdot \bar{n} \, dS$$

$$+ \oiint_{S(t)} \begin{bmatrix} 0 \\ p\bar{n} \\ p\bar{q} \cdot \bar{n} \end{bmatrix} dS = 0 \qquad (56)$$

The bounding surface S of the volume element has an outward normal $\bar{n}$ and moves with velocity $\bar{\lambda}$. This formulation differs from the usual Eulerian form in that the mesh velocity $\bar{\lambda}$ is included in the governing equations and the volume of every mesh element depends on time. Note that, for $\bar{\lambda} = 0$, Eq. (56) reduces to the usual Eulerian form, while for $\bar{\lambda} = \bar{q}$ it represents the Lagrangian formulation (i.e., the volume elements are fluid particles).

5.2   Mesh System

A typical mesh system is shown in Fig. 44 with the mesh locations given by integer counters j, k. The airfoil surface and shock correspond to $k = 1$ and $k = k_{max}$, respectively. For the case in which the surfaces S_j are fixed, Eq. (56) can be written in the integral form

$$\frac{\partial}{\partial t} (\text{vol} \cdot U) + \iint_{S_j} F \, dS + \iint_{S_k} H \, dS = 0 \tag{57}$$

where

$$U = \begin{bmatrix} \rho \\ \rho\bar{q} \cdot \hat{i}_x \\ \rho\bar{q} \cdot \hat{i}_y \\ e \end{bmatrix} \qquad F(U) = \begin{bmatrix} \rho(\bar{n} \cdot \bar{q}) \\ \rho(\bar{q} \cdot \bar{n})(\bar{q} \cdot \hat{i}_x) \\ \rho(\bar{q} \cdot \bar{n})(\bar{q} \cdot \hat{i}_y) \\ (e + p)(\bar{q} \cdot \bar{n}) \end{bmatrix}$$

$$H(U) = \begin{bmatrix} \rho(\bar{q} \cdot \bar{n} - \lambda) \\ \rho(\bar{q} \cdot \bar{n} - \lambda)(\bar{q} \cdot \hat{i}_x) + p(\bar{n} \cdot \hat{i}_x) \\ \rho(\bar{q} \cdot \bar{n} - \lambda)(\bar{q} \cdot \hat{i}_y) + p(\bar{n} \cdot \hat{i}_y) \\ e(\bar{q} \cdot \bar{n} - \lambda) + p(\bar{q} \cdot \bar{n}) \end{bmatrix}$$

and $\bar{\lambda}$ equals $\lambda\bar{n}$. The velocity is given by $\bar{q} = u\hat{i}_x + v\hat{i}_y$, where u and v are components in the x and y directions in the Cartesian coordinate system. The surfaces S_j and S_k are the surfaces $j = $ constant (D and B in Fig. 44) and $k = $ constant (A and C), respectively. The surface unit normals are given in terms of the corner locations of the volume element. For example,

$$\bar{n}_B = (-\Delta y \, \hat{i}_x + \Delta x \, \hat{i}_y)(\Delta x^2 + \Delta y^2)^{-1/2} \tag{58}$$

where
$$\Delta x = x_{j+1, k} - x_{j+1, k+1}$$

and
$$\Delta y = y_{j+1, k} - y_{j+1, k+1}$$

Note that the curvature of the volume elements is not considered in computing surface normals, since only projections of fluxes and forces in the x and y directions are used in Eq. (57). The curvature is important in computing the volumes of the elements. The U, $F(U)$, and $H(U)$ quantities are assumed uniform in each cell.

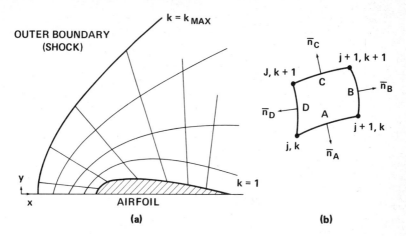

FIG. 44 Mesh network for supersonic free-stream case. (a) Mesh system; (b) typical volume element.

5.3 Treatment of Shock Waves

Before advancing the solution in time, the mesh speed $\bar{\lambda}$ must be specified. This is accomplished by first advancing the shock location by an iterative procedure using the Rankine-Hugoniot relations and characteristic data. The mesh system is then reconfigured such that the new location of the $k = k_{max}$ mesh line coincides with the shock. The other k = constant mesh lines are moved proportionately, keeping the body grid line $k = 1$ fixed. Note that, for the grid system shown in Fig. 44, $\bar{\lambda}$ is normal to the k = constant mesh lines. By adjusting the mesh in this manner, the exact shock solution is obtained; that is, the shock is essentially "fit" rather than "captured." If the $k = k_{max}$ mesh line deviates even slightly from the shock location, overshoots are obtained. This is illustrated in Fig. 45 (from [60]). Of course, the correct shock solution is obtained in the limit of vanishing cell volume whether the mesh is adjusted or not.

5.4 Finite-Difference Scheme

The solution at a new time level $(n + 1)\,\Delta t$ is obtained from the solution at the old time level $n\,\Delta t$ by the use of dimensionally split difference operators, that is,

$$U_{j,k}^{n+2} = [L_j(\Delta t)L_k(2\,\Delta t)L_j(\Delta t)]U_{j,k}^n \tag{59}$$

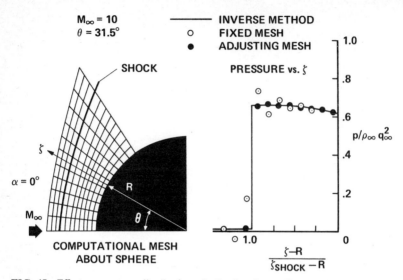

FIG. 45 Effect on pressure distribution of adjusting the mesh network to the motion of the shock wave. Three free-stream cells are differenced for the fixed mesh, none for the adjusting mesh.

where $L_j(\Delta t)$ has the two-step (predictor-corrector) form

$$\tilde{U}_{j,k}^{n+1} = \frac{(\text{vol} \cdot U)_{j,k}^n - \Delta t(F_{j,k}^n S_{j+1}^n + F_{j-1,k}^n S_j^n)}{\text{vol}_{j,k}^{n+1}}$$

$$U_{j,k}^{n+1} = \frac{1}{2} \frac{[(\text{vol} \cdot U)_{j,k}^n + \tilde{U}_{j,k}^{n+1} \cdot \text{vol}_{j,k}^{n+1} - \Delta t(\tilde{F}_{j+1,k}^{n+1} S_{j+1}^{n+1} + \tilde{F}_{j,k}^{n+1} S_j^{n+1})]}{\text{vol}_{j,k}^{n+1}} \quad (60)$$

and $L_k(\Delta t)$ has an analogous representation. The operators $L_j(\Delta t)$ and $L_k(\Delta t)$ generally do not commute, and hence the sequence

$$U_{j,k}^{n+1} = [L_k(\Delta t)L_j(\Delta t)]U_{j,k}^n$$

can be shown to be first-order accurate. Reversing the sequence at each step as in Eq. (59) cancels the first-order truncation errors introduced by the factorization (splitting), and second-order accuracy results.

5.5 Time-Step Restrictions and Stability

The time-step restriction for stability is

$$\Delta t \leq \min (\Delta t_j, \Delta t_k) \quad (61)$$

where
$$\Delta t_j \leq \min \left(\frac{l_j}{|\bar{q}_j| + a} \right)$$

$$\Delta t_k \leq \min \left(\frac{\frac{1}{2} l_k}{|\bar{q}_k - \bar{\lambda}| + a} \right)$$

and a is the speed of sound, l_j and l_k are the lengths of the volume sides $j = $ constant and $k = $ constant, and $\bar{q}_j$ and $\bar{q}_k$ are the velocities in the j and k mesh-line directions. Because of effects of mesh spacing, coordinate directions, flow velocity, and sound speed in Eq. (61), values of Δt_j and Δt_k may differ considerably. Splitting can reduce computation time by allowing the more restrictive operator to be applied several times for each application of the less restrictive operator. For example, if J is the next integer greater than $2\Delta t_k / \Delta t_j$, the sequence

$$U_{j,k}^{n+2} = \left[L_k(\Delta t) \prod_{j=1}^{J} L_j \frac{2\Delta t}{J} L_k(\Delta t) \right] U_{j,k}^n \tag{62}$$

is stable for $\Delta t \leq \max (\Delta t_j, 2\Delta t_k)$ and can advance the solution to a given time level in fewer operations than Eq. (59). An alternative approach (for the same step size) would be to use an implicit operator L_I in place of L_J to reduce the number of time levels at which a solution must be computed, that is,

$$U_{j,k}^{n+2} = [L_k(\Delta t) L_I(2\Delta t) L_k(\Delta t)] U_{j,k}^n \tag{63}$$

5.6 Computed Results

All three of these approaches, Eqs. (59), (62), and (63), were tested for efficiency in reaching a steady state in reference 61 for the case of $M_\infty = 1.25$ flow about an NACA 0012 airfoil; computed results for this and other Mach numbers are shown in Fig. 46. For each sequence, the flow was advanced to the same time level using a 10×20 mesh. Sequence (59), the fastest for a single cycle, is used as a reference. Although sequence (62) is slower per cycle, it can take larger time steps so that the overall effect is a 20% reduction in computer time. The mixed explicit-implicit sequence resulted in a 25% savings.

Results for a 6% thick parabolic-arc airfoil at $M_\infty = 0.909$ are compared with small-disturbance relaxation results and experiment in Fig. 47. The difference in computed shock locations is attributed to a deficiency in the small-disturbance procedure for strong shocks that has been alluded to earlier in this chapter. The shock Mach number in this case is 1.3.

Only preliminary computations using the finite-volume method have been performed thus far. The method is presently being extended to treat

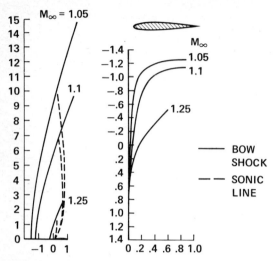

FIG. 46   Bow waves, sonic lines, and pressure coefficients for supersonic flows past an NACA 0012 airfoil.

lifting airfoils and should eventually prove useful in providing solutions to the complete Euler and Navier-Stokes equations for three-dimensional aerodynamic configurations.

6   AIRFOIL DESIGN BY NUMERICAL OPTIMIZATION

Some of the benefits of airfoil modifications to reduce drag are increased range, fuel conservation, and higher cruise Mach number. At transonic speeds, drag reduction can be achieved by reducing the strength and extent

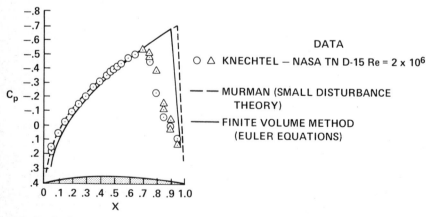

FIG. 47   Surface pressures for $M_\infty = 0.909$.

of embedded shocks, and several computational procedures have been developed with this in mind. For example, the hodograph method [62,63] produces airfoils that are shock-free for a particular design condition. This approach has proved useful in designing airfoils for some applications. Its main disadvantage is that the hodograph codes currently in use require extensive experience on the part of the user. Another design procedure, the inverse method [64,65], computes the airfoil shape corresponding to a specified pressure distribution. Although this procedure is simple to implement computationally, it requires a priori knowledge of what constitutes a desirable and achievable pressure distribution. Furthermore, constraints on lift, moment, airfoil thickness, and the like are not readily imposed.

A more recently developed approach, airfoil design by numerical optimization, is easy to apply and is capable of handling an almost limitless variety of specified constraints. This procedure has been developed and applied to both low-speed and transonic airfoil design by Hicks and his associates [66–69]. The numerical-optimization design code was constructed by linking two previously existing codes, an *optimization program* based on the method of feasible directions [70] and an *aerodynamic analysis program* that solves the full potential equation for flow about airfoils [71].

6.1 Design Method

Numerical optimization seeks to minimize some specified parameter (e.g., the drag coefficient C_d) for a set of design variables describing the airfoil geometry, while satisfying a number of specified constraints. These constraints may be aerodynamic (e.g., on the lift and moment coefficients C_l and C_m), geometric (e.g., on the airfoil thickness or volume), or flow field (e.g., on $C_{p_{min}}$ or the maximum surface-pressure gradient).

The design procedure for drag minimization is illustrated schematically in Fig. 48. The initial airfoil required to start each design problem is obtained by fitting a polynomial to the region of the airfoil to be modified, or to the entire airfoil if a complete modification is desired. The polynomial coefficients are the design variables perturbed by the optimization process to achieve the desired design improvement. In Fig. 48 the airfoil upper and lower surfaces are described by polynomials with coefficients $a_1, \ldots, a_7$ and $b_1, \ldots, b_7$, respectively. These coefficients, along with the Mach number, angle of attack, and specified constraints, are required to initiate the optimization. The optimization program perturbs each of the coefficients in turn, returning to the aerodynamic analysis program for evaluation of the drag coefficient after each perturbation. After all coefficients have been perturbed and the direction of change of C_d has been noted for each polynomial coefficient perturbation, the optimization program computes the partial derivatives (by one-sided finite differences) of C_d with respect to each coefficient,

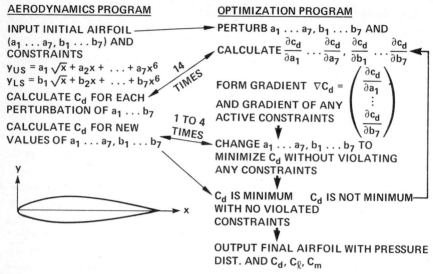

AERODYNAMICS PROGRAM

INPUT INITIAL AIRFOIL
$(a_1 \ldots a_7, b_1 \ldots b_7)$ AND
CONSTRAINTS
$y_{US} = a_1\sqrt{x} + a_2x + \ldots + a_7x^6$
$y_{LS} = b_1\sqrt{x} + b_2x + \ldots + b_7x^6$
CALCULATE C_d FOR EACH
PERTURBATION OF $a_1 \ldots b_7$
CALCULATE C_d FOR NEW
VALUES OF $a_1 \ldots a_7, b_1 \ldots b_7$

14 TIMES

1 TO 4 TIMES

OPTIMIZATION PROGRAM

PERTURB $a_1 \ldots a_7, b_1 \ldots b_7$ AND
CALCULATE $\dfrac{\partial c_d}{\partial a_1} \ldots \dfrac{\partial c_d}{\partial a_7}, \dfrac{\partial c_d}{\partial b_1} \ldots \dfrac{\partial c_d}{\partial b_7}$

FORM GRADIENT $\nabla C_d = \begin{pmatrix} \dfrac{\partial c_d}{\partial a_1} \\ \vdots \\ \dfrac{\partial c_d}{\partial b_7} \end{pmatrix}$
AND GRADIENT OF ANY
ACTIVE CONSTRAINTS

CHANGE $a_1 \ldots a_7, b_1 \ldots b_7$ TO
MINIMIZE C_d WITHOUT VIOLATING
ANY CONSTRAINTS

C_d IS MINIMUM C_d IS NOT MINIMUM
WITH NO VIOLATED
CONSTRAINTS

OUTPUT FINAL AIRFOIL WITH PRESSURE
DIST. AND C_d, C_ℓ, C_m

FIG. 48 Numerical optimization procedure for drag minimization.

thus forming the gradient of C_d (∇C_d). The optimization program then increments the polynomial coefficients one to four times, searching in the $-\nabla C_d$
direction until C_d begins to increase (owing to nonlinearity in the design
space) or until a constraint is encountered. If either of these conditions is
found, new gradients are computed, and a new direction is determined that
reduces C_d without violating any constraints. When a minimum value of C_d
is reached with no violated constraints, then forces, pressures, and airfoil
ordinates are output, and the process is terminated.

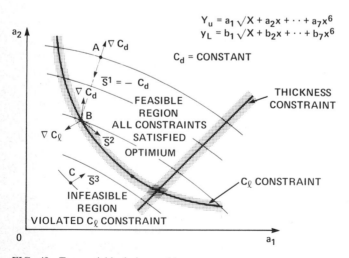

FIG. 49 Two-variable design problem.

Figure 49 illustrates a hypothetical optimization problem using only two design variables. The design is sought that minimizes drag with constraints on the lift coefficient C_l and airfoil thickness. The airfoil upper and lower surface geometry is given by y_u and y_L as shown, and, in this case, only a_1 and a_2 are perturbed in the optimization process. Note that the thickness constraint is a straight line; that is, thickness is a linear function of the polynomial coefficients a_1 and a_2.

The design space is divided into two regions, an infeasible region where one or both constraints are violated, and a feasible region where no constraints are violated. Minimum drag in the feasible region is sought. To begin with, assume that starting values of a_1 and a_2 correspond to point A. The gradient of C_d is computed for this point, giving the direction of change of a_1 and a_2 as $\bar{S}^1 = -\nabla C_d$. For this particular case, both a_1 and a_2 must be reduced to decrease C_d. The search in the $\bar{S}^1$ direction continues until the C_l constraint boundary is encountered at point B. At this point, both ∇C_d and ∇C_l are used to define a new search direction $\bar{S}^2$. Now a_1 must be increased and a_2 decreased to minimize C_d by moving along the C_l constraint boundary. The process continues until the design point reaches the optimum, that is, the point at which the line of constant C_d with the lowest value of C_d in the feasible region intersects the C_l constraint boundary.

Design conditions used to start the optimization need not correspond to points in the feasible region. For example, if the starting values of a_1 and a_2 correspond to point C, the initial design violates the C_l constraint. In such cases, an initial search direction $\bar{S}^3$ is determined that moves the design point toward the feasible region with a minimum increase in C_d. Such a move increases both a_1 and a_2. When the feasible region is reached, a new direction $\bar{S}$ is determined that moves the design point along the C_l constraint boundary toward the optimum.

In the design examples to be presented here, the upper surfaces of existing airfoils are redesigned to reduce the strength and extent of embedded shocks. The equation used to describe the portion of the airfoil to be modified is

$$y = a_1 \left(\frac{x}{c}\right)^{a_4} + k\frac{x}{c} + a_2 \left(\frac{x}{c}\right)^2 + a_3 \left(\frac{x}{c}\right)^3 \qquad (64)$$

where c is the airfoil chord. The coefficient k is used to match ordinates at the chordwise station where the modified and fixed regions of the airfoil surface meet. For cases in which the entire upper surface is modified, k is used to fix the trailing-edge bluntness. The design variables are a_1–a_4.

Two airfoil design examples from reference 69 are presented here, one that modifies only a portion of the airfoil upper surface, and another that redesigns the entire upper surface. The numerical optimization was performed using an inviscid aerodynamic analysis package. The final designs were analyzed using a transonic code with boundary-layer corrections provided using a theory described in reference 71. This method supplies a displacement thickness correction for the airfoil, which is smoothed and added to the airfoil contour to account for the presence of a turbulent boundary layer. The correction is computed as a part of the iterative solution process for the flow field.

6.2   NACA 23015 Section Modification

Results of a forward upper surface numerical-optimization redesign of the NACA 23015 airfoil at $M_\infty = 0.7$ and $\alpha = 0°$ are shown in Fig. 50. Note that the modified airfoil flow is nearly shock-free ($C_d = 0.0005$) and exhibits nearly the same pitching moment and lift coefficients as the original airfoil. The drag coefficient is obtained by a surface pressure integral and represents the drag due to shock losses in an inviscid flow.

The characteristics of the original and modified sections corrected for viscosity are shown in Fig. 51 for a Reynolds number of 10×10^6. The percent decrease in C_d is smaller than in the inviscid case because of the inclusion of skin friction drag. The lift and (nose-down) pitching-moment coefficients for both airfoils are also reduced by viscosity.

A plot of C_d versus M_∞ at $\alpha = 0°$ for both airfoils is shown in Fig. 52 for $Re_c = 20 \times 10^6$. Note that the modification increases the drag-rise Mach number, reduces the subsonic drag, and does not introduce a more rapid drag rise above the $M_\infty = 0.7$ design Mach number.

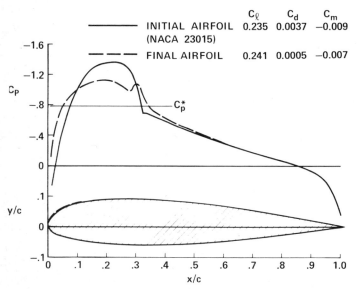

	C_ℓ	C_d	C_m
INITIAL AIRFOIL (NACA 23015)	0.235	0.0037	−0.009
FINAL AIRFOIL	0.241	0.0005	−0.007

FIG. 50   Inviscid drag minimization for an NACA 23015 airfoil at $M_\infty = 0.7$ and $\alpha = 0°$.

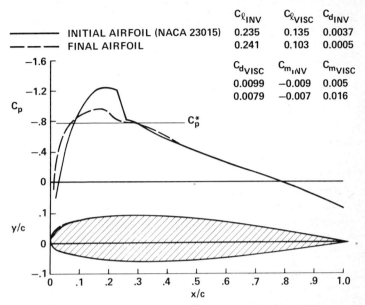

	$c_{\ell_{INV}}$	$c_{\ell_{VISC}}$	$c_{d_{INV}}$
INITIAL AIRFOIL (NACA 23015)	0.235	0.135	0.0037
FINAL AIRFOIL	0.241	0.103	0.0005

	$c_{d_{VISC}}$	$c_{m_{INV}}$	$c_{m_{VISC}}$
	0.0099	−0.009	0.005
	0.0079	−0.007	0.016

FIG. 51 Viscous results ($Re_c = 10 \times 10^6$) for $M_\infty = 0.7$ and $\alpha = 0°$.

In many cases, it is important to investigate the effect of transonic airfoil modification on low-speed high-lift requirements. This is especially important for the NACA 23015 airfoil, which exhibits one of the highest maximum lift coefficients of all NACA sections with similar design lift coefficients and, consequently is used on many general-aviation aircraft. Positive evaluation of the high-lift characteristics of the modified airfoil must be performed

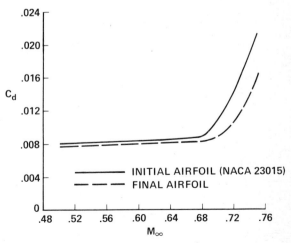

INITIAL AIRFOIL (NACA 23015)
FINAL AIRFOIL

FIG. 52 C_d versus M_∞ for $Re_c = 20 \times 10^6$.

experimentally. However, computational evidence is presented in reference 69 that suggests that the high-lift performance of the modified airfoil is equivalent to that of the original airfoil.

6.3 NACA 6-Series Section Modification

Results of a complete upper-surface redesign of a 13% thick NACA 6-series airfoil are shown in Fig. 53. Note that modification weakens the shock and reduces the wave drag to about one-quarter of the drag value for the initial profile. This was achieved by a redistribution of the area contained within the profile contour. A constraint on the profile cross-sectional area prevented further thinning of the section to reduce drag.

The forces and pressures for the initial and modified airfoil, corrected for viscosity, are shown at Reynolds numbers of 1.3×10^6 and 20×10^6 in Fig. 54a and b, respectively. The drag reduction is a smaller percentage than in the inviscid case but is still substantial.

A plot of C_d versus M_∞ at $\alpha = 0°$ is shown in Fig. 55 for both profiles for $Re_c = 20 \times 10^6$. The drag-rise Mach number is increased by about 0.04 by the redesign.

Low-speed $(M_\infty = 0.1)$ pressure distributions for $\alpha = 10°$ are shown in Fig. 56 for both airfoils. Note that the pressure suction peak near the leading edge on the upper surface is reduced by more than a factor of two by the

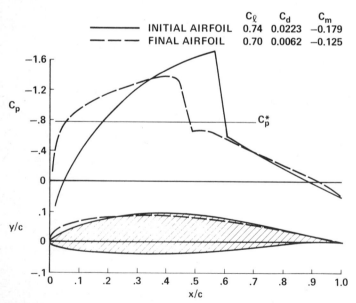

FIG. 53   Inviscid drag minimization for an NACA 6 series airfoil at $M_\infty = 0.7$ and $\alpha = 0°$.

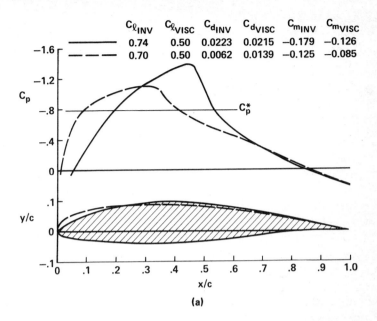

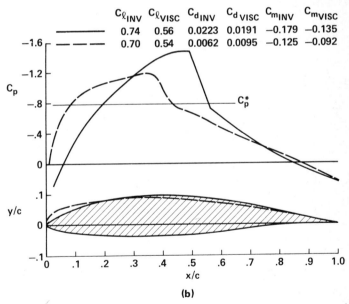

FIG. 54 Viscous results for $M_\infty = 0.7$ and $\alpha = 0°$. (a) $Re_c = 1.3 \times 10^6$; (b) $Re_c = 20 \times 10^6$.

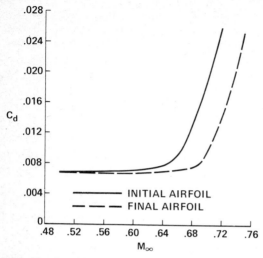

FIG. 55 C_d versus M_∞ for $Re_c = 20 \times 10^6$.

increased leading-edge bluntness of the modified airfoil. The adverse pressure gradient downstream of the leading-edge suction peak is substantially reduced, thus improving low-speed high-lift performance.

6.4   Extensions of the Numerical-Optimization Procedure

Several improvements to the numerical-optimization design procedure should be made. First, as Hicks [69] suggests, the polynomial approxima-

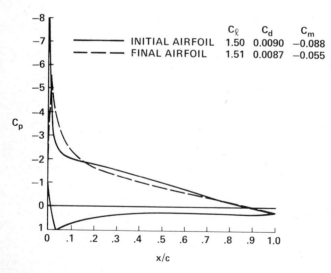

FIG. 56   Low-speed results at $M_\infty = 0.1$, $\alpha = 10°$, and $Re_c = 20 \times 10^6$.

tion with four design variables is probably too restrictive. A more general functional form expressing the airfoil geometry should be found so that a larger class of configurations can be considered in the optimization process. Second, it may be advantageous to treat angle of attack as a design variable and impose a constraint on the lift. Plotted results could then be given in terms of C_d versus M_∞ for a fixed lift. Third, it would be useful to include constraints for low-speed performance in the optimization procedure. This would offer a more direct approach for computing efficient airfoils for general-aviation and helicopter applications, that is, applications that have both transonic and low-speed performance requirements. Preliminary investigations by Hicks indicate that this would double the required computer time.

The numerical-optimization design approach is not restricted to airfoil design. For example, a three-dimensional transonic small-disturbance code could be linked with the optimization program to determine optimum spanwise variations of twist and camber, optimum wing-fuselage juncture design, etc.; such an effort is presently in progress.

7 CONCLUDING REMARKS

Remarkable progress has been made over the past few years in developing methods to handle each of the different aspects of transonic-flow analysis, including the effects of unsteadiness, three-dimensionality, and viscosity.

The development of suitable computational methods for unsteady flows has lagged the development of methods for steady flows by several years. However, efficient unsteady algorithms are now available, and the next few years should see them applied to a number of interesting engineering problems.

Small-disturbance approaches substantially simplify the governing equations and, especially, the treatment of boundary conditions in both three-dimensional and unsteady computational transonic aerodynamics. They provide solutions that are useful in qualitatively predicting the behavior of these complex flows. Furthermore, quantitatively good predictions can often be obtained by properly "tuning" the small-disturbance procedures.

Eventually, however, advances in algorithm development and computer technology will allow efficient solution of more complete systems of governing equations and boundary conditions, and aerodynamicists will no longer be satisfied with results provided by small-disturbance theories. The pacing item in providing more accurate alternatives to the small-disturbance procedures is the development of simple but adequate methods for satisfying surface boundary conditions. The finite-volume method, a good first step in this direction, was designed to accomplish this without the explicit use of coordinate transformations.

Viscous effects are often at least as important as non-small-disturbance effects in transonic flows. Hence, any less approximate procedure than small-disturbance theory should account for viscosity. Boundary-layer corrections with approximate treatment of shock–boundary-layer interaction regions and small regions of separated flow may be sufficient initially. However, eventually suitable turbulence models must be found to provide a complete treatment of separated turbulent flows.

As the accuracy and efficiency of computational simulations improves, they can be expected to assume an increasing percentage of the design burden now carried primarily by wind-tunnel simulations. New computational simulation methods can be incorporated in numerical-optimization design codes, like the airfoil design code reported on in Sec. 6, that automatically determine configurations that optimize performance while satisfying specified constraints.

REFERENCES

1 Nieuwland, G. Y., and B. M. Spee: Transonic Airfoils: Recent Developments in Theory, Experiment, and Design, *Annu. Rev. Fluid Mech.* **5**: 119–150, 1973.
2 Magnus, R., and H. Yoshihara: Inviscid Transonic Flow over Airfoils, AIAA paper 70–47, January, 1970.
3 Murman, E. M., and J. D. Cole: Calculation of Plane Steady Transonic Flows, *AIAA J.* **9**(1): 114–121, 1971.
4 Ballhaus, W. F., and F. R. Bailey: Numerical Calculation of Transonic Flow About Swept Wings, AIAA paper 72-677, June, 1972.
5 Bailey, F. R., and W. F. Ballhaus: Relaxation Methods for Transonic Flow about Wing-Cylinder Combinations and Lifting Swept Wings, in "Lecture Notes in Physics," vol. 19, pp. 2–9, Springer, New York, 1972.
6 Bailey, F. R.: On the Computation of Two- and Three-Dimensional Steady Transonic Flows by Relaxation Methods, in "Lecture Notes in Physics," vol. 41, pp. 1–77, Springer, New York, 1975.
7 Bailey, F. R., and W. F. Ballhaus: Comparisons of Computed and Experimental Pressures for Transonic Flows about Isolated Wings and Wing-Fuselage Configurations, NASA SP-347, pp. 1213–1231, 1975.
8 Bailey, F. R., and J. L. Steger: Relaxation Techniques for Three-Dimensional Transonic Flow About Wings, AIAA paper 72-189, January, 1972.
9 Newman, P. A., and E. B. Klunker: Computation of Transonic Flow about Finite Lifting Wings, *AIAA J.*, **10**(7), July, 1972.
10 Schmidt, W., S. Rohlfs, and R. Vanino: Some Results Using Relaxation Methods and Finite Volume Techniques for Two and Three-Dimensional Transonic Flows, in "Lecture Notes in Physics," vol. 35, pp. 364–372, Springer, New York, 1974.
11 Hall, M. G., and M. C. P. Firmin: Recent Developments in Methods for Calculating Transonic Flows Over Wings, ICAS paper 74, 1974.
12 Jameson, A.: Numerical Calculation of Three-Dimensional Transonic Flow over a Yawed Wing, *Proc. AIAA Comput. Fluid Dyn. Conf.* pp. 18–26, July, 1973.
13 Ashley, H., and M. Landahl: "Aerodynamics of Wings and Bodies," Addison-Wesley, Reading, Mass., 1965.
14 Jameson, A.: Iterative Solution of Transonic Flows over Airfoils and Wings, Including Flows at Mach 1, *Comm. Pure Appl. Math.*, **27**: 283–309, 1974.
15 Albone, C. M.: A Finite Difference Scheme for Computing Supercritical Flows in Arbitrary Coordinate Systems, RAE technical report 74090, 1974.

16 Lomax, H., F. R. Bailey, and W. F. Ballhaus: On the Numerical Simulation of Three-Dimensional Transonic Flow with Application to the C-141 Wing, NASA TN D-6933, 1973.

17 Monnerie, B., and F. Charpin: " Essais de buffeting d'une aile en fleche en trassonique," 10ᵉ Colloque D'Aérodynamique Appliquée, November, 1973.

18 Murman, E. M.: Analysis of Embedded Shock Waves Calculated by Relaxation Methods, *Proc. AIAA Comput. Fluid Dyn. Conf.*, pp. 27–40, July, 1973.

19 Murphy, W. D., and N. D. Malmuth: A Relaxation Solution for Transonic Flow over Three-Dimensional Jet-Flapped Wings, AIAA paper 76-98, January, 1976.

20 Krupp, J. A.: The Numerical Calculation of Plane Steady Transonic Flows Past Thin Lifting Airfoils, Boeing Scientific Research Laboratory Report D180-12958-1, June, 1971.

21 Ballhaus, W. F., R. Magnus, and H. Yoshihara: Some Examples of Unsteady Transonic Flows over Airfoils, *Proc. Symp. Unsteady Aerodyn.*, **II**: 769–791, 1975. Univ. of Ariz.

22 Albone, C. M., M. G. Hall, and G. Joyce: Numerical Solutions for Transonic Flows past Wing-body Combinations, presented at the Symposium Transsconicum II, September, 1975.

23 van der Vooren, J., J. W. Sloof, G. H. Huizing, and A. van Essen: Remarks on the Suitability of Various Transonic Small-Perturbation Equations to Describe Three-Dimensional Flow; Examples of Computations Using a Fully Conservative Rotated Difference Scheme, presented at the Symposium Transsonicum II, September, 1975.

24 Schmidt, W., and R. Vanino: The Analysis of Arbitrary Wing-Body Combinations in Transonic Flow Using a Relaxation Method, presented at the Symposium Transsonicum II, September, 1975.

25 Klunker, E. B., and P. A. Newman: Computation of Transonic Flow about Lifting Wing-Cylinder Combinations, *J. Aircr.*, **2**(4): 254–256, April, 1974.

26 Sedin, Y. C-J., and K. R. Karlsson: Some Numerical Results of a New Three-Dimensional Transonic Flow Method, presented at the Symposium Transsonicum II, September, 1975.

27 Bailey, F. R.: Numerical Calculation of Transonic Flow about Slender Bodies of Revolution, NASA TN D-6582, 1971.

28 Binion, T. W.: An Investigation of Three-Dimensional Wall Interference in a Variable Porosity Transonic Wind Tunnel, AEDC TR-74-76, 1973.

29 Loving, D. L., and B. B. Estabrooks: Transonic Wing Investigation in the Langley 8-Foot High-Speed Tunnel at High Subsonic Mach Numbers and at a Mach number of 1.2, NACA RM L51F07, 1951.

30 Tijdeman, H.: On the Motion of Shock Waves on an Airfoil with oscillating Flap, presented at the Symposium Transsonicum II, September, 1975.

31 Tijdeman, H.: High Subsonic and Transonic Effects in Unsteady Aerodynamics, NLR report TR 75079 U, October, 1975.

32 Garrick, I. E., and S. I. Rubinow: Flutter and Oscillating Air-Force Calculations for an Airfoil in a Two-Dimensional Supersonic Flow, NACA report 846, 1946.

33 Timman, R., A. I. van de Vooren, and J. H. Greidanus: Aerodynamic Coefficients of an Oscillating Airfoil in Two-Dimensional Subsonic Flow, *J. Aeronaut. Sci.* **21**(7): 499–500, 1954.

34 Ehlers, F. E.: A Finite Difference Method for the Solution of the Transonic Flow around Harmonically Oscillating Wings, NASA CR-2257, January, 1974.

35 Weatherill, W. H., J. D. Sabastian, and F. E. Ehlers: On the Computation of the Transonic Perturbation Flow Fields around Two- and Three-Dimensional Oscillating Wings, AIAA paper 76-99, January, 1976.

36 Traci, R. M., J. L. Farr, and E. Albano: Perturbation Method for Transonic Flow about Oscillating Airfoils, AIAA paper 75-877, June, 1975.

37 Traci, R. M., E. D. Albano, and J. L. Farr: Small-Disturbance Transonic Flows about Oscillating Airfoils and Planar Wings, AFFDL-TR-75-10, June, 1975.

38 Lomax, H., M. A. Heaslet, F. B. Fuller, and L. Sluder: Two- and Three-Dimensional Unsteady Lift Problems in High-Speed Flight, NACA report 1077, 1952.

39 Tobak, M.: On the Use of the Indicial Function Concept in the Analysis of Unsteady Motions of Wings and Wing-Tail Combinations, NACA report 1188, 1954.

40 Beam, R. M., and R. F. Warming: Numerical Calculations of Two-Dimensional, Unsteady Transonic Flows with Circulation, NASA TN D-7605, 1974.

41 van der Vooren, J., and J. W. Sloof: On Inviscid Isentropic Flow Models Used for Finite Difference Calculations of Two-Dimensional Transonic Flows with Embedded Shocks about Airfoils, NLR MP 73024U, October, 1973.

42 Landahl, M.: " Unsteady Transonic Flow," Pergamon, New York, 1961.

43 Magnus, R. J., and H. Yoshihara: Calculations of Transonic Flow over an Oscillating Airfoil, AIAA, paper 75-98, January, 1975.

44 Beam, R. M., and W. F. Ballhaus: Numerical Integration of the Small-Disturbance Potential and Euler Equations for Unsteady Transonic Flow, NASA SP-347, part II, pp. 789–809, March, 1975.

45 Beam, R. M., and R. F. Warming: An Implicit Finite-Difference Algorithm for Hyperbolic Systems in Conservation-Law Form, *J. Comp. Phys.*, **22**(1): 1976.

46 Ballhaus, W. F., and H. Lomax: The Numerical Simulation of Low Frequency Unsteady Transonic Flow Fields, in " Lecture Notes in Physics," vol. 35, pp. 57–63, Springer, New York, 1975.

47 Ballhaus, W. F., and J. L. Steger: Implicit Approximate-Factorization Schemes for the Low-Frequency Transonic Equation, NASA TM X-73,082, November, 1975.

48 Isom, M. P.: Unsteady Subsonic and Transonic Potential Flow over Helicopter Rotor Blades, NASA CR-2463, October, 1974.

49 Isom, M. P., and F. X. Caradonna: Subsonic and Transonic Potential Flow over Helicopter Rotor Blades, AIAA paper 72-39, January, 1972.

50 Ballhaus, W. F., and F. X. Caradonna: The Effect of Planform Shape on the Transonic Flow past Rotor Tips, AGARD CP-111, paper 17, September, 1972.

51 Ballhaus, W. F., and F. X. Caradonna: Numerical Simulation of Transonic Flow about Airplanes and Helicopter Rotors, *Proc. Ninth Army Sci. Conf.*, June, 1974.

52 Caradonna, F. X., and M. P. Isom: Numerical Calculation of Unsteady Potential Flow over Helicopter Rotor Blades, AIAA paper 75-168, January, 1975.

53 Schmitz, F. H., and D. A. Boxwell: In Flight Far Field Measurement of Helicopter Impulsive Noise, *J. Am. Helicopt. Soc.*, **21**(4): 1976.

54 Deiwert, G. S.: Numerical Simulation of High Reynolds Number Transonic Flows, AIAA paper 74-603, June, 1974.

55 Deiwert, G. S., J. B. McDevitt, and L. L. Levy: Simulation of Turbulent Transonic Separated Flow over an Airfoil, NASA SP-347, part I, pp. 419–436, March, 1975.

56 Deiwert, G. S.: Computation of Separated Transonic Turbulent Flows, AIAA paper 75-829, June, 1975.

57 McDevitt, J. B., L. L. Levy, and G. S. Deiwert: Transonic Flow about a Thick Circular-Arc Airfoil, AIAA paper 75-878, June, 1975.

58 McCormack, R. W., and A. J. Paullay: The influence of the Computational Mesh on Accuracy for Initial Value Problems with Discontinuities or Nonunique Solutions, *Comput. Fluids*, **2**: 339–361, 1974.

59 MacCormack, R. W., A. W. Rizzi, and M. Inouye: Steady Supersonic Flow Fields with Embedded Subsonic Regions, *Proc. Conf. Comput. Probl. Methods Aero. Fluid Dyn.*, University of Manchester, England, 1974.

60 Rizzi, A. W., and M. Inouye: Time Split Finite Volume Method for Three-Dimensional Blunt-Body Flow, *AIAA J.* **11**(11): 1478–1485, November, 1973.

61 Rizzi, A. W.: Transonic Solutions of the Euler Equations by the Finite Volume Method, presented at the Symposium Transsonicum II, September, 1975.

62 Bauer, F., P. Garabedian, and D. Korn: Supercritical Wing Sections, in " Lecture Notes in Economics and Mathematical Systems," vol. 66, Springer, New York, 1972.

63 Nieuwland, G. Y.: Transonic Potential Flow around a Family of Quasi-Elliptical Aerofoil Sections, NLR TR 172, 1967.

64 Arlinger, B.: An Exact Method of Two-Dimensional Airfoil Design, Saab TN 67, 1970.

65 Steger, J. L., and J. M. Klineberg: A Finite Difference Method for Transonic Airfoil Design, *AIAA J.* **11**(5), May, 1973.

66 Hicks, R. M., and G. N. Vanderplaats: Application of Numerical Optimization to the Design of Low-Speed Airfoils, NASA TM X-3213, March, 1975.

67 Vanderplaats, G. N., R. M. Hicks, and E. M. Murman: Application of Numerical Optimization Techniques to Airfoil Design, NASA SP-347, part II, pp. 749–768, March, 1975.

68 Hicks, R. M., and G. N. Vanderplaats: Design of Low-Speed Airfoils by Numerical Optimization, SAE paper 750524, 1975.

69 Hicks, R. M., G. N. Vanderplaats, E. M. Murman, and R. R. King: Airfoil Section Drag Reduction at Transonic Speeds by Numerical Optimization, NASA TM X-73,097, January, 1976.

70 Vanderplaats, G. N.: CONMIN—a Fortran Program for Constrained Function Minimization," NASA TM X-62,282, August, 1973.

71 Bauer, F., P. Garabedian, D. Korn, and A. Jameson: Supercritical Wing Sections II, in "Lecture Notes in Economics and Mathematical Systems," vol. 108, Springer, New York, 1975.

CHAPTER 4

Panel Methods in Aerodynamics

Werner Kraus

1 INTRODUCTION

The use of potential theory for the computation of flow fields around bodies of arbitrary shape in two- or three-dimensional flow has, for quite some time, been very well known. It is possible, for a number of simplified body geometries, to determine the exact or approximate solution of the resulting integral equation. A number of methods using singularity theory are applicable to two-dimensional profiles as well as to axisymmetric bodies. The positioning of singularities on the body contour, as well as on the profile chord line or body axis, has led to good results. However, the exact or approximate solution of the integral equation for a body with any given contour still poses a number of difficulties. The availability of modern computer systems to solve these problems resulted in new numerical methods. A breakthrough in solving the potential equation for bodies with any given contour was achieved by the so-called panel method.

Instead of a continuous distribution of singularities on the surface of the body, the singularity strength on a particular surface element (panel) is given by a function. The intent is to solve the integral equation for the function of each panel. In order to fulfill the boundary condition, each panel receives an assigned control point. By fulfilling all boundary conditions, the singularity strengths and thereby the flow field can be determined.

In the next section, this principle of panel methodology will be explained in more detail. The method has been successfully used at Messerschmitt-Bölkow-Blohm (MBB) to solve a number of problems. The MBB panel method [1–3] uses flat, rectangular panels on the outer contour of the body and works with a constant singularity strength as a distribution function.

2 BASIC THEORY, INTRODUCTION TO THE PANEL TECHNIQUE USING THE MBB METHOD

The basic equations of a singularity method placing the singularity distributions on the body contour are given in Fig. 1. For simplicity the lift-dependent terms have been eliminated. The starting point is the three-dimensional incompressible potential equation. Differentiation of the potential for the flow field gives the velocity vector **V**. Thus, the requirement for continuity and freedom of rotation for the flow field is satisfied. The onset flow plus the superimposed perturbation field contribute to the buildup of the potential of the flow field. A continuous source-sink distribution with intensity $\sigma(S)$ is then assigned to the body surface. The perturbation potential at a random point of the field and at a distance r can then be determined with the next equation. The boundary conditions for solving this equation are the disappearance of the perturbation potential at infinity and the kinematic flow condition at the contour (the velocity component perpendicular to the body surface has to be zero). **n** represents the normal vector to the body surface. Application of the kinematic flow condition to the perturbation-potential assumption results in the equation for determining the singularity strengths.

The computation then follows the arrows in the figure, taking into consideration the specific inputs (geometry and onset flow parameter) for the problem. The singularity sizes can be determined from the bottom equation in the figure. Thus, the perturbation potential and the total potential are set

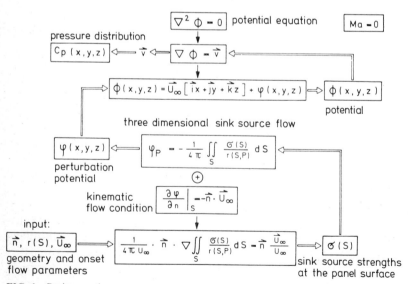

FIG. 1 Basic equations.

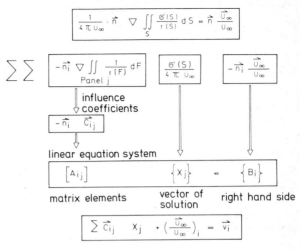

FIG. 2 Numerical solution for singularity strengths (panel method).

and, by differentiation, the velocity vector and the pressure distribution can be obtained.

Figure 2 shows the method used in the panel technique to determine the singularity strengths. A double summation is substituted for the double integral over the surface of an arbitrary three-dimensional body. The surface is split into a network of flat, rectangular areas (the panels). The double integral of the distribution function (constant value) over a single panel for an arbitrary point at a distance r can easily be solved. This double integral is identified as the influence coefficient C_{ij} of panel j on control point i in the next lower rectangle in the figure.

This procedure for determining the unknown singularity strengths for the given panels results in a set of linear equations. This set of equations can be solved easily, using "standard" mathematical procedures. Row i of the matrix satisfies the boundary condition at control point i. The sum of the influence coefficients of all panels at control point i in the direction n_i perpendicular to the surface, multiplied by the respective singularity sizes (equal to the sum of all perturbation velocities from the perturbation potential of the surface distribution), plus the component of the onset flow vector perpendicular to the surface of control point i (the velocity resulting from the potential of the onset flow) must be zero.

Figure 3 is a brief description of the panel method used at MBB to consider lift-dependent flows. The perturbation potential for the displacement effect [Eq. (1)] has already been described. To compute the flow about lift-generating bodies, additional singularities are required. The singularity strengths have to be determined by using the flow-off condition (Kutta

(1) $\varphi_S(P) = -\dfrac{1}{4\pi} \displaystyle\iint_S \dfrac{\sigma(S)}{r(S,P)}\, dS$

(2) $\varphi_D(P) = \dfrac{1}{4\pi} \displaystyle\iint_{S_i} \mu(S_i)\left[\dfrac{1}{r(S_i,P)}\right]_n dS_i$

(3) $\varphi(P) = \varphi_S(P) + \varphi_D(P)$

(4) $\left[\dfrac{\delta\varphi(P)}{\delta n}\right]_S = -\vec{n}\,\vec{U}_\infty$

(5) $\dfrac{1}{4\pi}\left[-\vec{n}\,\nabla \displaystyle\iint_S \dfrac{\sigma(S)}{r}\, dS + \vec{n}\,\nabla \displaystyle\iint_{S_i} \mu(S_i)\left(\dfrac{1}{r}\right)_n dS_i\right] =$

$\qquad = -\vec{n}\,\vec{U}_\infty$

FIG. 3   Basic equations for the lifting case.

condition). By differentiation of the source-sink distribution in the direction of the dipole axis, one gets the perturbation potential $\phi_D(P)$ of a three-dimensional dipole [Eq. (2)]. $\mu(S_i)$ represents the dipole strength on area S_i, which still has to be determined. The total potential, Eq. (3), results from the superposition of Eqs. (1) and (2). Equation (5) describes the solution of the boundary-value problem using the boundary condition [Eq. (4)]. The numerical solution is again obtained by using the panel technique.

The area S_i for the dipole distribution is determined in the following manner: An additional equation is obtained for lift-generating flows, in particular for flows around wings. A simple displacement flow around the wing profile does not represent the real situation, since there is a flow around the trailing edge (Fig. 4a). Therefore, an additional perturbation potential (circulatory flow) has to be added to prevent the flow around the trailing edge (Fig. 4b). However, this problem can also be interpreted as a boundary-value problem for a potential flow. The contour of the wing profile (representing a streamline in potential theory) is extended to provide a straight flow-off wake (Fig. 4c). The assumption of source-sink distribution converges to a dipole distribution on the extended contour (which is a streamline and infinitely thin), so that the dipole axes are perpendicular to the contour.

The singularity model thus contains the following parts:

A source-sink distribution on the contour
A dipole distribution on the extended contour

On the total contour, the perpendicular velocity components have to be zero (Fig. 5a).

Numerically, however, this problem still requires some manipulation. The source-sink distributions will show strong gradients at the trailing edge to compensate for the induced velocities from the dipole layer. This problem can be avoided if the dipole layer is extended into the inner part of the profile (on the camber surface), so that the distribution follows an arbitrary function, which has to be zero at the leading edge. This internal area, together with the wake, now represents the dipole distribution area S_i (Fig. 5b). The area S_i is then broken down into panels as well. In addition, the function for the dipole strength within the profile is defined (Fig. 5c).

A constant dipole strength is then assigned to each panel. For the three-dimensional case, a strip parallel to the flow is selected and treated as above (Fig. 6a). For numerical execution, the following analogy is used. The integral over a panel with constant dipole distribution for an arbitrary control point equals the integral of a concentric vortex around the edge of a dipole layer (Fig. 6b). This results in the well-known picture of the vortex lattice with spars and ribs (Fig. 6c).

The size of the total circulation separating from the trailing edge is determined from the flow-off condition in the usual way. The Kutta point (point satisfying the flow-off condition) is positioned slightly behind the

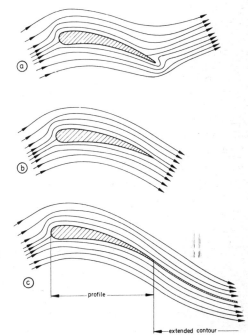

FIG. 4 Flow-off condition on wing profiles.

profile

extended contour

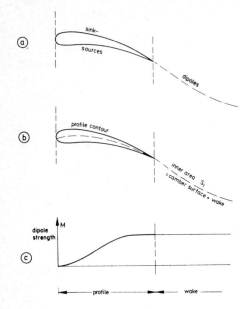

FIG. 5   Arrangement of singularities.

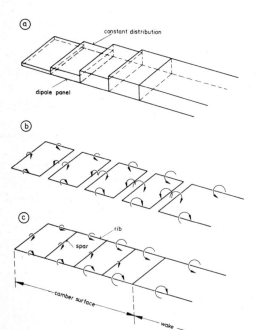

FIG. 6   Vortex lattice. (a) Dipole distribution of a strip; (b) vorticity rings instead of dipole panels; (c) vortex lattice.

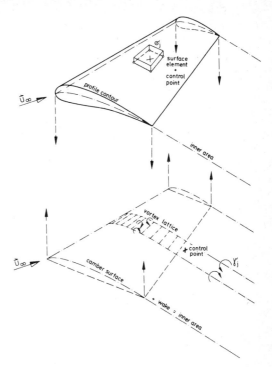

FIG. 7   Paneling of a wing.

trailing edge (Fig. 7) for numerical reasons. The effect of this control-point setback on the results is negligible, as has been shown by Loeve [4].

The control point of a surface panel (point satisfying the boundary condition) is determined from the fact that the source-sink distribution of the panel does not induce a velocity in the plane of the panel at this point. As a first approximation for the control point, the center of gravity of the panel area can be used.

Figure 8 shows the breakdown of the outer contour and the internal area S_i, with the wake.

The computation of the influence coefficients for the source, sink, and vortex distribution is shown in Fig. 9. Analysis of the double integral for the panel area (source-sink or parts of vortex lines) results in relatively simple algebraic equations. The matrix of the system of equations can then be determined as shown in Fig. 2.

$$[A_{ij}] \cdot \{X_j\} = \{B_i\}$$

$$A_{ij} = \mathbf{n}_i \cdot \mathbf{C}_{ij} \qquad B_i = -\mathbf{n}_i \frac{\mathbf{U}_\infty}{U_\infty}$$

$$X_j = \frac{\sigma_j}{4\pi U_\infty} \qquad \text{or} \qquad \frac{\mu_i}{4\pi U_\infty}$$

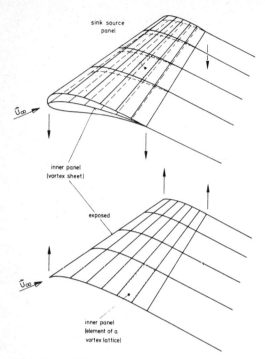

sink source
panel

$\tilde{U}_\infty$

inner panel
(vortex sheet)

exposed

$\tilde{U}_\infty$

inner panel
(element of a
vortex lattice)

FIG. 8   Paneling of a wing.

Even for modern computers, the number of matrix elements in general exceeds the available core storage. Therefore, the matrix has to be solved with an iterative procedure, whereby only one line of the matrix is put into core storage at a time. For the source-sink part, a simple Gauss-Seidel procedure is sufficient, since this part of the matrix usually shows a dominant main diagonal. To solve the vortex part, this section is computed by a direct procedure (Gauss-Jordan) after each iteration step of the source-sink portion. The approximation to the source-sink part is put on the right-hand side as a perturbation element. This procedure requires only three iterations for each significant decimal digit of the final result. Thus, the computer time required is very small, even for complicated configurations.

Usually, the Göthert rule is used to consider compressibility effects in three-dimensional flow. However, the accuracy of this rule is only sufficient for small disturbances, e.g., for bodies with high fineness ratios. Application of the Göthert rule to blunt bodies results in high suction peaks for the forebody. This is because the kinematic flow condition is satisfied by an equivalent body for which an incompressible computation is performed. Up to now, the application of semiempirical assumptions (e.g., nonlinear correction terms) resulted in significant improvements. MBB uses a modified form of the Göthert rule, whereby the boundary condition is satisfied by the real body (Fig. 10). This approach can be separated into the following steps:

1. Compute the geometry for the equivalent body using the Göthert transformation.
2. Set up the system of equations needed to compute the perturbation potential of the equivalent body.
3. Recompute the still unknown gradients of the perturbation potential $\nabla\varphi$ by transformation to the real-body geometry:

$$\frac{\partial\varphi}{\partial x} = \frac{\partial\varphi_a}{\partial x}\frac{1}{\beta^2} \qquad \frac{\partial\varphi}{\partial y} = \frac{\partial\varphi_a}{\partial y}\frac{1}{\beta} \qquad \frac{\partial\varphi}{\partial z} = \frac{\partial\varphi_a}{\partial z}\frac{1}{\beta}$$

4. Solve the system of equations while satisfying the boundary condition at the real body.
5. Compute the total velocity vector and the pressure coefficient C_p, using the isentropic equation.

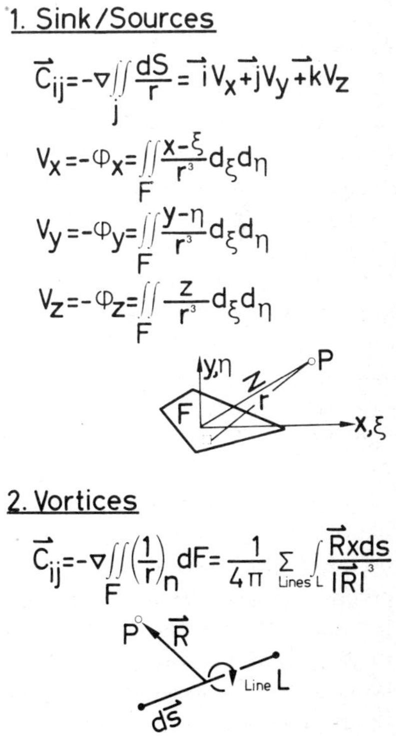

1. Sink/Sources

$$\vec{C}_{ij} = -\nabla\iint_j \frac{dS}{r} = \vec{i}V_x + \vec{j}V_y + \vec{k}V_z$$

$$V_x = -\Phi_x = \iint_F \frac{x-\xi}{r^3}d\xi d\eta$$

$$V_y = -\Phi_y = \iint_F \frac{y-\eta}{r^3}d\xi d\eta$$

$$V_z = -\Phi_z = \iint_F \frac{z}{r^3}d\xi d\eta$$

FIG. 9 Influence coefficients.

2. Vortices

$$\vec{C}_{ij} = -\nabla\iint_F \left(\frac{1}{r}\right)_n dF = \frac{1}{4\pi}\sum_{\text{Lines }L}\int_L \frac{\vec{R}\times ds}{|\vec{R}|^3}$$

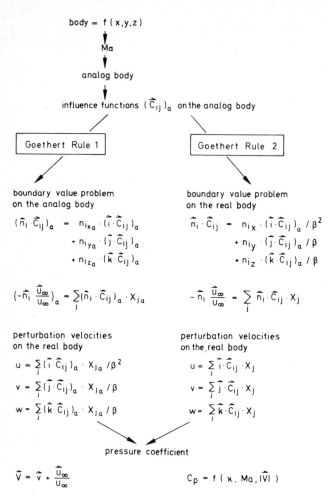

FIG. 10   Compressibility rules.

3   REVIEW OF EXISTING METHODS

A historic review (Fig. 11) reveals the fact that many panel programs existing in a number of countries around the world are based upon a very limited
number of basic panel programs. Classic panel programs were developed
without considering the lift problem; they are closely associated with the
names Smith [5], Hess [6], and Giesing [7]. The first approach was to solve
the two-dimensional problem; soon thereafter, consideration of lift effects
for this case was also possible. The circulation necessary for solving this
problem was generated by a number of single vortices within the profile
contour. A very elegant solution to the lift problem was found by A. M. O.
Smith [8] in 1966 (limited to the two-dimensional case). He used the

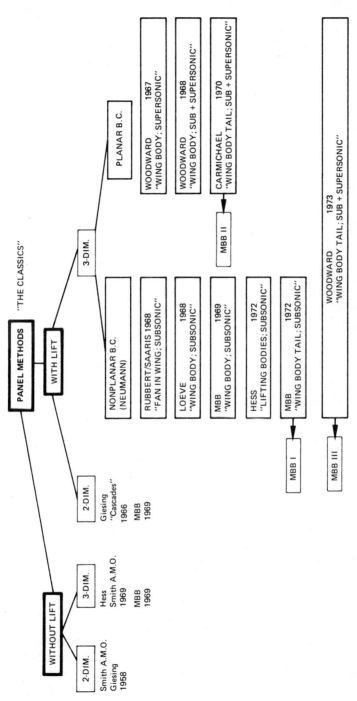

FIG. 11 A survey of panel methods.

influence functions of the source-sink panels located on the profile contour
and turned the velocity vector by 90°, with a subsequent summation. This
resulted in a continuous distribution of circulation on the total profile
contour. The circulation strength was determined by using the Kutta
flow-off condition. Each surface panel was then loaded with

1   A constant source-sink distribution to simulate the displacement effect,
    and
2   A constant circulation distribution to simulate the lift effect.

Two possibilities exist for satisfying the Kutta-Joukowski flow-off
condition:

1   Equal velocities for the last panels on the upper and lower surfaces.
2   The velocity vector at the trailing edge has to be parallel to the bisector.

In these methods, the displacement effect is verified only by using a surface
distribution and the Neumann boundary condition. This nonplanar boundary condition (NBC) was retained by Rubbert and Saaris in 1968 [9], when
they modified the three-dimensional method of Hess and Smith (1962) [5] to
include lift effects. As in the above-mentioned two-dimensional method, a
vortex distribution within the body contour was used (vortices positioned
on the wing camber plane). These vortices had to extend from the trailing
edge to infinity (free vortices). Two additional panel methods, with minor
differences in some details, were generated at NLR-Amsterdam in
1968 [10,11], and at MBB in 1969 [1]. A number of assumptions were introduced to allow the application of the wing vortex layer to the fuselage.
Finally, a method for computing a complete wing-body-tail combination,
considering all interference effects of the separating vortex wake, was available at MBB in 1972 [3].

A totally different method was used by Hess in 1972 [12] to extend his
method to include lift effects. Vortex distributions were placed on the upper
and lower surfaces of the wing, with both vortex layers flowing off at the
trailing edge.

A common feature of all these variants is the singularity distribution
(sink-source) on the surface to simulate the displacement effect (introduced
by Hess and Smith) and the limitation to subsonic flow. A second group of
panel programs emerged, using a totally different principle based upon the
vortex lattice method. This method has the advantage of allowing the computation of arbitrary wing planforms, which are, however, limited to flat
planforms (thickness zero). The classical two-dimensional profile methods
indicated a solution to the thickness problem. The general case of a
cambered profile with a given thickness and angle of attack was treated by:

Separation into thickness and camber (or angle of attack) problems

Linearization of the boundary condition [the boundary condition was not satisfied on the profile contour but on the wing chord line, by considering the respective contour slopes of the profile drop and camber; this is the planar boundary condition (PBC)]

A singularity distribution on the wing chord line

Woodward used the same approach in 1967 [13] in applying the panel technique to the three-dimensional case.

At first this method was designed for the supersonic region only. The program allowed consideration of the effect of axisymmetric bodies. Later, the program was extended to include the subsonic region. Various regions were reviewed and simplified by Carmichael in 1970 [14]. His program is the basis for most of the Woodward panel programs used all over the world. Such a program is available at MBB [15] and is used mainly for comparison purposes. In 1973, a new program developed by Woodward [16] appeared on the market, in which he tried to combine all the methods discussed above. This program is applicable to both the subsonic and supersonic regions, and includes the option of using either the NBC or the PBC for the wing. A version of this program was extended at MBB in 1974 to include the external flow [17].

As shown above, the classic panel programs are based upon three methods. Figure 12 compares these methods as to source data, singularity model used, boundary conditions, and limitations. Figures 13–16 illustrate these methods in more detail. Since the methods are fully described in the literature, the theory need not be discussed here.

Method 1 was described in detail in Sec. 2. Figure 13 shows a typical panel distribution for the surface of a wing-body combination. Figure 14 shows the panel arrangement for the internal area and the wake (separating vortex layer).

The singularity distribution in method 2 is somewhat different (Fig. 15). Chord planes are used for the wing, with source-sinks to simulate the thickness, and a vortex distribution to simulate camber and angle of attack. The method allows consideration of only axisymmetric bodies, since the fuselage is represented by linear elements (source-sink and dipole) on the body axis. The source-sinks are used to treat the axisymmetric flow, and the dipole to consider the angle-of-attack terms. As in the PBC method, for wings, the boundary condition is satisfied on the axis using the slope of the respective fuselage contour. For a wing-body combination, there are additional velocity components on the body contour that are induced by the perturbation field of the wing. These velocity components are compensated for by an additional surface panelization of the body (vortex distribution) within the interference region of the wing.

	METHOD I	METHOD II	METHOD III
PANELMETHODS	NONPLANAR B.C.	PLANAR B.C.	BOTH
Origin/Names	Hess/Smith/Rubbert/Loeve/Kraus	Woodward/Carmichael	Woodward 1973
Singularities	**Body:** constant sink/sources on surface panels **Wing:** constant sink/sources on surface panels. vortex lattice in the cambersurface, trailing vortices from the trailing edge of the wing	**Body:** constant line sinks/dipoles on the body axis. vortexshield in the influence region of the wing **Wing:** sink/source panels in the wing chord plane (for thickness) vortex panels in the wing chord plane (for camber)	**Body:** constant sink/sources on surface panels **Wing:** 1. planar B.C. linear varying source distribution in the chord plane (for thickness) linear varying vortex distribution in the chord plane (for camber) 2. nonplanar B.C. linear varying vortex distribution on panels lying on the wing surface (upper and lower side)
Restrictions of the methods	only subsonic	only axisymmetric bodies (but with camber) difficulties for wings with big dihedral or cruciform type. only cylindrical body in the influence region of the wing	none
Advantages	the most versatile method in respect to complicated body geometries	can be used for design purposes of wings	arbitrarily shaped bodies in supersonic

FIG. 12   The three methods used at MBB.

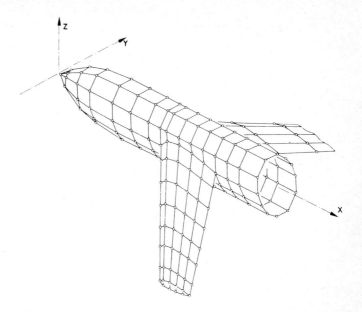

FIG. 13   Singularity distribution, method 1.

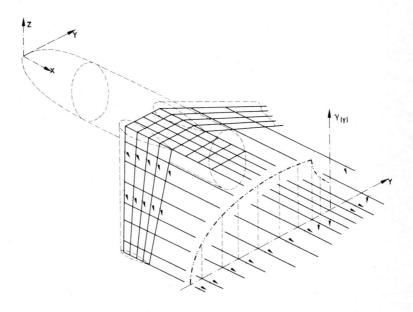

FIG. 14   Singularıty distribution, method 1.

Wing Body Combination

distribution of singularities

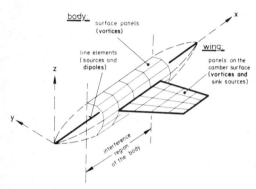

FIG. 15   Singularity distribution, method 2.

Figure 16 shows the singularity distribution for method 3 with the source-sink distribution on the body panels and a vortex distribution on the wing contour (NBC cases).

The classical methods have already been improved and refined. In particular, higher-order terms have been introduced. The improved methods use curved panel elements with varying singularity distributions. The main advantage of these methods is a reduction in the number of elements used to represent a body. This is equivalent to a reduction in computer time. In some cases, accuracy can be improved in spite of the fewer panel elements. However, the consistency of panel elements and singularity distributions has

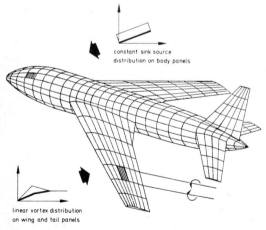

FIG. 16   Singularity distribution, method 3.

to be maintained. For example, for flat panels, a constant source-sink distribution is required; for parabolic panels, a linear distribution, and for cubic panels, a parabolic source-sink distribution.

Such a method for axisymmetric cases was developed and successfully used by Hess and Martin [19] in 1974. Particular improvements were achieved in the computation of inlet flow fields. A second method of this kind was published by Rubbert and Johnson [20] in 1975. However, to date, no test runs have been made with these program types at MBB.

4 BASIC EXAMPLES OF CALCULATIONS

We shall now look at some examples that demonstrate the wide range of applicability of panel methods. We begin with some simple examples of basic significance that make use of method 1.

A series of wing-body combinations were experimentally investigated by Körner [22]. His systematic measurements of pressure distributions gave an ideal basis for checking the MBB panel method. Figures 17 and 18 illustrate two geometries and the wing and body section planes where comparisons were made. Figure 17 shows the comparisons for a straight wing and body. Results using the method of Loeve are also shown. Agreement with experiment is very good. Figure 18 is an example of a swept-wing–body configuration. Agreement with experimental results is again very satisfactory.

5 WING-TAIL INTERFERENCE

Results were compared with wind-tunnel measurements made by Körner [24]. There was remarkably good agreement between analysis and measurement in the case of wing-body-tail combinations. Several parameters were tested, e.g., angle of attack, wing and tail incidence angles, and tail location. Here, we shall discuss some selected examples.

Figure 19 shows the positions of the sections where pressure distributions were compared with test data. Six wing, five tail, and three body streamwise sections were involved. Figure 20 gives the panel distributions in the three planes of view. Because of the symmetry, only half the body need be shown. The case presented in Fig. 20 is that with the smallest tail height and zero incidence of wing and tail. Comparisons of aerodynamic total coefficients are given in Figs. 21 and 22. Figure 21 compares tail-on and tail-off configurations, including the effects of the tail-height parameter. For the wing-body configuration, unstable longitudinal characteristics were found. Agreement between analysis and test data was very good for all cases. The same is true for the drag polars, if one takes into account the fact that there is no allowance for viscous effects (c_{D_0}) in the potential theory. Figure 22

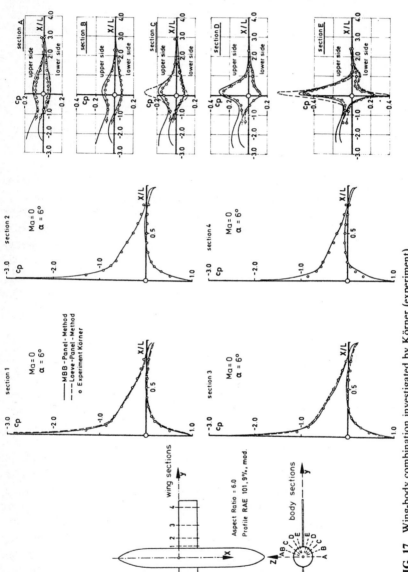

FIG. 17 Wing-body combination investigated by Körner (experiment).

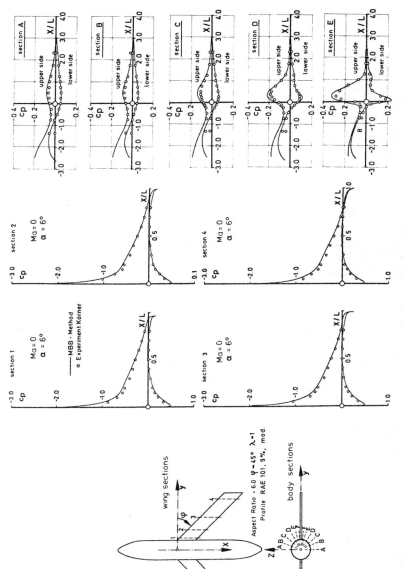

FIG. 18 Wing-body combination investigated by Körner (experiment).

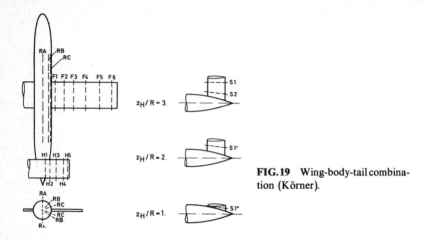

FIG. 19 Wing-body-tail combination (Körner).

illustrates the influence of different tail settings (trim conditions). The correlation between total zero-lift pitching moment and tail incidence shows up well.

Comparisons of the effects on wing and horizontal tail are presented in Figs. 23–26. Figure 23 shows the spanwise distribution for three different tail settings with zero angle of attack. Note the negative load distribution at the tail at zero angle of attack and zero incidence, caused by the body-area reduction at the tail (boat-tailing of the body). The equivalent results for three tail incidences but with 6° angle of attack are presented in Fig. 24. Lift

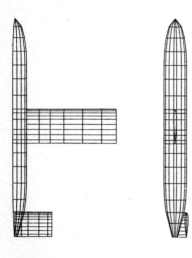

FIG. 20 Paneling of the configuration.

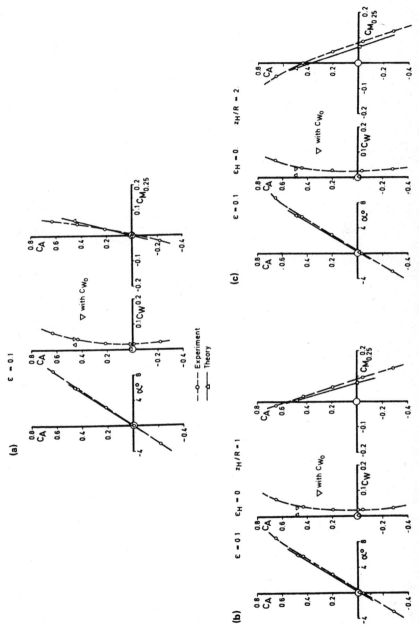

FIG. 21 Aerodynamic overall coefficients; influence of tail heights. (a) Wing-body combination; (b) and (c) total combination.

257

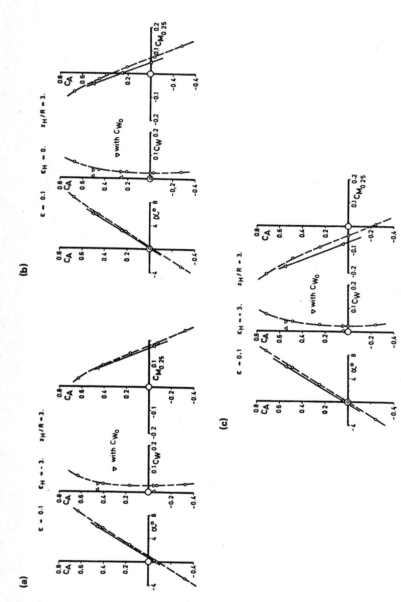

FIG. 22 Aerodynamic overall coefficients; influence of tail incidence. (a–c) Total combination.

258

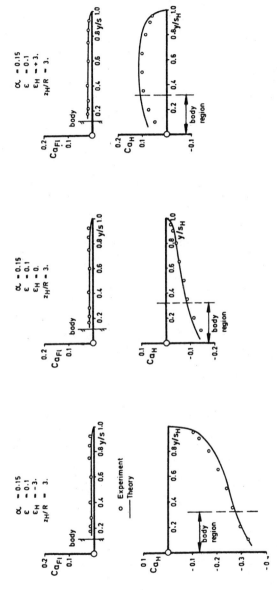

FIG. 23 Aerodynamic section coefficients; influence of tail incidence, $\alpha = 0.15$.

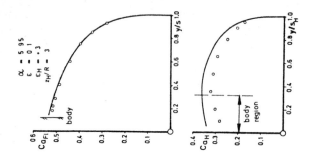

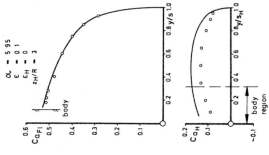

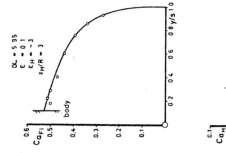

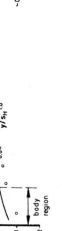

FIG. 24 Aerodynamic section coefficients; influence of tail incidence, $\alpha = 5.95$.

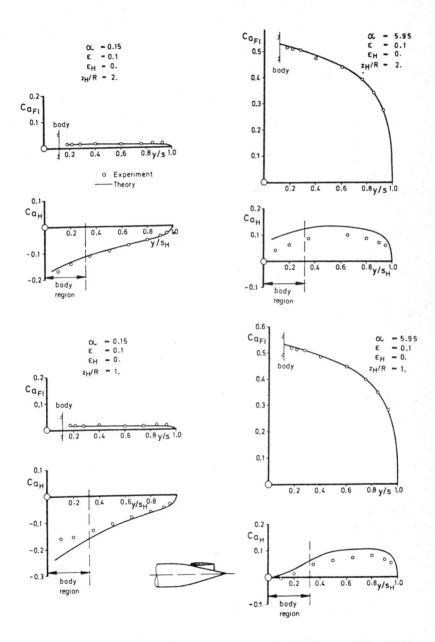

FIG. 25 Aerodynamic section coefficients; influence of tail heights, $\alpha = 0.15$ and 5.95.

is slightly overestimated. This is due to the position of the tail just below the body (the effect of body viscous wake), dependent on angle of attack.

Figure 25 demonstrates the influence of tail height, and Fig. 26 gives the results for different wing incidence angles. With 6° wing incidence, agreement between analysis and test results is pretty good at the horizontal tail, because in this special case the body is positioned at zero angle of attack, and there is no body wake interfering with the tail surfaces. Interference effects between wing and tail are well interpreted by the method, and wing wake influences at the tail are well predicted.

Only by an iterative procedure can one develop more accuracy in computing wing-wake rolling-up effects (see Sec. 6). The location and shape of the wing wake do, of course, influence the tail loads, but Fig. 4.8b definitely shows that the iteration procedure gives satisfactory results after two steps.

6 THE WAKES OF WINGS

Analysis of wing-wake characteristics is the basis for obtaining the wing-tail interference. The wake is determined by a procedure that computes streamlines in the flow field. This streamline procedure starts at the trailing edge of the wing. Calculated streamlines give the new vortex model for the free-trailing vortices. Then an additional computation of streamlines is performed for this new vortex model. In most cases, the iteration procedure for finding the final wake converges after two or three steps.

Figure 27 shows the initial assumption for the wake of two coplanar rectangular wings positioned at a distance of three wing chords. The final, iteratively determined wake of the first wing in the presence of the second, and the wake of the first wing alone, are also shown in the midchord plane.

Figure 28 illustrates streamlines in the vicinity of the wake, especially in the tip region of a rectangular wing. In Fig. 29 you can see the wake of this wing with and without consideration of the y component of the velocity vector.

The amount of lowering of the wake was also analyzed and compared, and Fig. 30 shows good agreement between experiment and theory. The comparison was performed for the wing-body-tail combination discussed above (see Chap. 4 of reference 24), at several spanwise positions behind the wing.

7 NONLINEAR CALCULATION METHOD FOR DISCRETE TRAILING VORTEX SHEETS

As we have seen that allowing for the trailing vortex sheet by means of potential theory gives reasonable results. This suggests that the nonlinear effects (lift versus incidence) of vortex sheets shed from the wing leading and side edges be included in the same way. A panel method is not required for

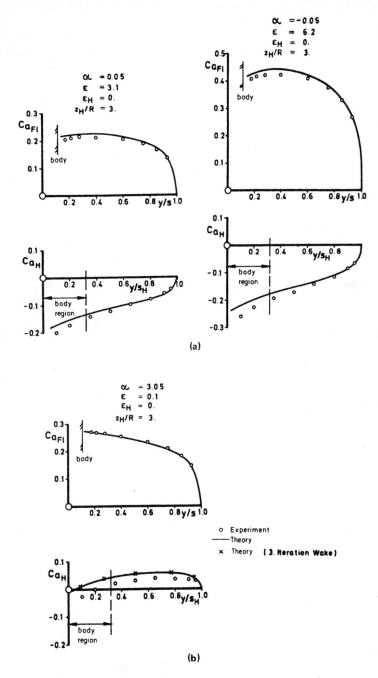

FIG. 26 Aerodynamic section coefficients. (a) Influence of several wing incidences; (b) influence of wake iteration.

263

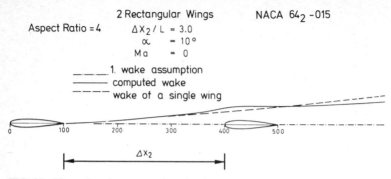

2 Rectangular Wings   NACA 64$_2$ -015

Aspect Ratio = 4 $\Delta X_2 / L$ = 3.0
                          α = 10 °
                          Ma   = 0

_____ 1. wake assumption
_____ computed wake
--------- wake of a single wing

0   100   200   300   400   500

ΔX_2

FIG. 27 The wake of a rectangular wing in the presence of a second rectangular wing.

this [26,27], but it was hoped that the universal applicability of panel methods would enhance the chances for success.

At MBB, several variants of these methods were checked [28]. In order to study the fundamental possibilities of these methods, it is sufficient to start with a method treating only thin wings. Three vortex models were studied (Fig. 31):

1   Bollay's model: Straight trailing vortex leaving the wing side edge with an incidence of $\alpha/2$ or $2/3\alpha$
2   Gersten's model: Straight vortices shed from the wing surface with $\alpha/2$
3   Belotserkovsky's model: Stepwise calculation of the trailing vortex sheet with a linear starting solution [29]

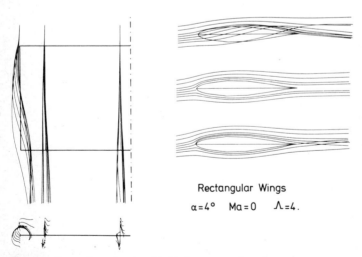

Rectangular Wings
$\alpha = 4°$ Ma = 0 $\Lambda = 4$.

FIG. 28 Streamlines about and behind a rectangular wing.

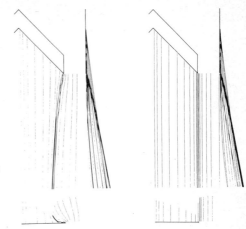

FIG. 29 The wake of a swept wing;
$\alpha = 10°$, M = 0.

Belotserkovsky's model should be explained in more detail. The calculation is started with a small incidence α. In this linear region, the streamlines leaving the wing side or leading edges (straight or swept wing) are calculated. The next calculation uses the streamlines as axes of vortices. This procedure is then repeated until the solution converges. The incidence is increased, and a new calculation starts with the converged solution for the initial small angle of attack. This process is rather expensive in time and money, owing to its slow convergence.

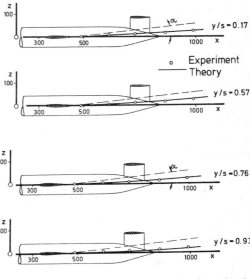

FIG. 30 The wake of a wing-body-tail configuration in comparison with experiment (Körner). Lowering of the wake; $\alpha = 5,95$, $\varepsilon = 0.1$, $\varepsilon_H = 0$, $z_H/R = 2$.

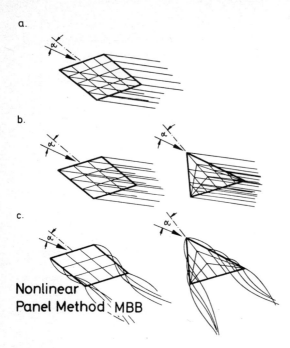

a.

b.

c.

Nonlinear
Panel Method MBB

FIG. 31 Several vortex-sheet models. (a) Bollay; (b)
Gersten; (c) Belotserkovsky.

There is, however, another way to use Belotserkovsky's vortex model, as
shown by Perrier [30]. At the desired incidence α, we start with a basic
distribution of vorticity and iterate for the streamlines. The starting distribution is provided by a nonlinear calculation with a Bollay or Gersten model.
In this calculation procedure, streamlines are also lines of constant vorticity.
On the streamline, the vortex elements should not be smaller than the disin-
tegration of the flow field. This is about the size of a panel. If they are
smaller, the streamlines will wind around the vortex lines emanating from
the trailing edge.

    Figure 32 presents the streamlines resulting from a linear calculation for
a slender rectangular wing at incidences of 2 and 20°. The angle of attack of
20° does not, of course, represent a realistic case. Figure 33 shows a nonlinear vortex sheet shedding from the side edge at α = 20°. It was calculated by
the procedure mentioned above, starting with the solution of α = 2°, and
iterating for the nonlinear vortex sheet in steps of Δα = 2°.

    The upper part of the figure shows the resulting normal force and
moment distribution as functions of incidence α. Experimental data are
shown for comparison. As this procedure is very laborious and expensive
(computer time-consuming), it was dropped at MBB.

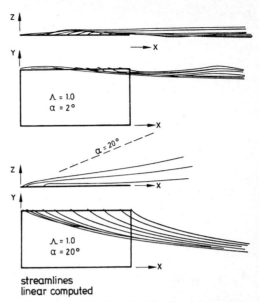

streamlines
linear computed

FIG. 32 Streamlines from the side edge of a rectangular wing. Streamlines are linearly computed.

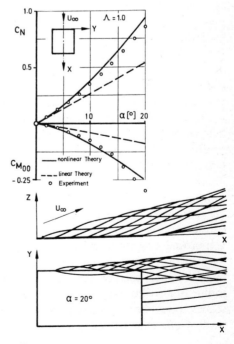

FIG. 33 Vortex sheet from the side edge; aerodynamic coefficients in comparison with experiment (Belotserkovsky vortex model).

267

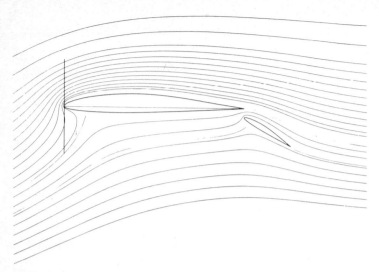

FIG. 34 Streamlines about a system of profiles. NACA 63A310: $\alpha = 10°$, $\vartheta = 30°$, M = 0.

8 GRAPHS OF FLOW FIELDS

In dealing with potential theory, it is clear that we need not confine ourselves to finding pressures and velocity vectors on body surfaces, but can extend the calculations to the complete flow field. We are especially interested in pictures of streamlines. Figure 34 shows a two-dimensional flow field around an airfoil system at an incidence of $\alpha = 10°$ and $\delta = 30°$ flap deflection. Figure 35 presents the flow field of an augmentor wing. The blown flow between two flaps has been accounted for by additional singularities. The influence function for an arbitrary point in the flow field is found in exactly the same way for a control point on the body surface. Thus, we obtain the velocity vector in the same way for body points and for flow field, and, by integration in space, we find the streamlines too.

The simple difference scheme for solving the differential equation for the two-dimensional case is given in Fig. 36. The upper index in the formulas indicates the approximate level. Figure 37 is a graphic display of the procedure. In most cases, the iteration converges within a few steps below the prescribed value ε. Further, the step width is controlled by maintaining a constant proportion of step size to radius of streamline curvature.

In Figure 38, the streamline around a circular cylinder as approximated by the difference scheme is compared with an exact solution. The deviation from the exact solution is shown for several step sizes Δs normalized by the radius of curvature.

It is possible to calculate the traces of water droplets in the air with a similar difference scheme. Thus, we can calculate the position and amount of

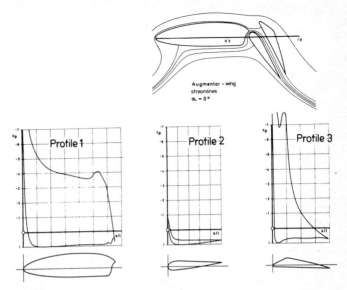

FIG. 35 Streamlines about an augmentor wing system.

ice forming per unit time on wings, bodies, and intakes. Using this method, it is possible to predict the shape and height of ice buildups for different temperatures and moisture contents.

For different sizes of droplets, traces are calculated from the differential equation giving the equilibrium between the forces of inertia and drag for a droplet. This equation is solved by a simple difference scheme. Figure 39

$$\frac{V_x}{dx} = \frac{V_y}{dy}$$

in stream line direction:

$$\frac{V_x}{|\vec{V}|} = \frac{dx}{ds} \qquad \frac{V_y}{|\vec{V}|} = \frac{dy}{ds}$$

iteration procedure to determine the coordinates of a stream line:

$$x_{n+1}^{(v+1)} = x_n + \Delta s \frac{V_{xn}}{|\vec{V}_n|} + \Delta x_n^{(v)}$$

$$y_{n+1}^{(v+1)} = y_n + \Delta s \frac{V_{yn}}{|\vec{V}_n|} + \Delta y_n^{(v)}$$

FIG. 36 Iteration procedure to determine the streamlines.

with $\Delta x_n^{(v)} = \frac{\Delta s}{2}\left(\frac{V_{xn+1}^{(v)}}{|\vec{V}_{n+1}^{(v)}|} - \frac{V_{xn}}{|\vec{V}_n|}\right)$

$$\Delta y_n^{(v)} = \frac{\Delta s}{2}\left(\frac{V_{yn+1}^{(v)}}{|\vec{V}_{n+1}^{(v)}|} - \frac{V_{yn}}{|\vec{V}_n|}\right)$$

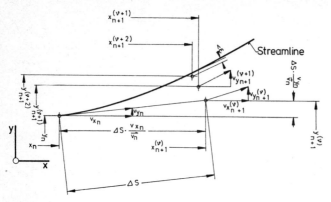

FIG. 37   Streamline iteration.

shows this scheme as applied to the velocities of the droplets. Having found these velocities, it is easy to determine the traces using the scheme shown in Fig. 36. Figure 40 shows the droplet path lines around a circular cylinder for different inertia parameters C.

Figure 41 illustrates an application of the theory. The ice formation at the empennage of a commercial airplane is clearly shown. The results are compared with measurements of the traces around the airfoils and the height of the ice buildup along the wing span [32].

9   APPLICATION TO EXTERNAL STORES

The effects of external stores are, generally speaking, hard to treat. Panel methods can be applied to this problem, to calculate:

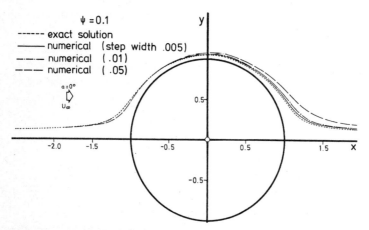

FIG. 38   Streamline around a circular cylinder: comparison with exact solution.

$$\frac{d\vec{V}}{dt} = K_1 C(\vec{U} - \vec{V})$$

$\vec{V}$ = velocity of waterdrops

$\vec{u}$ = " " air

$K_1 = \frac{1}{24} C_D Re$ Stokes factor

$C = \frac{9}{2} \frac{\mu_L l}{\rho w(d/2)^2 u_\infty}$ droplet inertia parameter

Re = Re-number of droplet

C_D = dragcoefficient of d

ν_L = viscosity of air = $\frac{\mu_L}{\rho}$

l = characteristic length of body

d = droplet diameter

ρw = density of water

FIG. 39 Differential equation for water droplets (two-dimensional) in Eulerian frame and in streamline direction.

$$\frac{\delta V_x}{\delta s} = K_1 C \frac{U_x - V_x}{\vec{V}} \qquad \frac{\delta V_y}{\delta s} = K_1 C \frac{U_y - V_y}{\vec{V}}$$

difference scheme:

$$V_{xn+1}^{(v+1)} = V_{xn} + \Delta s K_1 C \frac{u_x - V_{xn}}{\vec{V}} + \Delta V_{xn}^{(v)}$$

$$V_{yn+1}^{(v+1)} = V_{yn} + \Delta s K_1 C \frac{u_y - V_{yn}}{\vec{V}} + \Delta V_{yn}^{(v)}$$

$$\Delta V_{xn}^{(v)} = \frac{\Delta s}{2} K_1 C \left(\frac{u_x - V_{xn+1}^{(v)}}{|\vec{V}_{n+1}|} - \frac{u_x - V_{xn}}{|\vec{V}_n|} \right)$$

$$\Delta V_{yn}^{(v)} = \frac{\Delta s}{2} K_1 C \left(\frac{u_y - V_{yn+1}^{(v)}}{|\vec{V}_{n+1}|} - \frac{u_y - V_{yn}}{|\vec{V}_n|} \right)$$

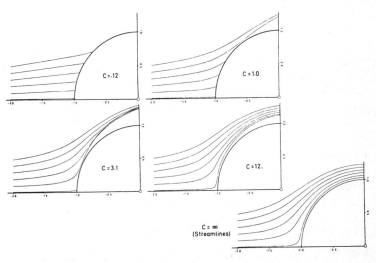

FIG. 40 Water-drop trajectories around a cylinder for several droplet parameters.

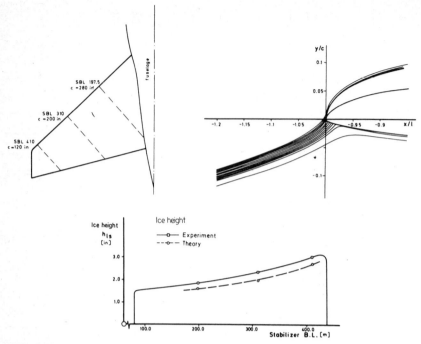

FIG. 41 Water-drop trajectories and ice heights for a horizontal stabilizer.

1   The aerodynamic loads on carriers and stores or weapons (estimation of
    interference effects)
2   The forces and moments on separating stores in different positions rela-
    tive to the airplane (interference effects), to determine the trajectories of
    the separating stores.

Aerodynamic loads can be treated with panel methods; they hold for any
desired geometry. Figure 42 illustrates the simplest possible way to find the
aerodynamic forces and moments on external stores in different positions
relative to the airplane. Of course, a solution has to be found for the complete configuration (airplane plus external store) each time. This procedure
becomes very expensive with respect to calculation time. For this reason,
MBB uses the method shown in Fig. 43. Velocity vectors of nonuniform flow
fields are calculated at distinct points (three-dimensional flow grid). The
external store, in different positions relative to the aircraft, can now be taken
into account. The three-dimensional grid is used to interpolate velocity
vectors at the control points of the store. A calculation is then necessary in
order to find the forces and moments on the store, with the store in the
previously determined curved flow.

In order to obtain trajectories with the given aerodynamic forces, one has

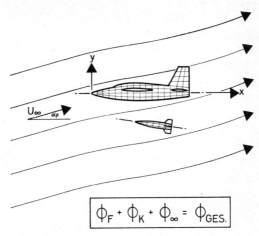

$$\phi_F + \phi_K + \phi_\infty = \phi_{GES.}$$

solution for the complete Configuration
(aircraft clean and external Store)

FIG. 42 Calculation of aerodynamic coefficients of external stores in different positions relative to the airplane.

Panel - Method - Aircraft Clean

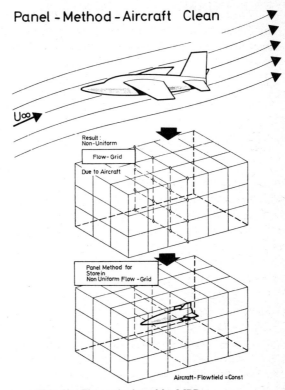

FIG. 43 The method used by MBB.

to establish and solve an equation of motion for the external store. This is done with a method similar to that used for the trajectories of water drops (Sec. 8). Now not only the drag, but all three aerodynamic forces and moments must be taken into account as driving forces. Moreover, the external store cannot be regarded as a mass point. We shall not discuss the problem further here; more specific data can be found in the report of Deslandes [33].

Figure 44 shows a paneling of the well-known Phantom F4. Two different kinds of weapons are attached. The aerodynamic forces (lift and side force) and moments (pitching, rolling, and yawing) of those weapons were calculated and compared with experimental results. Figures 45 and 46 give two comparisons for the first external store (with and without pylon loads). Figure 47 refers to the second store. Agreement is satisfactory.

In Fig. 48, the jettison characteristics (trajectories) of a usually retarded flying external store are plotted. As one can see, the store will loop and turn without a parachute. Figure 49 illustrates the jettison characteristics of a ballistic store. A well-controlled jettison from the carrier is evident.

10 INTAKES

There was close collaboration between MBB and the Dornier Company, Friedrichshafen, as there was between MBB and DFVLR-Braunschweig. Dornier extended the standard version of the MBB panel method by simply introducing further (sink-source-doublet) singularity arrangements re-

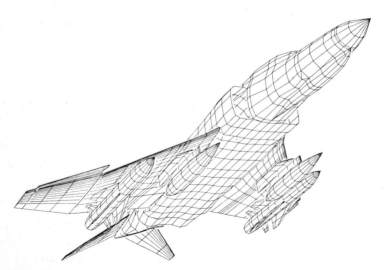

FIG. 44 F4F Phantom with stores.

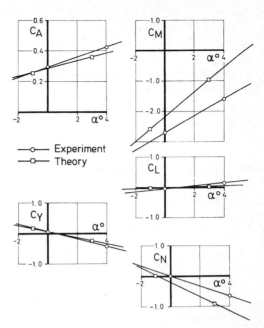

FIG. 45 Loads on store 1 (without pylon loads); $M = 0.7$, $\beta = 0°$.

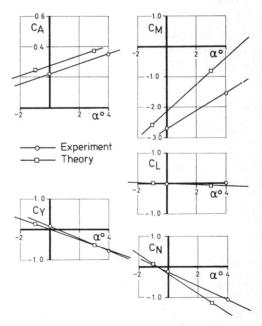

FIG. 46 Loads on store 1 (pylon included); $M = 0.7$, $\beta = 0°$.

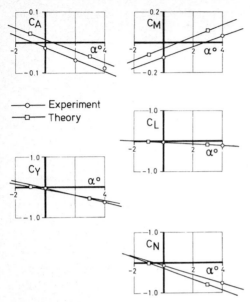

—o— Experiment
—□— Theory

FIG. 47   Loads on store 2; $M = 0.7$, $\beta = 0$.

presenting flows through intakes [35]. The results of this work were first published by H. Zimmer during a symposium at DGLR, Friedrichshafen, in 1972.

At first, attempts were made to simulate the mass flow by using a disk of sink-sources with given boundary conditions for normal velocity components (Fig. 50). Speed ratios less than 1 and ground-run cases should be

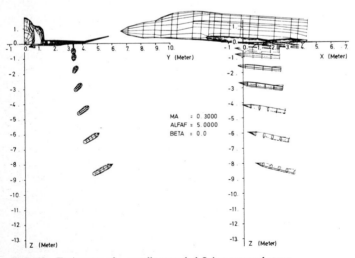

FIG. 48   Trajectory of a usually retarded flying external store.

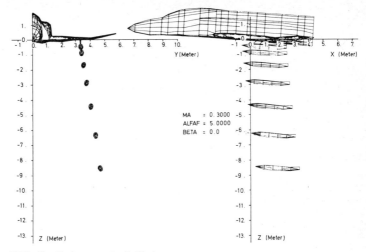

FIG. 49 Trajectory of a ballistic store.

analyzed in this way. However, in addition to the problem of leakage (this is a well-known problem in panel methods when computing internal flows), there were problems of convergence in solving the system of linear equations. Leakage problems can be avoided by moving the singularity sheet away from the intake contour. Both difficulties are minimized by moving this sheet to the most forward position. If an assumed distribution of normal velocities across this area is used, the pressure distribution on the inlet contour is changed in an arbitrary manner, and reasonable results cannot be realized along the internal intake. This singularity model may, however, be applicable to and useful for the calculation of total coefficients, including the interference effects of intakes.

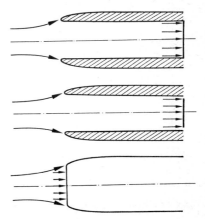

FIG. 50 Simulation of the intake with sink-source panels.

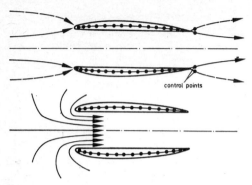

FIG. 51 Simulation of the intake with an internal vortex system.

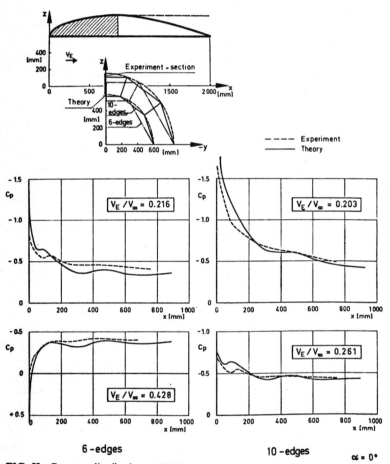

6 -edges   10 -edges

$\alpha = 0°$

FIG. 52 Pressure distribution at NACA 1–70–50 intake.

The use of other singularities (vortices) finally gave more success. These vortex singularities curve the flow in such a way that mass flow is varied. The inlet is represented as an annular wing with a sharp trailing edge (Fig. 51). Proper selection of flow directions at the trailing edge (Kutta condition) gives any desired mass-flow ratio. There is a second method of solution: prescribing the circulation of the vortex system. The advantage of this procedure is that it includes the ground running case. In both methods, the mass flow through the engine is evaluated simply by integrating over a control area directly behind the intake lips.

Comparative calculations were performed for the NACA 1-70-50 intake (Fig. 52). The intake is "transformed" into an annular wing via a supplementary connected part. Even for a small number of panels, the results are in good agreement with measurements. Vortex model B was used; that is, the vortex circulation was prescribed.

Figure 53 compares the two vortex models (solution methods). All results are in fair agreement with the test data and, in this particular case, the models are equivalent.

Finally, some results derived from the computation of the intake of Alpha Jet are shown in Fig. 54. Panel elements are defined on parts of the body side contour for partial representation of the total forebody. The intake is considered as half an annular wing mounted directly on the body side. For

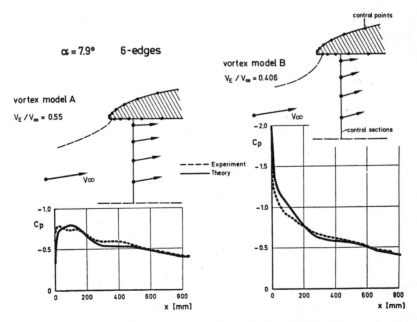

FIG. 53 Comparison of the two vortex models at NACA 1-70-50 intake.

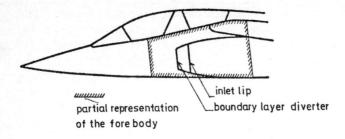

partial representation
of the fore body

inlet lip

boundary layer diverter

Paneling

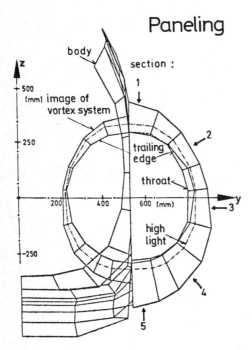

FIG. 54   Engine inlet.

numerical reasons, the vortex system does not stop at the wall; its image extends inside the body in order to avoid the effects of strong shed vortices.

Some results of analysis and experiment are compared in Fig. 55. Pressure distributions along the outside contour of the intake fairings and isobars at the body side wall are shown. Agreement is satisfactory; only the negative peaks at the intake lips show a low level of accuracy in this method of analysis.

11 SUPERSONIC PANEL METHODS

In this section we discuss some results of calculations using Woodward's supersonic panel method (method 3 of Fig. 12). This method was extended by MBB, among other reasons, to account for flow fields. Figure 56 shows such a flow at the center section of a straight, untapered wing of aspect ratio 6. For a Mach number of 1.4 and angle of attack of 15°, velocity vectors were

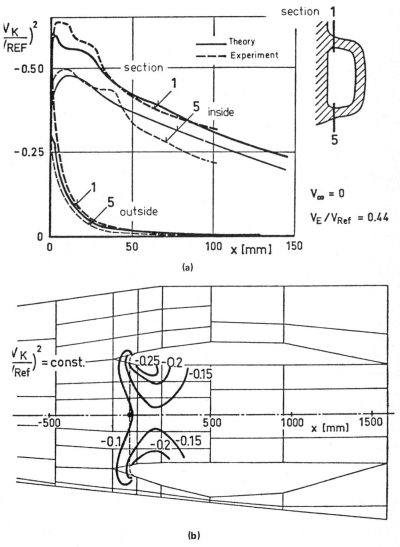

(a)

(b)

FIG. 55 Comparison of analysis and experiment. (a) Pressure distribution around inlet; (b) isobar plot on body side.

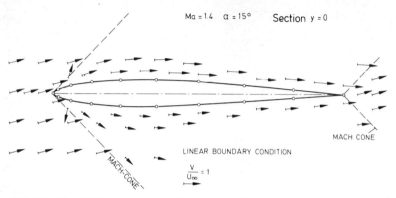

Ma = 1.4   α = 15°   Section y = 0

LINEAR BOUNDARY CONDITION

$$\frac{V}{U_\infty} = 1$$

MACH CONE

FIG. 56. Flow field in the center section of a rectangular wing, M = 1.4.

calculated at several control points. It is easily seen in Fig. 56 that there is an undisturbed translatory flow beyond the Mach cone. Absurd values are produced by the method for points positioned directly on the Mach cone, for numerical reasons. Figure 57 illustrates that a reasonable amount of upwash is found near the wing leading edge in the subsonic case.

One additional advantage of Woodward's second method, relative to earlier methods, is its applicability to arbitrary body shapes. An elliptical cone, for which experimental results already existed, was recalculated to compare results. The distribution of surface elements is drawn in Fig. 58. Figures 59 and 60 compare theory and measurements for two angles of attack and a Mach number of 1.89.

Finally, Fig. 61 compares calculated results using Woodward's method with experimental results. The graphs show results calculated with the re-programmed MBB version in addition to Woodward's originally calculated results. The agreement between calculated and experimental results can be regarded as satisfactory.

Ma = 0   α = 15°   Section y = 0

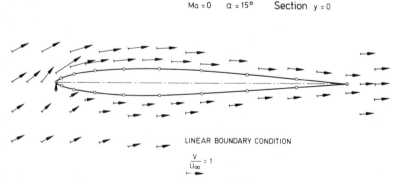

LINEAR BOUNDARY CONDITION

$$\frac{V}{U_\infty} = 1$$

FIG. 57 Flow field in the center section of a rectangular wing, subsonic case.

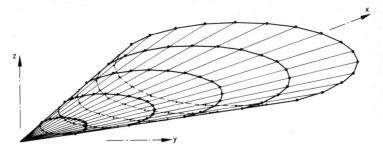

FIG. 58 Elliptical cone.

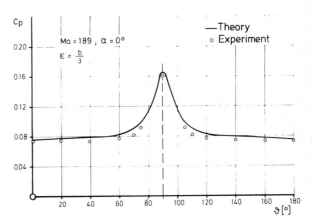

FIG. 59 Theory versus measurement for the elliptical cone at $\alpha = 0°$.

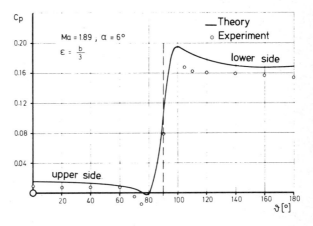

FIG. 60 Theory versus measurement for the elliptical cone at $\alpha = 6°$.

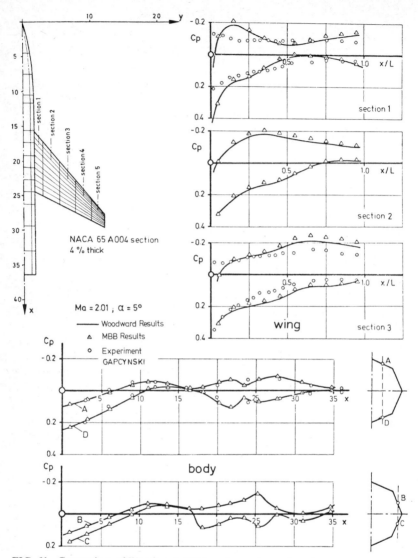

FIG. 61 Comparison of Woodward, MBB, and experimental results.

12 PANEL METHODS FOR THE CALCULATION OF COMPRESSIBLE FLOWS ABOUT AIRFOILS

12.1 Analysis of Airfoils in Purely Subsonic Flow

We usually apply panel methods to the determination of pressures on three-dimensional configurations immersed in an incompressible fluid flow distorted throughout the flow field by the constant Prandtl factor. In

(6) $(\rho\Phi_x)_x + (\rho\Phi_y)_y = 0$

(7) $\rho = \rho_0 \left[1 - \frac{\kappa-1}{2}(\Phi_x^2 + \Phi_y^2)\right]^{\frac{1}{\kappa-1}}$

(8) $\Phi_{xx} + \Phi_{yy} = \frac{-1}{\rho}(\rho_x\Phi_x + \rho_y\Phi_y) = g$

(9) $\Phi_p = u_\infty x + v_\infty y + \Gamma\phi\left[(\ln r)_n\right]_p ds$

$\qquad + \frac{1}{2\pi}(\phi\mu\ln r_p ds + \iint g\ln r_p dx dy)$

FIG. 62   The equations.

contrast, panel methods can be developed that allow for the correct density relation in space, at least for subsonic flow. We shall discuss such a method for compressible potential flows about airfoils [37,38].

Making use of the well-known assumptions of potential theory, we describe a two-dimensional flow by the continuity equation (6) in Fig. 62, where the density ρ is a function of ϕx and ϕy [Eq. (7)]. Explicit differentiation of the continuation equation leads immediately to the elliptic equation (8). By splitting the potential ϕ into the free-stream potential and a disturbance potential φ, this equation can be transformed into an integral relation using Green's theorem and the well-known parametric singularity of two-dimensional potential theory, $\ln r$ [Eq. (9)], where r is the distance between field points and control point P. The terms have the meanings given in Fig. 63.

For evaluating the singularity strengths Γ, μ, and g of the integral equation, a classic constant-strength panel technique is used. This is achieved by

$$u_\infty x + v_\infty y =$$

freestream potential

$$\phi(\ln r)_n ds =$$

unit vortex distribution on airfoil contour

$$\phi\mu\ln r ds =$$

source distribution on airfoil contour of yet unknown strength μ

FIG. 63   Meanings of the terms.

$$\iint g\ln r\, dx dy =$$

source distribution throughout flowfield representing compressibility

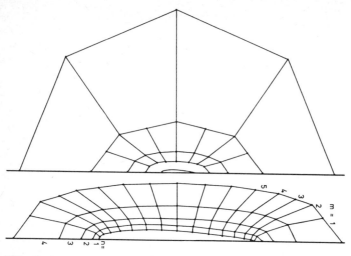

FIG. 64   Paneling of profile and flow field.

subdividing the airfoil contour into straight-line elements, and the flow field into appropriate quadrilaterals (Fig. 64). As μ and g are assumed to be constant over each element, the influence on a fixed control point P of each source can be determined by a priori integration, thus decomposing the integral relation to a sum formula per control point (Figs. 65 and 66).

In order to find the singularity strengths of the boundary sources, the sum formula is solved directly for μ_i for each boundary control point i by applying the boundary condition $\phi_{ni} = 0$, where it is assumed that all singularity strengths are set equal zero before the first iteration (Fig. 67a). The evaluated strengths are used immediately at each sweep through the matrix (Gauss-Seidel procedure). The compressibility strengths g_i of the field points are determined by immediate updating of the relation for g_i (Fig. 67c). This process has to be repeated until the boundary condition, including the Kutta condition (Fig. 67b), is satisfied to an acceptable degree.

The final summing of source influences on boundary control points leads to the velocity distribution along the airfoil contour. Some results calculated by Stricker are shown in Figs. 68 and 69.

Basically, this method can be extended to three-dimensional cases by replacing

$\ln r$ by r

The line integrals by area integrals over the surface of the configuration
The area integral by a volume integral over the flow field, where g becomes a
   three-dimensional function

$$g = \frac{1}{\rho} (\rho_x \phi_x + \rho_y \phi_y + \rho_z \phi_z)$$

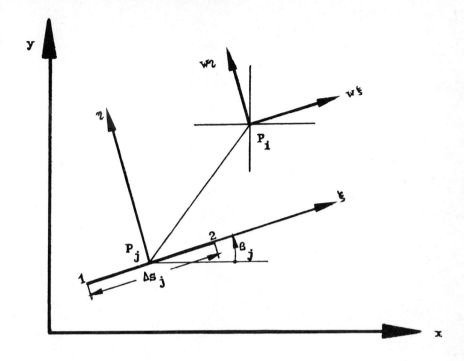

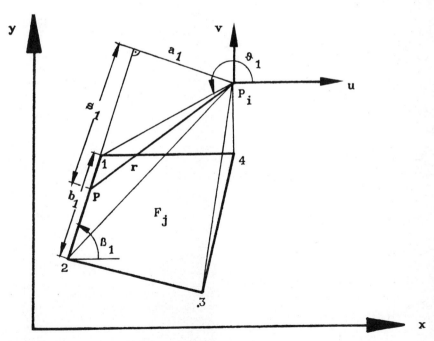

FIG. 65   Computation of influence coefficients.

$$\Phi_{xp} \quad = \quad u_\infty + \Gamma \sum_i \int_i (\ln r)_{nxp} ds + \Big\}$$
$$\left. + \sum_i \mu_i \int_i (\ln r)_{xp} ds + \right\} \quad \text{boundary}$$
$$+ \sum_i g_i \iint_i (\ln r)_{xp} dx dy \Big\} \quad \text{field}$$

$$\Phi_{yp} \quad = \quad v_\infty + \Gamma \sum_i \cdots yp + \sum_i \mu_i \cdots yp + \cdots$$

$$\Phi_{xyp} \quad = \quad \Gamma \sum_i \cdots xpyp + \sum_i \mu_i \cdots xpyp + \cdots$$

$$\Phi_{xxp} \quad = \cdots$$

$$\Phi_{yyp} \quad = \quad g_p(\text{„old value"}) - \Phi_{xxp}$$

FIG. 66 Sum formula per control point.

12.2 Design of Supercritical Airfoils Using a Two-Dimensional Panel Method

A method similar to the one described above can be applied to the transonic airfoil design problem in hodograph theory [38]. To begin, the boundary streamline, in terms of local Mach number and flow angle, is defined in the

a) boundary condition:

$$\Phi_{ni} = (\Phi_x x_n)_i + (\Phi_y y_n)_i = 0$$
$$\longrightarrow \mu_i$$

b) Kutta-condition:

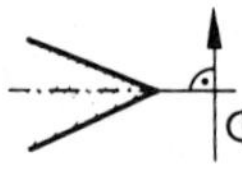

$$\Phi_n = 0$$

c) compressibility:

$$g_i = -\frac{1}{\rho_i}(\rho_{xi}\Phi_{xi} + \rho_{yi}\Phi_{yi})$$

iteration a,b,c repeated until

$$\left| \Phi_{n\,\text{boundary max}} \right| < \varepsilon$$

FIG.67 Iteration process to determine the singularity strength.

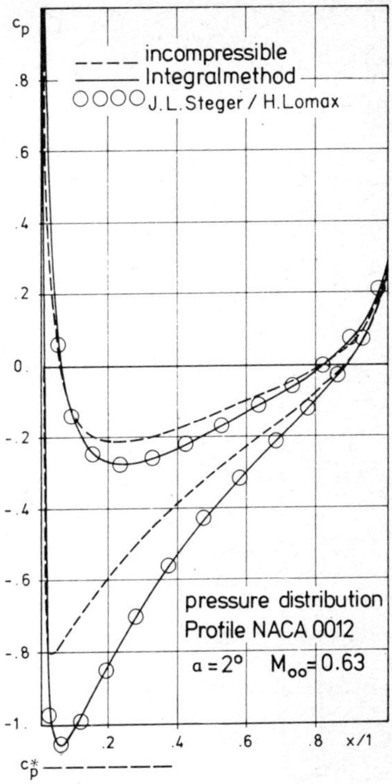

FIG. 68 Comparison with other theories.

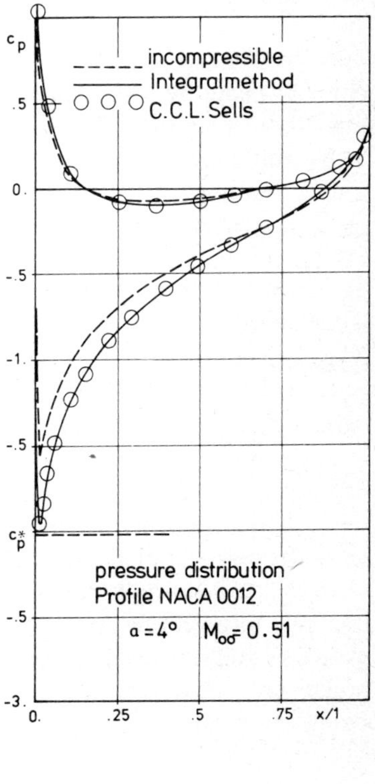

FIG. 69 Comparison with other theories.

$$dz = dx + idy = \frac{d\phi + i\frac{g_0}{g}d\psi}{\overline{w}} \qquad \overline{w} = we^{-i\vartheta}$$

$$z_{w\vartheta} = z_{\vartheta w}$$

$$\phi_w = \frac{g_0}{gw}\left(M^2 - 1\right)\psi_\vartheta$$

$$\phi_\vartheta = \frac{g_0 w}{g}\,\dot{\psi}_w$$

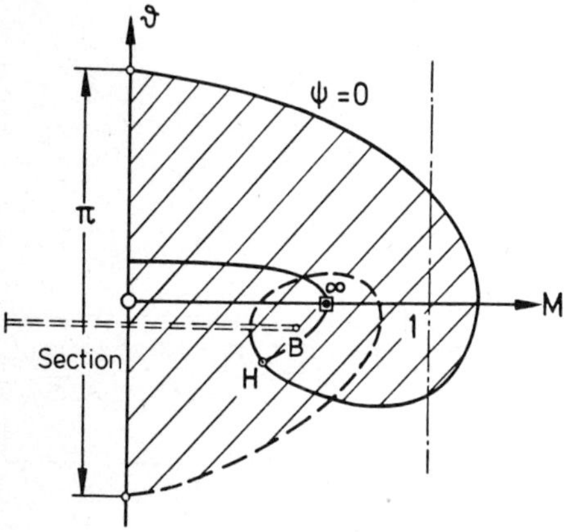

FIG. 70   Hodograph transformation.

velocity plane (Fig. 70). Our goal is to map the zero streamline into the physical plane by means of the well-known integral relation at the top of Fig. 70, thus getting the airfoil plus the associated pressure distribution.

By suitable conformal mapping in addition to Sobieczky's rheograph transformation (which need not be discussed in detail here), the velocity plane is distorted to a single-valued elliptic working plane ξ, η, the given hodograph boundary assuming the shape of a cylinder immersed in a uniform free-stream flow of known direction (Fig. 71). Thus, the problem is reduced to solving a normalized potential equation of the Poisson type (Fig. 72).

In contrast to the continuity equation in the physical plane, the right-hand side is now a linear function of the potential derivatives, since in a hodograph method such as the present one, all gas dynamic and mapping functions depend directly on the coordinates. For the solution of the potential equation, again the well-known integral relations holds; Fig. 72 indicates the physical meanings of its terms.

A constant-strength panel method is applied in order to "discretize" the integrals to algebraic sums of influence coefficients multiplied by the associated singularity strengths (Fig. 73). To evaluate the latter, these sum formulae are converted into a system of linear equations by fulfilling the boundary conditions, inclusive of the stagnation-point condition replacing

$$\zeta = \xi + i\eta = \frac{1}{\pm \sqrt{-e^{\sigma + i(\theta - \theta_B)} - \zeta_{3B} - \zeta_{6\infty}}}$$

$$\phi_\xi = K\psi_\eta$$

$$\phi_\eta = -K\psi_\xi$$

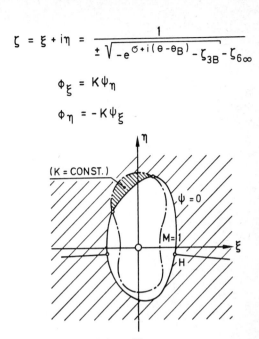

FIG. 71   The elliptic problem.

$$\nabla\phi = -\frac{\varkappa + 1}{2} \frac{M^4}{(1 - M^2)^{3/2}} \left(\sigma_n \phi_n\right)_{M = const} = g$$

Solution

$\phi_P =$	$\xi + a\eta$	
$+$	$\Gamma \delta$	
$+$	$\dfrac{1}{2\pi} \oint \varphi_n \ln r\, ds$	
$+$	$\dfrac{1}{2\pi} \iint g \ln r\, dF$	

FIG. 72   The elliptic problem.

291

Discretisation :

Controlpoint i

Singularity j

$$\phi_{ni} = \xi_{ni} + a\eta_{ni} + \Gamma\delta_{ni} + \sum_{1}^{N} h_j a_{ij}$$

Boundary Condition $\phi_{ni,} = 0 \ (i = 1, NR)$

Compressibility $h_i = \left(\dfrac{K_G}{K} G_n\right)_i \phi_{ni} \ (i = NR + 1, N)$

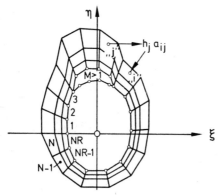

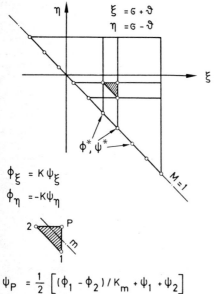

FIG. 73 The elliptic problem.

$$\xi = G + \vartheta$$
$$\eta = G - \vartheta$$

$$\phi_\xi = K\psi_\xi$$
$$\phi_\eta = -K\psi_\eta$$

FIG. 74 The hyperbolic problem.

$$\psi_P = \frac{1}{2}\left[(\phi_1 - \phi_2)/K_m + \psi_1 + \psi_2\right]$$

$$\phi_P = \frac{1}{2}\left[K_m(\psi_1 - \psi_2) + \phi_1 + \phi_2\right]$$

292

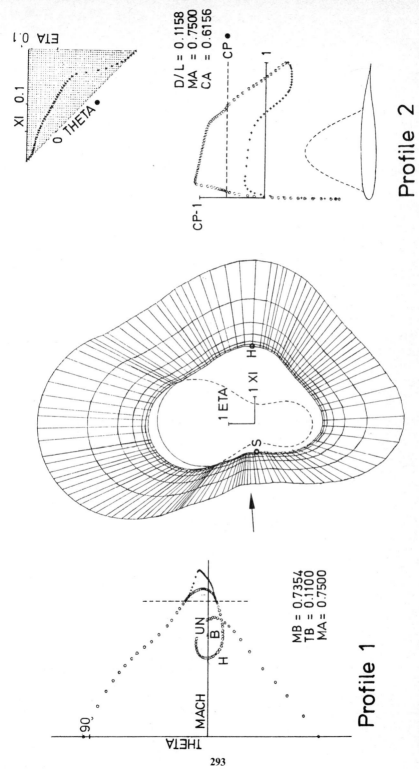

FIG. 75 Typical results.

293

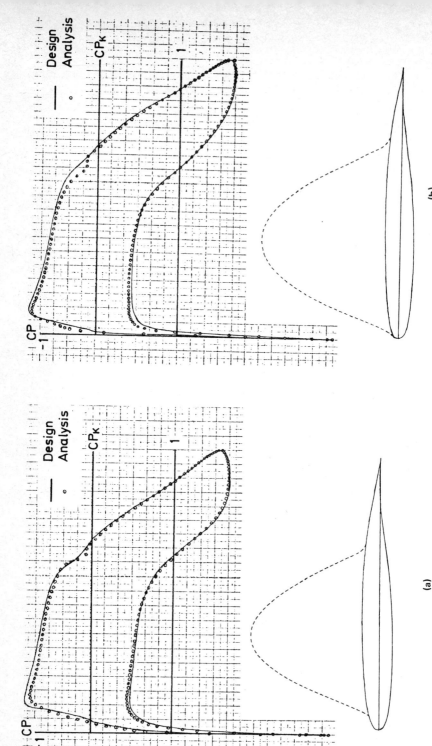

FIG. 76 Typical results. (a) Profile 1, $M = 75$, $\alpha = 0°$; (b) profile 3, $M = 75$, $\alpha = 0°$.

the Kutta condition in hodograph theory, and direct evaluation of the right-hand side by immediate updating, which is a sort of Gauss-Seidel procedure.

In doing so, we treat the embedded supersonic region as an elliptical, in our case, even as an incompressible flow regime, which does not make sense, of course, for physical flow. So we transfer the distribution of the potential and the stream function along the sonic line to the characteristic plane, where the supersonic region is recalculated in a straightforward manner, in accordance with the physically relevant equations (Fig. 74).

The subsonic and supersonic solutions for the potential versus boundary streamline, matched together, allow for the evaluation of the airfoil contour integral. Figures 75 and 76 show typical results. Comparisons between design and analysis show the usefulness of the present method for engineering purposes. Better accuracy can be achieved, of course, with additional rows of panels, which is only a matter of computer time cost.

NOMENCLATURE

General

$[A_{ij}]$	matrix of influence coefficients (components normal to surface) = $\mathbf{n}_i \cdot \mathbf{C}_{ij}$
$\{B_i\}$	Right-hand side of matrix (boundary condition)
$\mathbf{C}_{ij}$	vector of influence coefficients (influence of the singularity i on the control point j)
$\mathbf{i}, \mathbf{j}, \mathbf{k}$	unit vectors of x, y, z
Ma	mach number
$\mathbf{n}$	vector normal to body surface
r	distance between two points
S	surface
s	half span of wing
$\mathbf{U}_\infty$	velocity vector of the onset flow
$\mathbf{V}$	velocity vector
x, y, z	coordinate system
$\{X_j\}$	vector of solution of the equation system (singularity strength)
α	angle of attack
β	Prandtl-Glauert factor
γ	strength of vortex singularities
ε	inclination of wing (or tail) to body
κ	isentropic exponent
Λ	aspect ratio
μ	strength of dipole singularities
ρ	density factor
σ	strength of sink-source singularities

φ perturbation potential of velocity field
ϕ potential of velocity field
ψ stream function

Aerodynamic Coefficients

C_A lift coefficient
C_L coefficient of rolling moment
C_M coefficient of pitching moment
C_N coefficient of yawing moment
C_W drag coefficient
C_Y coefficient of side force
C_a local lift coefficient
C_p pressure coefficient

REFERENCES

1   Kraus, W.: Das MBB-Unterschall Panelverfahren, Teil 2: Das auftriebsbehaftete Verdrängungsproblem in kompressibler Strömung, MBB UFE-633-70, 1970.

2   Kraus, W., and P. Sacher: Das Panelverfahren zur Berechnung der Druckverteilung von Flugkörpern in Unterschall, *Z. Flugwiss.* 21(9), 1973.

3   Kraus, W.: Weiterentwicklung des Panelverfahrens, Teil 2: Erweiterung des Panelverfahrens auf die Flügelleitwerksinterferenz (mit Programmliste), MBB UFE-1017 (ö), 1973.

4   Labrujere, T. E., W. Loeve, and J. W. Slooff: An Approximate Method for the Calculation of the Pressure Distribution on Wing-Body Combination at Subcritical Speeds, AGARD-CP-71, 1971.

5   Smith, A. M. O., and J. L. Hess: Calculation of the Nonlifting Potential Flow about Arbitrary 3-dimensional Bodies, Douglas report no. E.S. 40622, 1962.

6   Hess, J. L.: Calculation of Nonlifting Potential Flow about Arbitrary 3-dimensional Bodies, *J. Ship Res.*, 8, 1964.

7   Giesing, J. P.: Potential Flow about 2-dimensional Airfoils, Douglas Aircraft Division report no. LB31946, 1965.

8   Smith, A. M. O., and W. C. Schliesser: Two-dimensional Cascades, Douglas memo C1-210-TM-22Y, 1966.

9   Rubbert, P. E., and G. R. Saaris: A General Three-dimensional Potential-Flow Method Applied to V/STOL Aerodynamics, SAE paper no. 680304, 1968.

10   Loeve, W.: Computer Programmes in Use at NLR for the Calculation of Stationary Subsonic Flow around Wing-Body Combinations, NLR AT-69-02, 1969.

11   Loeve, W., and J. W. Sloof: On the Use of "Panel Methods" for Predicting Subsonic Flow about Airfoils and Aircraft Configurations, NLR MP-71018-U, 1971.

12   Hess, J. L.: Calculation of Potential Flow about Arbitrary Three-dimensional Lifting Bodies, MDC J5679-01, final report, 1972.

13   Woodward, F. A., E. N. Tinoco, and J. W. Larsen: Analysis and Design of Supersonic Wing-Body Combinations, Including Properties in the Near Field, NASA CR-73106/Boeing D6-15044-1, 1967.

14   Carmichael, R. L.: A Computer Program for the Estimation of the Aerodynamics of Wing-Body Combinations, NASA Ames Program Description, 1971.

15   Kraus, W.: Ein Computer-Programm zur Berechnung der Druckverteilung und der Beiwerte von Flügeln, Rümpfen, Flügelrumpfkombinationen und Flügelleitwerkskombinationen im Unter-, Über- and Hyperschall (mit Programmliste), MBB UFE-1014 (ö), 1973.

16 Woodward, F. A.: An Improved Method for the Aerodynamic Analysis of Wing-Body-Tail Configurations in Subsonic and Supersonic Flow, NASA CR-2228, 1973.

17 Kraus, W.: Ein allgemeines Panelverfahren zur Berechnung der dreidimensionalen Potentialströmung um beliebige Flügel-Rumpf-Leitwerkskombinationen im Unter- und Überschall (mit Programmbeschreibung), MBB UFE-1136, 1974.

18 Maskew, B.: Calculation of the Three dimensional Potential Flow around Lifting Nonpanar Wings and Wing Bodies Using a Surface Distribution of Quadrilateral Vortex Rings. N71-20115 (Loughborough TT 7009), 1970.

19 Hess, J. L., and R. P. Martin: Improved Solution for Potential Flow about Arbitrary Axisymmetric Bodies by the Use of Higher Order Surface Source Method, NASA CR134694/MDC J6627-01, 1974.

20 Rubbert, P. E., and F. T. Johnson: Advanced Panel-Type Influence Coefficient Methods Applied to Subsonic Flows, AIAA paper no. 75-50, 1975.

21 Besigk, G.: Halbempirische Theorie zur Berechnung des Heckwiderstandes, MBB Bericht 437-69, 1969.

22 Körner, H.: Untersuchungen zur Bestimmung der Druckverteilung an Flügel-Rumpf-Kombinationen, DFVLR-Bericht, 1969; DFVLR-Bericht, 1970.

23 Schneider, W.: Druckverteilungsmessungen an Pfeilflügel-Rumpfanordnungen bei hohen Unterschall Machzahlen, AVA-Bericht 70A23, 1970.

24 Körner, H., and W. Schröder: Druckverteilungs- und Kraftmessungen an einer Flügel-Rumpf-Leitwerksanordnung, DFVLR IB-080-72/13, 1972.

25 Labrujere, T. E.: A Numerical Method for the Determination of the Vortex Sheet Location behind a Wing in Incompressible Flow, NLR TR-72091-U, 1972.

26 Gersten, K.: Nichtlineare Tragflächen Theorie für Rechteckflügel bei inkompressibler Strömung, *Z. Flugwiss.* 5, 1957.

27 Bollay, W.: A Theory of Rectangular Wings of Small Aspect Ratio, *J. Aero. Sci.*, 4, 1937.

28 Kraus, W., and W. Sonnleitner.: Weiterentwicklung des Panelverfahrens, Teil I: Nichtlineares Panelverfahren unter Berücksichtigung diskreter, abgelöster Wirbelschichten an gepfeilten, schlanken Tragflügelformen, MBB UFE-1070 (ö), 1973.

29 Belotserkovskiy, S. M.: Calculation of the Flow around Wings of Arbitrary Planform in a Wide Range of Angle of Attack, NASA TT-F-12/291, 1969.

30 Perrier, P., and W. Vitte: Elements de calcul d'aérodynamique tridimensionelle en fluide parfait, AFITAE 7, Col. Aero. Appl., 1970.

31 Rubbert, P. W., T. J. Johnson, *et al.*: A Three-Dimensional Solution of Flows over Wings with Leading-Edge Vortex Separation, AIAA paper no. 75 866, 1975.

32 Kraus, W.: Untersuchungen zur Leitwerksvereisung der A300B, MBB UFE-874-72, 1972.

33 Deslandes, R. M.: Load Prediction for External Stores on a Fighter Airplane, MBB UFE-1226, 1976.

34 Ahmed, S. R.: Prediction of the Optimum Location of a Nacelle-Shaped Body on the Wing of a Wing-Body Configuration by Inviscid Flow Analysis, AGARD CPP-150, 1974.

35 Zimmer, H.: Berechnung der Druckverteilung an dreidimensionalen Unterschalltriebwerkseinläufen mit Quellpanelverfahren, DGLR Fachausschuß December, 1972.

36 Ahmed, S. R.: Berechnung des reibungslosen Strömungsfelds von dreidimensionalen auftriebsbehalteten Tragflügeln, Rümpfen und Flügel-Rumpf-Kombinationen nach dem Panelverfahren, DLR FB-73-102, 1973.

37 Stricker, R.: Zur Berechnung der stationären unterkritischen Potentialströmung um ebene Profile beliebiger Form, MBB UD-135-74 (ö), 1974.

38 Eberle, A.: Eine exakte Hodographenmethode zum Entwurf überkritischer Profile, MBB UFE-1168 (ö), 1975.

CHAPTER 5

Progress in Transonic Flow Computations: Analysis and Design Methods for Three-dimensional Flows

Wolfgang Schmidt

1 INTRODUCTION

Transonic flows are characterized by the presence of adjacent regions of subsonic and supersonic flow, generally terminated by weak shock waves. The development of the transonic flow pattern over a lifting airfoil for increasing free-stream Mach number is schematically shown in Fig. 1. The flow pattern progresses from a predominantly subsonic flow with embedded supersonic regions to a predominantly supersonic flow with embedded subsonic regions.

Let us first consider the main reasons for aerodynamicists to be interested in a speed range with such a complicated flow pattern. Modern commercial airplanes must have improved cruise performance. Based on Breguet's range formula for aircraft with jet engines,

$$R = K \underbrace{\left(\frac{1}{F_{\text{spec}}}\right)}_{\text{Propulsion}} \underbrace{\left(\frac{W_{\text{payload}}}{W_{\text{empty}}}\right)}_{\text{Structures}} \underbrace{\left(M_\infty \frac{L}{D}\right)}_{\text{Aerodynamics}}$$

This work was sponsored by the German Ministry of Defense under ZTL contract T/R 720/R 7600/42009 and by the German Ministry of Research and Technology under contract LFF 28.

I would like to thank my colleagues at FFA, Mrs. Agrell, A. Gustavsson, and S. Hedman, and at Dornier, W. Fritz, S. Rohlfs, H. W. Stock, and R. Vanino, for participating in the program development and performing most of the numerical calculations.

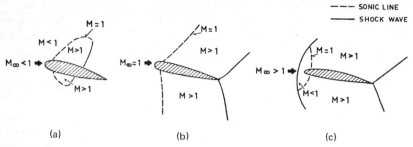

FIG. 1   Transonic flow patterns. (a) Supercritical flow; (b) sonic flow; (c) supersonic flow.

the maximum range, for a given propulsion system and ratio of payload to weight given mainly by the structures, is found for optimum $M_\infty L/D$. As shown in Fig. 2, the optimum cruise Mach number with respect to the maximum range is given by the drag rise behind the drag divergence Mach number, which is caused by shocks.

Moreover, for a shock-free supercritical design, the wing thickness can be increased (or wing sweep decreased) compared with conventional wings. This generally pays off in structural weight. Fighter-type aircraft are designed not for one point such as cruise, but for the whole flight range (limited by maneuver boundaries such as buffet). The main objectives for fighter design are high lift at a low drag level ("dogfighting"), high thrust-to-drag ratio for acceleration, and high load factors for maneuvers, limited by the buffet boundary.

To improve these factors limiting the performance of aircraft, detailed studies, comprising both wind-tunnel testing and fluid-dynamic computations, have to be performed. The goal of such studies is to design for decreased strength of the shock waves terminating the supersonic regions and shock–boundary-layer interaction without separation.

Numerical techniques for the analysis of inviscid transonic potential flow have become highly developed. Mathematically, the description of steady transonic flows requires the solution of "mixed" equations, which are ellip-

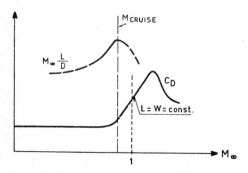

FIG. 2   Optimum cruise Mach number.

tic in subsonic regions. The problem is essentially nonlinear, and solutions generally contain discontinuities representing shock waves. The basic numerical procedure used is that first introduced by Murman and Cole [1]. It accounts for the mixed elliptic-hyperbolic character of the governing equations by using a mixed finite-difference scheme. The general procedure is to employ centered differences when the flow is locally subsonic and one-sided differences when it is locally supersonic.

Using line relaxation, different methods have been developed to solve the full potential equation as well as the small-perturbation form. Three-dimensional results have been given for lifting wings [2–17] and for lifting wing-body combinations. First attempts have also been made to take in account wind-tunnel wall effects [10].

This chapter presents some details of the relaxation methods used at Dornier [18,19]. The fundamentals of the relaxation method and difference schemes are given in Chaps. 1 and 3; here, differences between the Dornier methods and other methods are the main topics of discussion. Special attention is given to the assumptions required by the small-perturbation form of the equations. Detailed evaluations of numerical results show the limitations of different formulations. The chapter concludes with some applications and comparisons with other methods, as well as wind-tunnel results for wings and wing-body combinations.

2 TRANSONIC SMALL-DISTURBANCE THEORY

We begin the discussion of three-dimensional transonic relaxation methods by considering the basic steady transonic equations and boundary conditions and proving the different assumptions by error estimates based on calculated results.

2.1 Basic Equations

By expanding the three-dimensional steady velocity potential as

$$\Phi(x, y, z) = U_\infty[x + \varepsilon\phi(x, y, z)] \tag{1}$$

and assuming $\phi_x{}^2$, $\phi_y{}^2$, and $\phi_z{}^2 \ll 1$ if $\varepsilon = O(1)$, or $\varepsilon^2 \ll 1$ if $\phi = O(1)$, which is true for the Krupp scaling $\varepsilon = \delta^{2/3}/\mathrm{M}_\infty$, the perturbation form of the full potential equation is simplified to

$$\left[1 - \mathrm{M}_\infty{}^2 - (\kappa + 1)\mathrm{M}_\infty{}^2\varepsilon\left(1 + \frac{\varepsilon}{2\phi_x}\right)\phi_x - \frac{\kappa - 1}{2}\mathrm{M}_\infty{}^2\varepsilon^2(\phi_y{}^2 + \phi_z{}^2)\right]\phi_{xx}$$
$$+ [1 - (\kappa - 1)\mathrm{M}_\infty{}^2\varepsilon\phi_x]\phi_{yy} + [1 - (\kappa - 1)\mathrm{M}_\infty{}^2\varepsilon\phi_x]\phi_{zz}$$
$$- 2\mathrm{M}_\infty{}^2\varepsilon(\phi_y\phi_{xy} + \phi_z\phi_{xz}) = 0 \tag{2}$$

Furthermore, if we assume that ϕ_x, ϕ_y, and $\phi_z \ll 1$ if $\varepsilon = O(1)$, or $\varepsilon \ll 1$ if $\phi = O(1)$, Eq. (2) reduces to

$$[1 - M_\infty^2 - (\kappa + 1)M_\infty^2 \varepsilon \phi_x]\phi_{xx} + \phi_{yy} + \phi_{zz} - 2M_\infty^2 \varepsilon(\phi_y \phi_{xy} + \phi_z \phi_{xz})$$
$$= 0 \tag{3}$$

If, finally, the last term on the left side of Eq. (3) is neglected, we obtain the generally used classical small-perturbation form

$$[1 - M_\infty^2 - (\kappa + 1)M_\infty^2 \varepsilon \phi_x]\phi_{xx} + \phi_{yy} + \phi_{zz} = 0 \tag{4}$$

which gives the Spreiter form for $\varepsilon = 1$

$$[1 - M_\infty^2 - (\kappa + 1)M_\infty^2 \phi_x]\phi_{xx} + \phi_{yy} + \phi_{zz} = 0 \tag{4a}$$

and the Murman-Krupp form for $\varepsilon = \delta^{2/3}/M_\infty$

$$[1 - M_\infty^2 - (\kappa + 1)M_\infty \delta^{2/3}\phi_x]\phi_{xx} + \phi_{yy} + \phi_{zz} = 0 \tag{4b}$$

The form of the equation generally used in the Dornier methods is the Murman-Krupp form, Eq. (4b). All the results given in this chapter were obtained with this equation.

We shall now show that these small-disturbance assumptions are satisfied, using computational results from a test case that will be discussed in detail later. Figure 3 shows the numerical results for the three velocity components at two different spanwise stations.

The assumptions that led from the full potential equation to the final small-perturbation form, eq. (4), are

1 $\varepsilon^2 \phi_x^2$, $\varepsilon^2 \phi_y^2$, and $\varepsilon^2 \phi_z^2 \ll 1$
2 $\varepsilon \phi_x$, $\varepsilon \phi_y$, and $\varepsilon \phi_z \ll 1$
3 $\varepsilon \phi_y \phi_{xy}$ and $\varepsilon \phi_z \phi_{xz} \ll \phi_{yy}$, ϕ_{zz}, $[\Box]\phi_{xx}$

The results in Fig. 3 show that

Assumption 1 is satisfied at wing stations.
Assumption 2 is hardly satisfied, especially not on highly loaded wings, but it can be assumed that $(\kappa - 1)M_\infty^2 \varepsilon \phi_x \ll 1$. Only some distance away from the wing is assumption 2 satisfied.
Assumption 3 is far from satisfied. Especially for wings with pressure plateaus ($\phi_{xx} \ll 1$) and sections close to the tip, which are generally not highly loaded but have a strong cross flow, this assumption is not consistent with the physical problem being treated.

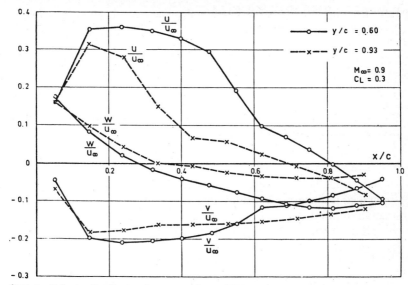

FIG. 3 Velocity distribution for two sections, PT7Q wing.

From this error estimate it can be concluded that for wings of practical interest, the potential equation can only be simplified to the form of Eq. (3). Due to the high subsonic Mach numbers at high loading, an even less simplified form, similar to that of Eq. (2), should be used:

$$\left[1 - M_\infty^2 - (\kappa + 1)M_\infty^2 \varepsilon \left(1 + \frac{\varepsilon}{2} \phi_x\right) \phi_x - \frac{\kappa - 1}{2} M_\infty^2 \varepsilon^2 (\phi_y^2 + \phi_z^2)\right] \phi_{xx}$$

$$+ \phi_{yy} + \phi_{zz} - 2M_\infty^2 \varepsilon (\phi_y \phi_{xy} + \phi_z \phi_{xz}) = 0 \qquad (5)$$

Since this form is not consistent in the sense of the usual first- or second-order expansions, it should be kept in mind that even if the ϕ_x^2 term in the ϕ_{xx} expression is of higher order than the ϕ_x term in the ϕ_{yy} or φ_{zz} expression of Eq. (2), generally $\phi_x(\kappa - 1)$ is much smaller compared to 1 than the ϕ_x^2 term compared to $1 - M_\infty^2$. Furthermore, it must still be shown how well this form of the equation can approximate the Rankine-Hugoniot conditions for shocks.

2.2 Boundary Conditions

The flow-tangency condition at the surface, with its normal vector **n** including the angle of attack, is expressed as

$$(1 + \varepsilon\phi_x)n_x + \varepsilon\phi_y n_y + \varepsilon\phi_z n_z = 0 \qquad (6)$$

2.2.1 Wing Boundary Conditions

In two-dimensional flow, this condition is generally linearized to

$$n_x + \varepsilon\phi_z n_z = 0 \tag{7}$$

Based on thin-wing theory, this basically two-dimensional form is also used for three-dimensional wings with large sweep. From the results in Fig. 3 we know that our wings are highly loaded, that is, $\varepsilon\phi_x$ is not $\ll 1$. Figure 4 indicates the orders of the different terms in Eq. (6) for a supercritical transonic wing with 8–12% thickness variation along the span and a leading-edge sweep of almost 40°.

Indications are that the full boundary condition should be used, even if we apply it in a plane $z = $ constant, i.e., do not allow for any movement of the stagnation point. The Dornier method is now being improved in this sense; however, all the results reported in this chapter correspond to the wing boundary condition (7).

2.2.2 Body Boundary Condition

For infinitely long bodies with no changes in cross section, Eq. (6) can be simplified to

$$\phi_y n_y + \phi_z n_z = 0 \tag{8}$$

For all other cases, the full form, Eq. (6), should be used.

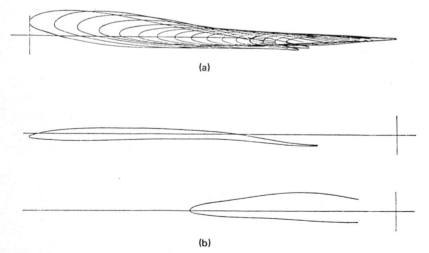

(a)

(b)

FIG. 4 PT7Q wing. (a) Spanwise distribution of wing sections, $Y = $ constant; (b) wing sections, $X = $ constant for two chordwise stations.

2.2.3 Wake Boundary Condition

For lifting wings and wing-body combinations, the Kutta condition is applied on the wing, thus forcing the flow to leave all subsonic trailing edges smoothly. In the small-disturbance theory, the Kutta condition is satisfied by requiring that the pressure ϕ_x be continuous across the trailing edge. The vortex sheet is assumed to be straight and to lie in the wing mean plane $z = 0$, with the conditions that ϕ_x and ϕ_z be continuous and ϕ_y be discontinuous through it. This jump in potential is independent of x at any span station y and is equal to the circulation about the wing section defined by

$$\phi(x, y, +0) - \phi(x, y, -0) = \sigma(x, y) \qquad (9)$$

While Bailey [20] uses $\sigma(x, y) = \sigma(y) = \Gamma_{TE}(y) = \Gamma_{FF}(y)$ as constant along x from the trailing edge to the far-field boundary, we use the three-dimensional form of the procedure introduced by Krupp [21] for two-dimensional airfoil flow.

$$\sigma(x, y) = \Gamma_{TE}(y) + [\Gamma_{FF}(y) - \Gamma_{TE}(y)] \frac{x - x_{TE}(y)}{x_{FF} - x_{TE}(y)} \qquad (10)$$

Figure 5 shows some results for a wing of aspect ratio 4, leading-edge sweep 30°, and taper ratio 1.0 having a biconvex airfoil at 1° angle of attack and $M_\infty = 0.908$. As can be seen, the change of potential of upper and lower wake surfaces along x giving the same jump $\Delta\phi$ seems to be better indicated by Eq. (10), although there is little difference in terms of lift and pressure distribution. It could be stated that Eq. (10) gives a slightly better rate of convergence and allows for a trailing far-field plane closer to the configuration.

2.2.4 Far-Field Solutions

Since we do not transform infinity to a finite domain, we must introduce far-field solutions, along the boundaries of the domain, in which the flow field has to be calculated from the potential equation. These far-field solutions have to satisfy the flow field between the finite domain and infinity. Since it is some distance away from the configuration, the far flow field is subsonic, and it can be described by the linear potential equation. Therefore, classical source and vortex methods can be used.

The Dornier method for wings now uses a vortex distribution with the correct spanwise lift distribution and one straight leading filament, similar to that suggested by Klunker [22]. The influence of the body on the far field can be simulated by use of a line-source and line-doublet distribution. As shown in Fig. 6, a source distribution representing thickness effects for wings has very little influence on the pressure distribution, and these effects can be

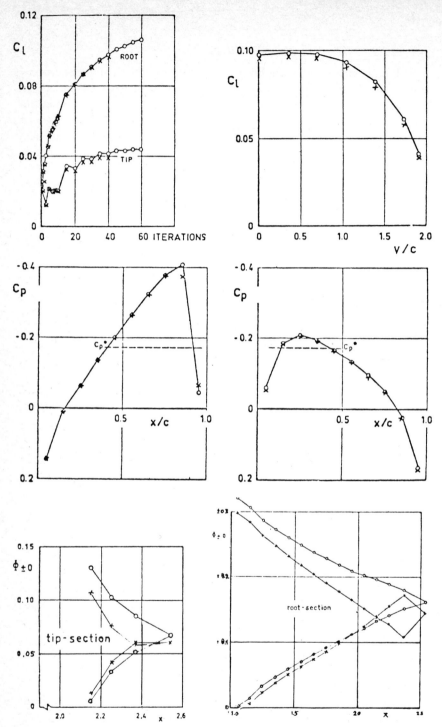

FIG. 5 Wake circulation algorithm. (x) Bailey; (o) modified Krupp.

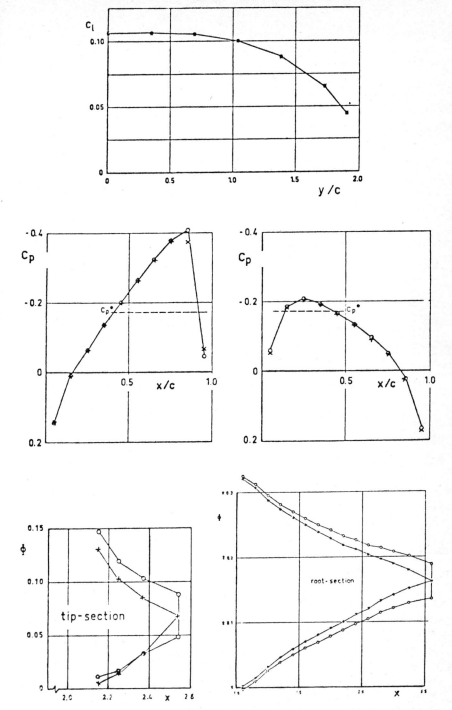

FIG. 6 Far field, with and without thickness term. (o) With thickness; (×) without thickness.

neglected. Nevertheless, it should be mentioned that, owing to the asymmetry of the potential in the wake, including thickness terms, the far field can be closer to the configuration. The entrance plane $x = 0$ and the exit plane $x = x_{FF}$ are not calculated with far field expressions, but by use of the Laplace equation.

As shown by P. E. Rubbert [23], the use of far-field expressions generally fits the potential ϕ at the finite domain, but not the normal derivatives, that is, the normal velocity. Therefore, mass can be produced in the domain. To avoid this problem, a new far-field treatment is being developed at Dornier, to fit ϕ as well as its normal derivatives and fulfill continuity requirements.

Wind-tunnel wall boundary conditions can be introduced in the same manner.

2.2.5 Pressure Calculation

For isentropic potential flow, the pressure coefficient C_p is given by the relation

$$C_p = \frac{2}{\kappa M_\infty^{\,2}} \left(\left\{ 1 - \frac{\kappa - 1}{2} M_\infty^{\,2} [2\varepsilon\phi_x + \varepsilon^2(\phi_x^{\,2} + \phi_y^{\,2} + \phi_z^{\,2})] \right\}^{(\kappa-1)/\kappa} - 1 \right) \tag{11}$$

the second-order series expansion of which is

$$C_p = -[2\varepsilon\phi_x + (1 - M_\infty^{\,2})\varepsilon^2\phi_x^{\,2} + \varepsilon^2(\phi_x^{\,2} + \phi_z^{\,2})] \tag{12}$$

For wings, the first-order form

$$C_p = -2\varepsilon\phi_x \tag{13}$$

is generally used. This form has been empirically modified by Krupp [21] for two-dimensional flow, to give better agreement with experimental results and with the full potential method using exact boundary conditions:

$$C_p = -2\frac{\varepsilon}{M_\infty^{\,1/4}}\phi_x \tag{13a}$$

Equation (12) is generally recommended for pressure calculations for wings and bodies; it is consistent with the results given in Sec. 2.1. In addition, it is well known from linearized potential flow that the second-order pressure formula is consistent with the first-order Euler equations [24,25].

2.2.6 Force and Moment Calculations

Lift on wings can either be calculated by spanwise integration of the circulation Γ, which is the jump in potential at the trailing edge, or by integration of the local pressures over the wing surface. Pitching moment is calculated by integration of the pressure times the x distance from the center of moment over the wing surface. Body surfaces are divided into many quadrilaterals and triangles, depending on the body representation by grid points. The vector components of these surfaces times the mean of the pressures in the corners of the elementary surfaces form the lift, drag, side force, and pitching moment when summed over all elements.

As discussed in detail by Murman [26] and Bailey [20], an additional integration along the shock surface has to be performed to get the correct total transonic pressure drag. It should be mentioned that the correct drag level, as well as a good theoretical prediction of drag rise, can be calculated only if viscous effects are included.

2.3 Design Procedure

When a new wing is designed, generally the planform, design-point lift coefficient, and Mach number are given. From two-dimensional studies, a desired design pressure distribution for the three-dimensional wing is also given. The core of the design process now is to find the wing-section shapes that produce such a pressure distribution. While this can be done by trial and error using an analysis procedure as described in the previous section, it is very helpful to have a design method available, in which the pressure distribution is the given boundary condition while the shape, i.e., normal velocity component, is unknown.

This classical Dirichlet boundary problem is generally solved by prescribing the tangential velocity $u = \phi_x$, instead of the normal velocity component, as the boundary condition. In two-dimensional studies it was found that in a relaxation procedure this method seems to fail [27], the reason being the a priori unknown potential distribution in front of the wing. Instead, in the manner of Steger and Klineberg [28] and Langley [29] for airfoil and wing design the vorticity equation can be used as an intermediate step that must be identically fulfilled by the potential flow. The steps in such a design procedure are

1 Prescribe the pressure C_p.
2 From similar solutions or iteratively, suggest the velocities ϕ_y and ϕ_z.
3 Calculate the tangential velocity ϕ_x from the given C_p [Eq. (12)] using ϕ_y and ϕ_z from step 2.
4 From the vorticity equation

$$\phi_{xz} = \phi_{zx} \tag{14}$$

calculate the cross derivative by differentiation in z.

5 Now integrate in the x direction to get a boundary condition in ϕ_z

$$\phi_z = \int \phi_{xz}\, dx \qquad (15)$$

to have a Neumann-type direct boundary condition.

These steps reduce the inverse problem to a direct one that can be solved by the standard procedure.

3 NUMERICAL PROCEDURE

3.1 Orthogonal Coordinate System

3.1.1 *Difference Equation*

The basic feature of the numerical method is that it solves the transonic perturbation equation in a rectangular grid box, as shown in Fig. 7, with variable mesh size. As in the two-dimensional method of Murman and Cole [1] we account for the mixed elliptic-hyperbolic nature of the equation by using central differences for the streamwise derivatives when the equation is elliptic, and backward when the equation is locally hyperbolic. The y and z derivatives are replaced everywhere by second-order central differences.

Unlike Jameson's rotated scheme, this difference scheme is only allowed if the mean velocity direction is close to the free-stream direction; otherwise information from outside the domain of dependence is taken in supersonic regions. Compared with Bailey's method there exists only the slight differ-

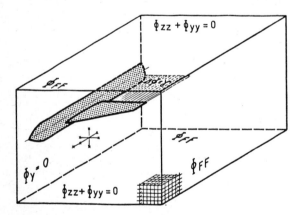

FIG. 7 Relaxation method to solve the transonic potential equation

$$[(1 - M_\infty^2) - (K + 1)\,\varepsilon M_\infty^2 \phi_x]\phi_{xx} + \phi_{yy} + \phi_{zz} = 0.$$

ence that in his method the y derivatives are taken along percent lines as central differences, while in our method $x =$ constant lines are taken. This might make a difference for highly swept wings and supersonic regions.

The resulting set of nonlinear algebraic equations is, for constant spacing and the small-perturbation form [Eq. (4)] for elliptic points,

$$\frac{1}{\Delta x^2} [1 - M_\infty^2 - (\kappa + 1)\varepsilon M_\infty^2 (\phi_{i+1,j,k}^{(v)} - \phi_{i-1,j,k}^{(v+1)})]$$

$$(\phi_{i+1,j,k}^{(v)} - 2\phi_{i,j,k}^{(v+1/2)} + \phi_{i-1,j,k}^{(v+1)})$$

$$+ \frac{1}{\Delta y^2} (\phi_{i,j,k+1}^{(v)} - 2\phi_{i,j,k}^{(v+1/2)} + \phi_{i,j,k-1}^{(v+1)})$$

$$+ \frac{1}{\Delta z^2} (\phi_{i,j+1,k}^{(v+1/2)} - 2\phi_{i,j,k}^{(v+1/2)} + \phi_{i,j-1,k}^{(v+1/2)}) = 0$$

$$(16)$$

which is solved iteratively by line relaxation. Each line forms an equation with a tridiagonal matrix for which fast solution procedures exist. The relaxation used is underrelaxation for hyperbolic points ($\omega = 0.7$–0.8) and overrelaxation for elliptic points ($\omega = 1.7$–1.9):

$$\phi^{(v+1)} = \phi^{(v)} + \omega(\phi^{(v+1/2)} - \phi^{(v)}) \qquad (17)$$

This scheme is fully conservative if, in addition, for shock points Murman's shock-point operator is used—that is, the sum of the elliptic plus the hyperbolic operator.

3.1.2 Boundary Conditions

The numerical formulation of the wing boundary conditions is essentially the same as that of Krupp [21] for airfoils. The wing boundary condition on ϕ_z enters the scheme in the first plane above (or below) the wing plane J ($z = 0$). Using a series expansion with minimized truncation error for mesh distances $t = s\sqrt{3}$, the resulting difference equation for the upper surface is

$$\phi_{zz_{i,J+1,k}} = \frac{2}{t(t+2s)} (\phi_{i,J+2,k} - \phi_{i,J+1,k}) - \frac{2}{t+2s} \phi_{z_{i,k}\,\text{boundary}} \qquad (18)$$

s and t being the mesh sizes between J and $J + 1$ (or $J + 1$ and $J + 2$). ϕ_z itself is estimated either by Eq. (6) or by Eq. (7).

For the design problem, the numerical formulation of the boundary condition according to Sec. 2.3 is

$$\phi^{v+1/2}_{xz_{i,J,k}} = -\frac{2s+t}{s(s+t)}\phi^{v+1/2}_{x_{i,J,k}} + \frac{s+t}{st}\phi^{v}_{x_{i,J+1,k}} - \frac{s}{t(s+t)}\phi^{v}_{x_{i,J+2,k}} \quad (19)$$

where
$$\phi^{v+1/2}_{x_{i,J,k}} = \omega_1 u_{i,k} + (1-\omega_1)\phi_{x_{i,J,k}} \quad (20)$$

using ω_1 as the relaxation facor for better convergence, and where $u_{i,k}$ is the velocity estimated via Eq. (12) or Eq. (13) from the prescribed pressure $C_{p_{i,k}}$. From Eq. (19), via numerical integration,

$$\phi^{v+1/2}_{z_{i,J,k}} = \int_{x_{i-1}}^{x_i} \phi^{v+1/2}_{xz_{i,J,k}} \, dx + \phi^{v+1}_{z_{i-1,J,k}} \quad (21)$$

and the final updated value is

$$\phi^{v+1}_{z_{i,J,k}} = \omega_2\, \phi^{v+1/2}_{z_{i,J,k}} + (1-\omega_2)\phi^{v}_{z_{i,J,k}} \quad (22)$$

using ω_2 as a second relaxation factor. This updated value then enters the general boundary condition, Eq. (18), for the analysis problem.

Figure 8 shows some details of the implementation of the body boundary

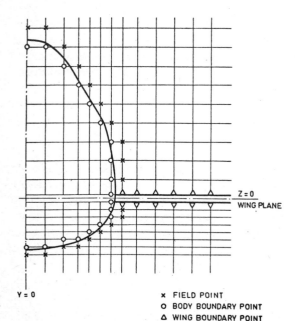

Z = 0
WING PLANE

Y = 0

x   FIELD POINT
o   BODY BOUNDARY POINT
Δ   WING BOUNDARY POINT

FIG. 8   Body boundary conditions.

conditions. For body points marked with a circle, instead of the potential equation, the body boundary equation (6) is used directly to calculate the potential $\phi_{i, J, k}$ on (or inside) the body surface:

$$\phi_{i, J, k} = \frac{1}{\varepsilon} n_x + \frac{n_x}{\Delta_x} \phi^v_{i-1, J, k} + \frac{n_y}{\Delta_y} \phi^v_{i, J \pm 1, k} + \frac{(n_z/\Delta_z)\phi^v_{i, J, k+1}}{n_x/\Delta_x + n_y/\Delta_y + n_z/\Delta_z} \quad (23)$$

This potential does not enter the line relaxation but is used only to calculate the derivatives at the neighboring points.

Instead of the first-order differences for the boundary condition, second-order formulas including the potential in $i \pm 2, J \pm 2. k + 2$ can be used, but they should not result in an increase in accuracy.

3.1.3 Grid Optimization

The core of a numerical method using a Cartesian coordinate system is its accuracy in the nose region of swept wings. Detailed studies [11] have shown that this problem can be solved by special grid arrangements.

Figure 9 shows the strong influence of mesh arrangement on the computed results for the highly swept and tapered RAE wing " C." The accuracy of the results depends perhaps more on the proper arrangement than on the total number of mesh points. Grid 1, in which the last point in front of the wing and the first point on the wing are percent lines and the wingtip is midway between two stations, gives reasonable results, at least for the center part. Such a grid (about 15,000 points) is recommended for use in the very first iterations, to get a good initial solution for the field. Grid 2 is an arrangement that cannot be recommended. Owing to the way the grid points are located on the wing along the span with respect to the percent line, the computed pressure distribution is wavy. The total number of points in grid 3 is only slightly higher than in grid 2, but grid 3 uses the proper spacing rule.

Figure 10 shows converged solutions for three different grids for the ONERA M6 wing. The nose region and the rear shock position change very much from grid 1 to grid 3, that is, from 24,570 to 90,000 points. Grid 1 has only 10 points on the upper and lower surfaces in the tip section, whereas at least 20 points are needed to represent the details of a section shape in the numerical method. The very large number of grid points required in all sections is a second disadvantage of the orthogonal Cartesian grid system.

For swept wings, dense spacing occurs not only on the wing but also in front of the wing, unless $x < x_{\text{leading edge}}$ in the center section and behind the wing, and $x > x_{\text{trailing edge}}$ in the tip section.

To avoid very large numbers of grid points, the use of different cuts for different meshes is recommended. That is, calculate converged solutions in a medium grid with far-field boundaries; then " place " a grid box inside the first box, with its edges closer to the aircraft configuration, and solve with a

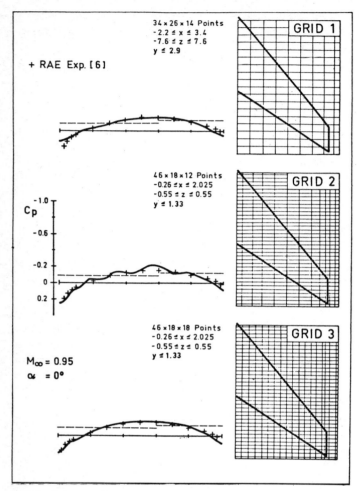

FIG. 9 Influence of grid arrangement on the pressure in the root section. Swept and tapered RAE wing "C"; $\lambda = 1/3$, $\vartheta_{25} = 47.6°$.

fine mesh and with fixed potential values interpolated from the previous calculations. By repeating this procedure, the potential on the cut edges can be updated. The mesh can be further refined by using an even smaller cut box. We summarize the basic rules for using a Cartesian grid:

Generate a mesh having first points on the wing along percent lines.
To reproduce the influence of the nose shape, the first points should have a distance from the leading edge of the order of the nose radius.
The mesh must be fine enough to resolve the pressure shape close to the nose as well as in the shock regions, if it is to give the proper shock position and strength.

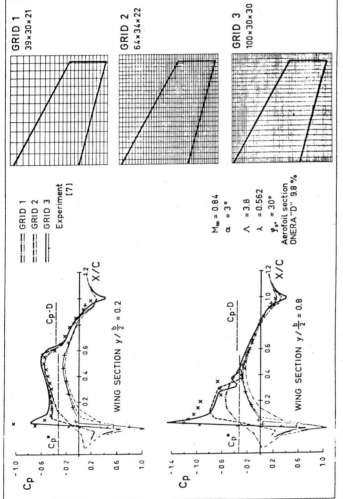

FIG. 10 ONERA M6: wing influence of number of grid points.

To increase the point density, cut boxes are recommended that have the potential distribution given by previous converged solutions in a coarse mesh with almost the correct lift distribution.

3.2 Nonorthogonal Coordinate System

Paralleling the work with Cartesian coordinates, some work was started as a joint effort between FFA and Dornier on a relaxation method [18,30] for the small-perturbation equation ($4b$) using a nonorthogonal coordinate system. The basic idea of this coordinate system can be discussed in conjunction with Fig. 11.

By constituting the mesh of i lines as percent lines and k lines as the usual lines y = constant a very efficient grid system can be generated that concentrates points where fine resolution is needed. This is very similar to the methods used at Ames, RAE, and NLR. However, the differential equation is not transformed but rather the finite-difference operator along lines i = constant. Since the sweep angle of the percent lines is known, the usual ϕ_{yy} derivatives can be calculated from the derivatives ϕ_{xx} in the stream direction, and ϕ_{nn} along the percent lines. These derivatives can be estimated either by exact transformation or by numerical interpolation, as shown in Fig. 11.

Figures 12 and 13 show some results with this approach, compared with experimental results and orthogonal-mesh results. The swept-wing case in Fig. 12 shows almost perfect agreement, while in Fig. 13 both computed results differ from the experimental values to about the same degree. Detailed studies on the finite-difference star used when the ϕ_{yy} derivatives are calculated by numerical interpolation show that information is taken

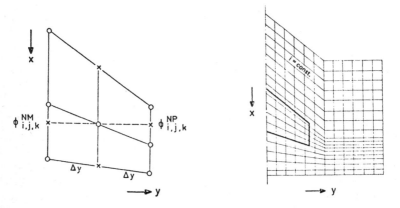

FIG. 11 The FFA-Dornier nonorthogonal coordinate system.

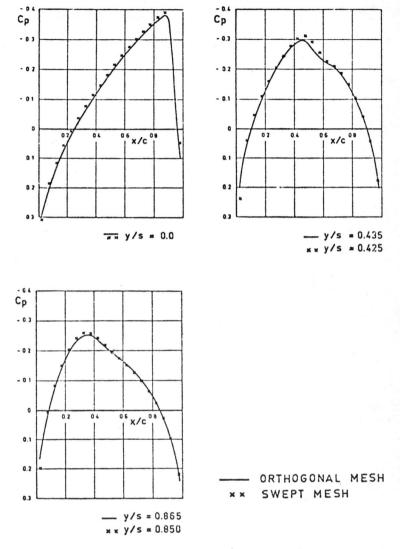

FIG. 12   Comparison of results for a swept wing; $\Lambda = 4$, $\lambda = 1$, $M_\infty = 0.908$, $\alpha = 0°$.

from outside the domain of dependence, which would explain the differences.

It is difficult to decide which of the two coordinate systems to use. The first generally is faster and easier to extend to complex configurations, while the second pays off if a very large number of wing points is needed. One recommendation is to use the swept mesh as a final, very fine grid to get very detailed information on pressures as well as velocities.

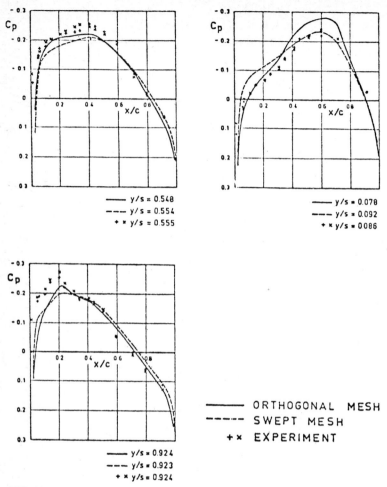

—— y/s = 0.548
––– y/s = 0.554
+ × y/s = 0.555

—— y/s = 0.078
––– y/s = 0.092
+ × y/s = 0.086

—— ORTHOGONAL MESH
– – – SWEPT MESH
+ × EXPERIMENT

—— y/s = 0.924
––– y/s = 0.923
+ × y/s = 0.924

FIG. 13 Comparison of results for the NACA RM A51G31 wing.

4. APPLICATIONS

4.1 Wing Analysis

The first set of calculations to be discussed is that for the swept and tapered RAE wing C having a 5.4% thick RAE 101 airfoil section. Figure 14 shows the calculated pressure distributions and isobars for 0 and 3° angle of attack at $M_\infty = 0.95$. Figure 15 is a comparison with RAE experimental values [31] as well as with theoretical results. It shows a quite good agreement.

Figure 16 shows a detailed comparison between the numerical results and wind-tunnel results [32] for the ONERA M6 wing having a 9.8% thick

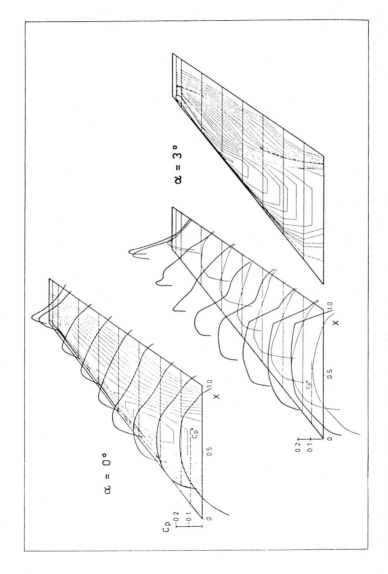

FIG. 14 RAE wing C, $M_\infty = 0.95$.

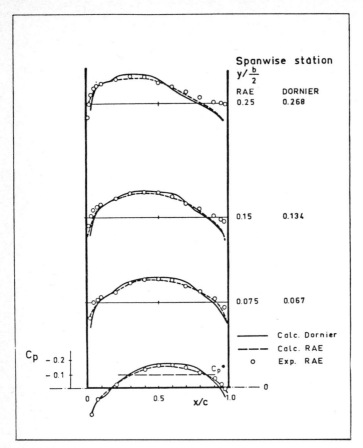

FIG. 15   RAE wing C: comparison of theory and experiment; $M_\infty = 0.95$, $\alpha = 0°$.

ONERA D airfoil. The calculated pressures are in fairly good agreement with the experimental, even in the nose region with its very high gradients. The rear shock position is quite well located, as is the front shock, which might only be a strong isentropic recompression. Differences among the results increase with span station. This might be due to the omission of the cross-derivative terms in the perturbation equation and the cross-flow terms in the boundary condition.

4.2   Wing-Body Analysis

Figures 17 and 18 show numerical results that prove the method for wing-cylinder combinations. Figure 17 shows the slightly decreased pressure on the upper and lower surfaces of a 30° swept wing mounted in midposition on

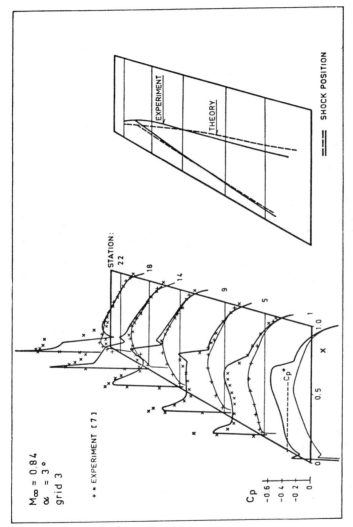

FIG. 16 ONERA M6 wing: pressure distribution and shock position.

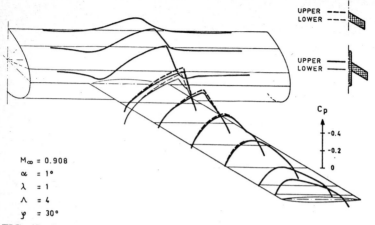

FIG. 17 Swept wing mounted in midposition on a circular body. Section: biconvex.

a circular body as compared with those of the exposed wing. Figure 18 indicates the asymmetric influence of the rectangular body on the same wing as in Fig. 17, but high-mounted. Differences due to the body on the upper and lower surfaces are noticeable only within the first 30 % half span. Figure 19 gives a comparison between the present method and that of Newman and Klunker [9,10], obtained for the same grid system (except for additional points close to the wing in z direction). Using Krupp scaling on the wing as well as on the body gives slightly higher pressures. The shock position is only one grid point ahead of that obtained by Newman and Klunker; the difference is mainly due to the additional term they use in their small-perturbation form.

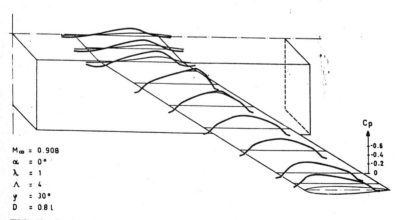

FIG. 18 Swept wing high-mounted on a rectangular body. Section: biconvex.

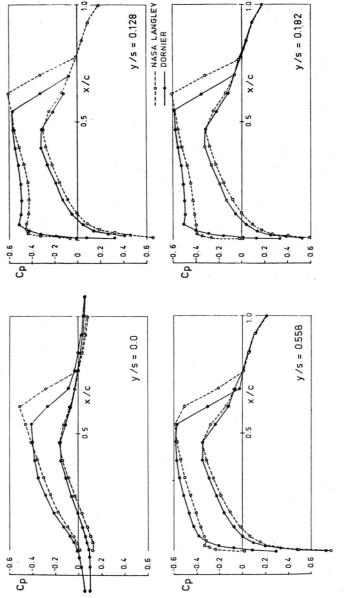

FIG. 19 NASA Langley wing-body; $M_\infty = 0.9$, $\alpha = 2°$.

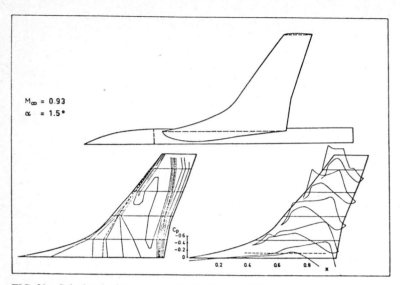

FIG. 20 Calculated wing pressure distribution for PT3 wing-body model.

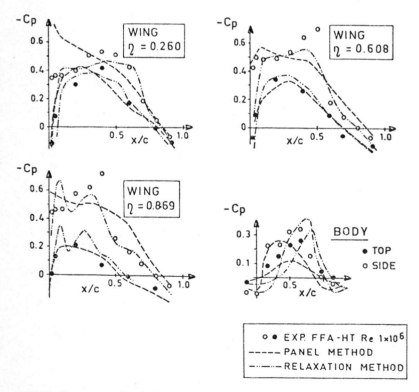

FIG. 21 Some results for the PT2 wing-body combination.

Figure 20 shows calculated pressure distributions for the PT3 wing, one of the wing-body combinations designed by Dornier and FFA and tested at FFA [33]. No experimental pressure distributions are available, but oil-flow pictures give the same qualitative results. In Fig. 21 some results are given for the PT2 wing-body combination designed and tested at FFA [34]. This comparison also includes results using a well-known subsonic panel method. The present relaxation method gives better agreement. The differences at the $\eta = 0.869$ station are due mainly to the tip fairing, which is not included in the computation. Furthermore, it should be mentioned that only about 30,000 mesh points were used.

Figures 22–24 show some of the ways in which the present method is used at Dornier in the development of supercritical wings for fighter-type aircraft as well as transport aircraft. The SKF design will lead to an experimental aircraft, based on the Alpha Jet, for evaluating the handling qualities and maneuvering boundaries of a supercritical wing. The transport design is mainly for a study of future aircraft with improved cruise performance. Detailed comparisons with experimental pressure distributions again show reasonably good agreement. Finer grid systems and the inclusion of boundary-layer effects will give improved results.

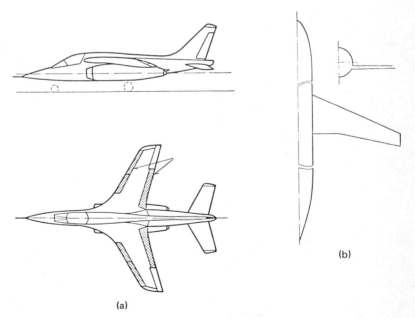

(a)

(b)

FIG. 22   (a) SKF transonic wing on alpha jet; (b) transport aircraft with supercritical wing.

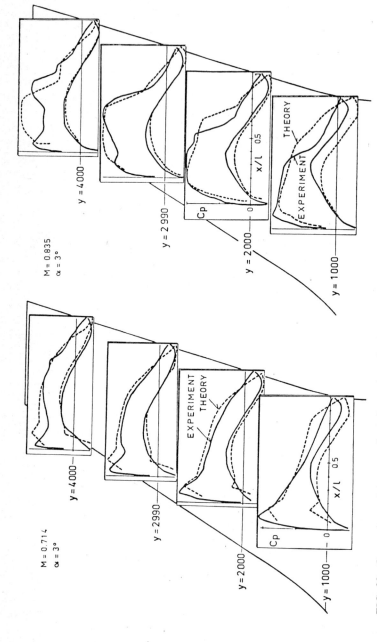

FIG. 23 Comparison of theory and wind-tunnel results for SKF wing; $\Lambda = 4.8$, $\lambda = 0.44$, $\varphi_{LE} = 31.7°$.

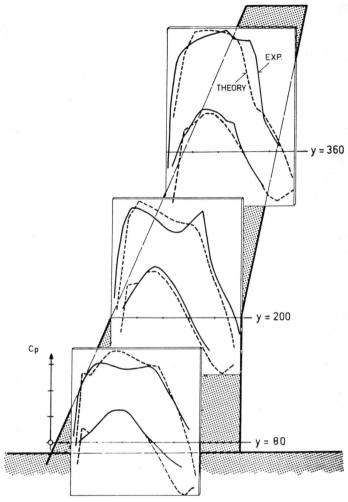

FIG. 24   Comparison of theory and wind-tunnel results for transport aircraft; $M_\infty = 0.8$, $\alpha = 5.4°$.

4.3   Wing and Wing-Body Design

The analysis and design capabilities of the present relaxation method have been tested in a study of a swept wing in the FFA PT series [37]. The wing was to be designed to have straight upper-surface isobars with a maximum local Mach number of 1.07 normal to the isobars. The study was performed with a very coarse grid, using only 20 iterations between the different steps shown in Figs. 25–27.

The basic wing with constant section geometry was the input for design J1 and $M_\infty = 0.90$ and $\alpha = 0.5°$. The characteristic aft shift of the inboard

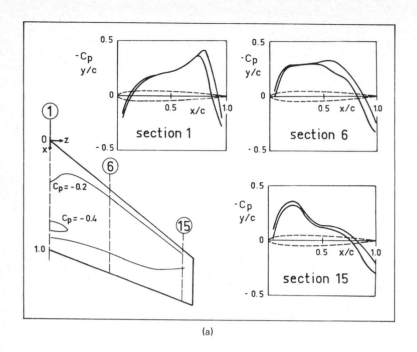

(a)

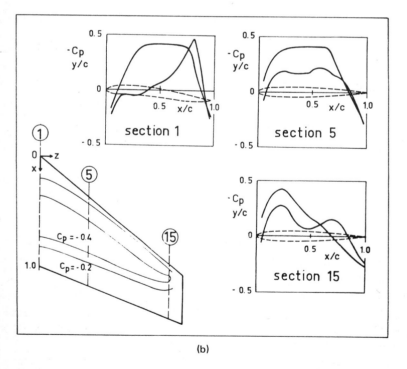

(b)

FIG. 25 (a) Wing design J1; (b) wing design J3.

pressure distribution and forward shift of the outboard distribution are clear in Fig. 25. The J3 wing was obtained from the basic wing through the requirements that the upper surface have a constant pressure plateau $C_p = -0.4$ with linear trailing-edge compression and that the basic wing thickness be retained. Figure 25 shows that these goals were reached but also that the inboard section gets a very steep lower-surface compression. $C_p = -0.4$ corresponds to a Mach-number component of 1.07 perpendicular to the trailing-edge generator.

To relieve the steep lower-surface compression, arbitrary pressure distributions were drawn for the lower surface at the three innermost sections (shown in Fig. 26 for the root section). Again the requirement on the upper-surface pressures was imposed. The resulting J4 wing, shown in Fig. 26, has overlapping upper and lower surface contours.

A new wing (J6) was obtained through rotation of the root-section lower surface contour of wing J4 around the leading edge until zero wing thickness was obtained at the trailing edge. For the rest of the wing, the profile slopes of wing J4 were maintained. The upper-surface pressure pattern was enforced. The new pressure distributions appear in Fig. 26. An acceptable design was obtained, but an attempt to extend the plateau forward would be interesting.

The J7 wing surface slopes have been increased over the J6 values by an amount $dy/dx = 0.05\{1 - \sin [\pi y/y(17)]\}$ for the inboard sections in the airfoil nose region, forward of the second grid points. The same pressure level on the plateau was maintained. The forward limit of the pressure plateau was shifted slightly forward, as can be seen in Fig. 27.

To pull the plateau at the root section further forward, one other twist increment was added to the inboard leading-edge part of the J8 wing, $dy/dx = 0.1\{1 - \sin [\pi y/y(6)]\}$. The wing reference angle of attack was increased to 1.5°. Here, too, a small change forward of the plateau resulted, as can be seen in Fig. 27. It is possible that an increase in root-section nose radius would have the effect of straightening the leading-edge isobars.

To obtain a new supercritical wing, more advanced design philosophies have to be used, in a manner similar to the design procedures outlined.

5 BOUNDARY-LAYER EFFECTS

From the various comparisons shown in this chapter, we may conclude that the present method is well suited for use in transonic wing development. To improve the results, boundary-layer effects must be considered along with the previously discussed modifications in the equation and boundary conditions. While a lot of work has been done in two-dimensional flow by using displacement thickness or the equivalent-source concept for the iterative treatment of potential flows and boundary layers to get results for the weak interaction, nothing has been published yet for three-dimensional flow.

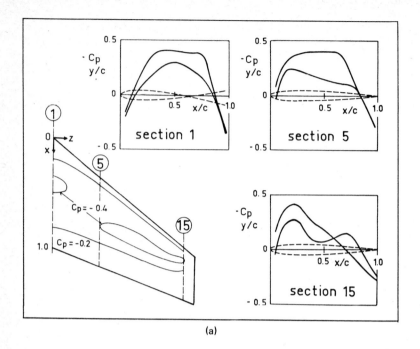

(a)

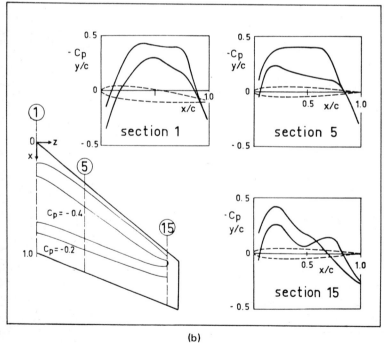

(b)

FIG. 26 (a) Wing design J4; (b) wing design J6.

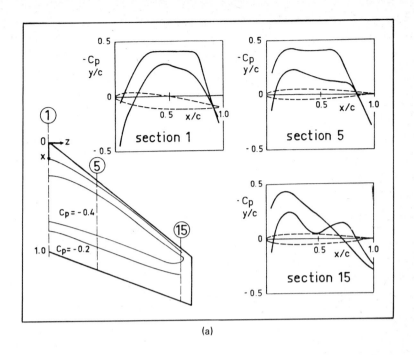

(a)

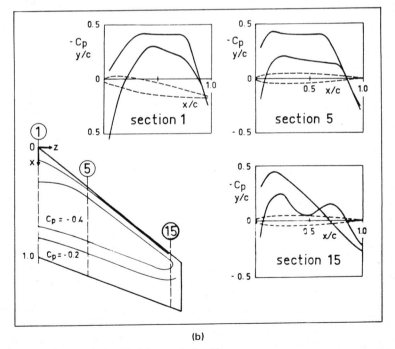

(b)

FIG. 27 (a) Wing design J7; (b) wing design J8.

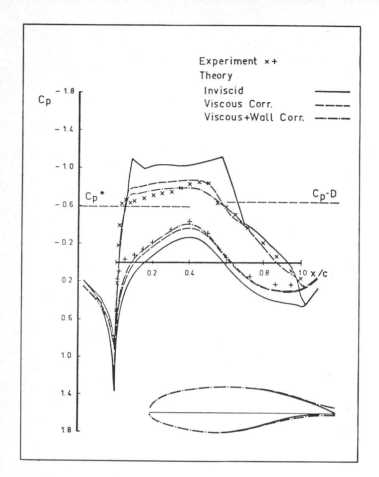

FIG. 28   Influence of viscosity and wind-tunnel corrections.

Figure 28 shows the strong two-dimensional influence and its treatment using the two-dimensional relaxation method and a two-dimensional boundary-layer integral method iteratively [38].

Recently some work was initiated at Dornier, based on the three-dimensional boundary-layer-integral method of P. D. Smith. First results are shown in Fig. 29; they were calculated by H. W. Stock [19] for the test case of the NASA Langley wing-body combination. The boundary layer is, of course, calculated only for the wing, but using the wing-body velocity distribution. For the two sections close to the body and close to the tip, the boundary-layer values show the expected behavior on the upper and lower surfaces. The calculation was started at 1% chord using the values for a yawed cylinder as initial conditions. In the region of strong pressure rise on

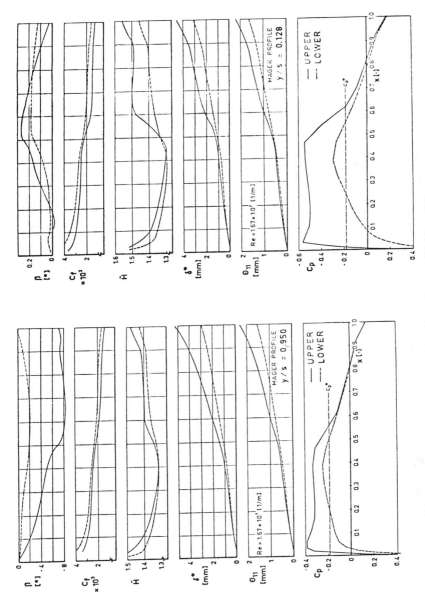

FIG. 29 Boundary-layer development for NASA Langley wing; $\alpha = 2°$, $M_\infty = 0.9$.

333

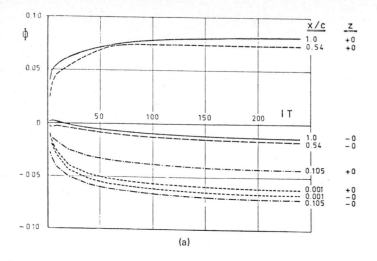

(a)

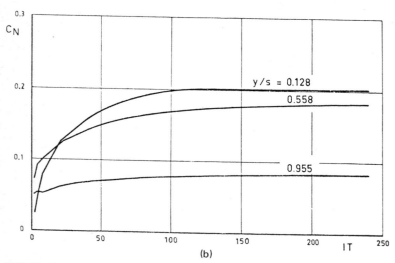

(b)

FIG. 30 NASA Langley wing body; $M_\infty = 0.9$, $\alpha = 2°$. (a) Convergence of potential for different points along the station $Y/S = 0.558$; (b) convergence of normal force coefficient for different sections.

the upper surface, the variation of the boundary-layer characteristics is maximum.

Displacement and momentum thickness show a strong increase, while wall-friction coefficient decreases. More detailed results and iterative potential–boundary-layer solutions showing the boundary-layer influence on shock position will be published soon.

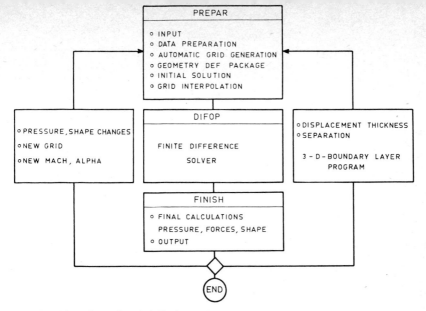

FIG. 31 Flow chart of analysis/design system.

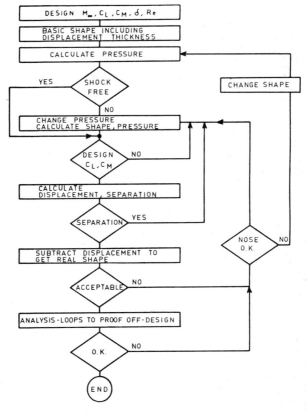

FIG. 32 Transonic design loop.

1. ANALYSIS   NORMAL VELOCITY COMPONENT SPECIFIED ;
                  PRESSURE CALCULATED

             o OPTIONALLY CAMBER OR THICKNESS CAN BE CHANGED
               ON PORTIONS

2. DESIGN   TANGENTIAL VELOCITY COMPONENT SPECIFIED ;
                  NORMAL COMPONENT COMPUTED ⟶ SHAPE

             o ARBITRARY PORTIONS ON UPPER AND/OR LOWER SURFACE
               CAN BE CHANGED

3. MIXED   ON SOME PORTIONS TANGENTIAL, ON OTHERS NORMAL VELOCITY
    PROBLEMS   COMPONENTS SPECIFIED
                  FINAL SHAPE AND PRESSURE COMPUTED

             o OVER SOME CHORDRANGE PRESSURE IS PRESCRIBED ON UPPER AND/OR LOWER
               SURFACE SPAN-SECTIONS WHILE FOR THE REST SHAPE IS GIVEN

             o SHAPE ON SOME STATIONS IS CHANGED

             o PRESSURE ON UPPER OR LOWER SURFACE IS CHANGED WITH THICKNESS OR
               CAMBER CONSTRAINT

FIG. 33   Analysis and design options

6   CONVERGENCE AND COMPUTER TIME

Figure 30 shows the extremely good behavior of the relaxation method, for solutions and iterations. The results converge nicely, for both the potential in the field and the section normal forces. The same is true for the pressure distributions.

Numerous numerical experiments have proved the present relaxation method to be very stable. Computer time for the analysis and design method could be reduced to only 0.14 ms per point per iteration on the IBM 370/158 computer, that is, 210 cpu-s for 100 iterations on a wing-body combination like the NASA-Langley one with 15,048 mesh points.

o SIMPLE INPUT BY USING GEOMETRY DEFINITION PROGRAMS

o AUTOMATIC GRID DESIGN INDEPENDENT OF USER

o SOLVES ANALYSIS AND DESIGN PROBLEM INCLUDING COMBINATIONS
   OF BOTH CLOSE TO THE PHYSICAL PROBLEM

o NO RESTRICTION TO WING-BODY-SHAPE EXCEPT DUE TO TPE

o VISCOUS EFFECTS INCLUDED BY USE OF DISPLACEMENT THICKNESS
   CONCEPT

FIG. 34   Payoff.

7 CONCLUSION

The methods described in this chapter are the basic tools for the evaluation of supercritical transonic wings and wing-body combinations at Dornier for future transport and fighter-type aircraft. An improved transonic perturbation equation and boundary conditions are being developed. Further experience has to be gained in the use of three-dimensional boundary-layer calculations.

In summary, Fig. 31 shows the flow chart of the Dornier analysis and design system. The use of the different blocks minimizes the input information needed. Figure 32 illustrates a possible procedure for designing new configurations, including the proof of off-design behavior and a check on possible constraints. Figures 33 and 34 show the capabilities and advantages of such a program.

REFERENCES

1 Murman, E. M., and J. D. Cole: Calculation of Plane Steady Transonic Flows, *AIAA J.*, 9(1): 114–121, 1971.

2 Bailey, F. R., and W. F. Ballhaus: Relaxation Methods for Transonic Flow about Wing-Cylinder Combinations and Lifting Swept wings, "Lecture Notes in Physics," vol. 19, pp. 2–9, Springer, New York, 1972.

3 Ballhaus, W. F., and F. R. Bailey: Numerical Calculation of Transonic Flow about Swept Wings, AIAA paper 72-677, June, 1972.

4 Bailey, F. R., and J. L. Stelger: Relaxation Techniques for Three-Dimensional Transonic Flow about Wings, *AIAA J.*, 11(3), 318–325, 1973.

5 Hall, M. G., and M. C. P. Firmin: Recent Developments in Methods for Calculating Transonic Flows over Wings, ICAS paper 74, 1974.

6 Jameson, A.: Numerical Calculation of the Three-dimensional Transonic Flow over Yawed Wings, *Proc. AIAA Comput. Fluid Dyn. Conf.*, 1973.

7 Jameson, A.: Iterative Solution of the Transonic Flows over Airfoils and Wings, Including Flows at $M = 1$.

8 Lomax, H., F. R. Bailey, and W. F. Ballhaus: On the Numerical Simulation of Three-dimensional Transonic Flow with Application to the C-141 Wing, NASA TN-D 6933, 1973.

9 Klunker, E. B., and P. A. Newman: Computation of Transonic Flow about Lifting Wing-Cylinder Combinations, *J. Aircr.*, 11(4), 1974.

10 Newman, P. A., and E. B. Klunker: Numerical Modeling of Tunnel-Wall and Body-Shape Effects of Transonic Flows over Finite Lifting Wings, NASA SP-347, pp. 1189–1212, 1975.

11 Rohlfs, S., and R. Vanino: A Relaxation Method for Two- and Three-dimensional Transonic Flows, Euromech 40, Saltjöbaden, 1973.

12 Schmidt, W., S. Rohlfs, and R. Vanino: Some Results Using Relaxation Methods for Two- and Three-dimensional Transonic Flows, in "Lecture Notes in Physics," vol. 35, pp. 364–372, Springer, New York, 1975.

13 Rohlfs, S., and R. Vanino: Berechnung der dreidimensionalen transsonischen Potentialströmung, *Z. Flugwiss.*, 23(7/8): 239–245, 1975.

14 Bailey, F. R., and W. F. Ballhaus: Comparisons of Computed and Experimental Pressures for Transonic Flows about Isolated Wings and Wing-Fuselage Configurations, NASA SP-347, pp. 1213–1232, 1975.

15 Albone, C. M., M. C. Hall, and Joyce Gaynor: Numerical Solutions for Transonic Flows past Wing-Body Combinations, Symposium Transsonicum II, Göttingen, 1975.

16 van de Vooren, J., H. W. Slooff, G. H. Huizing, and A. van Essen: Remarks on the Suitability of Various Transonic Small Perturbation Equations to Describe Three-dimensional Transonic Flow, Symposium Transsonicum II, Göttingen, 1975.

17 Schmidt, W., and R. Vanino: The Analysis of Arbitrary Wing-body Combinations in Transonic Flow Using a Relaxation Method, Symposium Transsonicum II, Göttingen, 1975.

18 Schmidt, W., and R. Vanino: Berechnung der dreidimensionalen transsonischen Strömung um Tragflügel-Rumpf-Kombinationen, Dorner FB 74/43 B, 1974.

19 Hedman, S., H. W. Stock, W. Fritz., and W. Schmidt: Entwurfsverfahren für dreidimensionale reibungsbehaftete transsonische Strömung, Dornier FB 75/45 B, 1975.

20 Bailey, F. R.: On the Computation of Two and Three-dimensional Steady Transonic Flows by Relaxation Methods, VKI Lecture Series 63, February, 1974.

21 Krupp, J. A.: The Numerical Calculation of Plane Steady Transonic Flows past Thin Lifting Airfoils, BSRL Report D 180-12958-1, 1971.

22 Klunker, E. B.: Contribution to Methods for Calculating the Flow about Thin Lifting Wings at Transonic Speed, NASA TN-D-6530, 1971.

23 Rubbert, P. E.: Private communication, Workshop of Transonic Flows, Friedrichshafen, 1975.

24 Wellman, J.: Der Einfluß der Machzahl auf wellenwiderstandsoptimierte Rümpfe und Profile bei überschallströmung, Vortrag 108, DGLR Tagung, 1972.

25 Schmidt, W.: Potentialtheoretische Berechnungsverfahren für Unter- und Überschallströmungen—Ein kritischer Überblick, Dornier FB 74/54 B, 1974.

26 Murman, E. M., and J. D. Cole: Inviscid Drag at Transonic Speeds, AIAA paper 74-540, 1974.

27 Kühl, P., and Zimmer, H.: Die Entwicklung von Tragflügelprofilen für Verkehrsflugzeuge mit verbesserten Schnellflugeigenschaften, Dornier FB 74/16, 1974.

28 Steger, J. L., and J. M. Klineberg: A Finite-Difference Method for Transonic Airfoil Design, AIAA paper 72-679, 1972.

29 Langley, M. J.: Numerical Methods for Two-dimensional and Axisymmetric Transonic Flows, ARA memo 143, 1973.

30 Agrell, N.: Some Results on Transonic Wings Using a Swept Mesh, to be published as FFA note, Sweden.

31 Lock, R. C.: Theoretical and Experimental Results on RAE Wing C, private communication, 1973.

32 Monnerie, B.: Experimental Study of a Swept-Back Wing in the Transonic Range, Euromech 40, Saltjörbaden, 1973.

33 Gustavsson, S. A. L., and R. Vanino: Entwurf und Windkanalmessung einer Flügel-Rumpf-Kombination mit überkritischem Profil, Z. Flugwiss., 23(7/8): 257–262, 1975.

34 Gustavsson, S. A. L., and S. G. Hedman: Design and Test of a Sonic Roof Top Pressure Distribution Wing, Symposium Transsonicum II, Göttingen, 1975.

35 Lotz, M., B. Moeken, and D. Welte: Experimental programm Superkritischer Flügel, Phase I, Dornier report, November, 1975.

36 Rohlfs, S.: Tragflügelentwicklung für Verkehrsflugzeuge, Dornier FB, to be published, 1976.

37 Hedman, S.: Wings in Transonic Flow, Calculation of Pressure or Surface Slope Distribution, FFA AU-1043, February, 1975.

38 Zimmer, H., and Stanewsky, E.: Entwicklung und Windkanalerprobung von drei überkritischen Tragflügelprofilen für Verkehrsflugzeuge, Z. Flugwiss., 23(7/8): 246–256, 1975.

CHAPTER 6

Computational Experiment: Direct Numerical Simulation of Complex Gas-Dynamics Flows on the Basis of Euler, Navier-Stokes, and Boltzmann Models

Oleg M. Belotserkovskii

1 INTRODUCTION

1.1 The Use of Numerical Methods

At present, specialists in the applied sciences are confronted with various kinds of practical problems whose successful and accurate solution, in most cases, may be attained only by numerical methods with the aid of computers. Certainly, this does not mean that analytical methods that permit us to find the solution in "closed" form will not be developed. Nevertheless, it is absolutely clear that the range of problems permitting such a solution is rather narrow; therefore, the development of general numerical algorithms for the investigation of problems of mathematical physics is important. It is especially urgent in continuous mechanics (gas dynamics, theory of elasticity, etc.) for at least two basic reasons.

1 *The difficulty of carrying out the experiment.* In studying the phenomena taking place, for example, at hypersonic flight velocities, the resulting high temperatures give rise to the effects of dissociation, ionization in the flow, and, in a number of cases, even to "luminescence" of a gas. In these cases it is enormously difficult to simulate the experiment in the laboratory, since, for similarity between the natural environment and the conditions of the experiment, it is not sufficient to satisfy the classical criteria of similarity, i.e., the equality of the Mach and Reynolds numbers. The equality of absolute pressures and absolute temperatures is also required, and this is possible only if the sizes of the model and the real object are equal. All this involves numerous technical difficulties which lead to high

experimental costs. In addition, in many cases the results of the tests are rather scarce. However, the importance of experimentation, in principle, must not be underestimated. Experimentation is always the basis of investigations confirming (or rejecting) the scheme and the solution according to some theoretical approach.

2 *The complexity of the equations considered.* Deep penetration of numerical methods into the mechanics of continua is also explained by the fact that the equations of aerodynamics, gas dynamics, and the theory of elasticity represent the most complicated system of partial differential equations (as compared to other branches of mathematics). In a general case, this is a nonlinear system of mixed type with the unknown form of the conversion surface (where the equations change their type) and with "movable boundaries"; i.e., the boundary conditions are given on surfaces or lines that, in turn, are determined by calculations. Moreover, the range of the unknown functions is so wide that ordinary methods of analytical investigation (linearization of equations, series expansion, separation of a small parameter, etc.) cannot be used to derive the full solution of the problem for this general case. It should be noted that, in solving complicated problems on electronic computers, the preliminary analytical investigation of a problem may be of great aid; sometimes, this investigation is decisive in the successful realization of the numerical algorithm.

Let us dwell on one more peculiarity of the algorithms used for solving concrete problems of the mechanics of continua. As is known, numerical methods have found wide practical application in design offices and research institutes. Substantial progress in the exploration of the cosmos, the optimum control of vehicles, the choice of rational configurations of vehicles, and so on is, to a considerable extent, due to serial calculations and the use of scientific information obtained in this way. The volume of information obtained by means of the calculation is far more complete and substantially cheaper than the corresponding experimental investigations if the problem is correctly formulated, well simulated, and algorithmically rational. However, wide application of numerical methods for practical purposes requires sufficient simplicity and reliability.

Thus, on the one hand, one has to deal with rather complicated mathematical problems. On the other hand, it is necessary to develop rather simple and reliable numerical methods permitting us to carry out serial calculations at project institutes and design offices.

Note that, for most problems in gas dynamics, the mathematical theorems of existence and uniqueness have not been proved, and very often there is no confidence that such theorems can be derived. As a rule, the very mathematical formulation of the problem is not strictly given, and only the physical treatment is presented—which are far from being one and the same thing. The mathematical difficulties of investigating such types of problems are related to the nonlinearity of the equations, as well as to the great number of independent variables.

The state of affairs with regard to the methods of solving gas-dynamics equations is no better. So far, investigations related to the possibility of realization of the algorithm, its convergence to the unknown solution, and its stability have been rigorously performed only for linear systems, and, in a number of cases, only for equations with constant coefficients. When confronted with the necessity of solving a problem, the mathematician has to use the known algorithms and to develop new methods without rigorous mathematical basis for their applicability.

This does not mean that the situation is very much different from that in any other new field. In science, as well as in mathematics, one can find many examples of new ideas and concepts that originated and were successfully used without a solid basis, which appeared later. Of course, this is not to imply that, when developing new numerical algorithms, one must act at random, without thinking about the accurate formulation of the problem or without comprehending its physical meaning. That path inevitably leads to numerous mistakes and to the waste of time—and the experience, without being theoretically interpreted, does not lay the foundation for further development of the method.

We draw your attention to this rather clear question only because there is still an opinion that the main thing is to write down differential equations, and all the rest reduces to a trivial substitution of differences for derivatives and to programming (to which too much importance is sometimes attached). In this connection, it is reasonable to formulate the main stages of the numerical solution of a mechanical or physical problem with the aid of computers in the following way:

1 Construction of a physical model and the mathematical statement of a problem
2 Development of a numerical algorithm and its theoretical interpretation
3 Programming (manual or automatic) and the formal adjustment of the program
4 Methodical adjustment of the algorithm, i.e., a test of its operation with concrete problems, elimination of the drawbacks found, and experimental investigation of the algorithm
5 Serial calculations, accumulation of experience, and estimation of the effectiveness and the range of applicability of the algorithm

At all stages, the mathematical theory and physical and numerical experimentation (with the aid of computers) are used jointly and consistently. The realization at each stage may be illustrated by solving concrete problems, and this will be done later in this chapter. At this point, we make only some common observations.

The main principle in using mathematical results is that the conditions providing the solution of a problem for simple and special cases must be fulfilled for more common and more complicated cases. Parallel to this, consideration of the physical meaning of the phenomenon gives a qualitative picture with the help of which the statement of the problem is checked and defined more exactly. Ultimately, the final experimental test allows us to judge the correctness of the assumption and to estimate the accuracy of the algorithm and the derived solution. It should be noted that the accuracy of the numerical solution of the formulated differential problem must be estimated purely mathematically, without using the results of the physical experiment. The latter may be used for qualitative comparison, but quantitative comparison between the calculation and the experiment must provide information on how closely the physical model approaches the natural environment.

1.2   Numerical Methods Applicable to Gas-Dynamics Problems

In the exact sciences there may arise many important problems whose investigation is tied to the solution of a system of nonlinear partial differential equations. Gas dynamics is one of these sciences, and, furthermore, it includes many problems with discontinuous solutions.

The construction of reasonably accurate solutions of the exact equations of gas dynamics in the general case has become possible only with the aid of numerical methods that exploit the advantages of high-speed electronic digital computing machines. Technology has called for an intensive development of numerical methods and their application to the solution of a wide variety of gas-dynamics problems. Scientists and research engineers in the area of gas dynamics have contributed significantly to the development of modern numerical methods for solving systems of nonlinear partial differential equations.

Four universal numerical methods are applicable to the solution of the nonlinear partial differential equations of gas-dynamics problems.

1.2.1   Method of Finite Differences

This method is the most highly developed of the four at the present time and is widely applied to the solution of both linear and nonlinear equations of the hyperbolic, elliptic, and parabolic types. The region of integration is subdivided into a network of computational cells by a generally fixed orthogonal mesh. Derivatives of functions in the various directions are replaced by finite differences of one form or another; usually, a so-called implicit difference scheme is applied to the integration of the equations. This results in the solution, at each step of the procedure, of a system of linear algebraic equations involving perhaps several hundred unknowns.

Finite-difference schemes are often used for solving steady and unsteady gas-dynamics equations. Lagrangian and Eulerian approaches are widely used here. In the first case, where the coordinate network is related to the liquid particles, the structure of the flow is better defined and one succeeds in constructing rather accurate numerical schemes for flows with comparatively small relative displacements. In the second case, when the calculational network is fixed over space, the schemes are used for constructing flows with large deformation. Recently, the approaches mentioned here have also found wide application in the calculation of steady flows.

1.2.2 Method of Integral Relations

In this method, which is a generalization of the well-known method of straight lines, the region of integration is subdivided into strips by a series of curves whose shape is determined by the form of the boundaries of the region. The system of partial differential equations written in divergence form is integrated across these strips, the functions occurring in the integrands being replaced by known interpolation functions. The resulting approximate system of ordinary differential equations is integrated numerically. The method of integral relations, like the method of finite differences, is applicable to equations of various types.

1.2.3 Method of Characteristics

This method is applied only to the solution of equations of the hyperbolic type. The solution, in this case, is computed with the aid of a grid of characteristic lines, which is constructed in the course of the computation. Actually, the method of characteristics is a difference method for integrating systems of hyperbolic equations on the characteristic calculational network, and it is used mainly for the detailed description of flows. Its distinguishing feature, as compared to other difference methods, is the minimum utilization of interpolation operators and associated maximum proximity of the region of influence of the difference scheme as well as the region of influence of the system of differential equations. The smoothing of the profiles inherent in the difference schemes with fixed network is minimal here, since the calculational network used in the method of characteristics is constructed exactly with the region of influence of the system taken into account.

Irregularity (nonconservativeness) of the calculational network should be mentioned as a drawback of the method of characteristics. It is possible to develop a technique, based on this method, in which the calculations are carried out in layers bounded by fixed lines. The method of characteristics permits one to determine accurately the point of origin of secondary shock waves within the field of flow, as the result of the intersection of the characteristics of one family. However, if a large number of such shock waves are developed, difficulties are encountered in their calculation. Accordingly, the method of characteristics is most expediently applied to hyperbolic problems in which the number of discontinuities is small (for example, problems concerning steady supersonic gas flow).

1.2.4 Particle-in-Cell (PIC) Method

In certain respects, the PIC method incorporates the advantages of the Lagrangian and Eulerian approaches. The range of solution here is separated by the fixed (Eulerian) calculation network. However, the continuous medium is interpreted by a discrete model; i.e., the population of particles of fixed mass (Lagrangian network of particles) that move across the Eulerian network of cells is considered. The particles are used for determining parameters of the liquid itself (mass, energy, velocity), whereas the Eulerian network is employed for determining parameters of the field (pressure, density, temperature).

The PIC method allows the investigation of complex phenomena of multicomponent media in dynamics, because particles carefully "watch" free surfaces, lines of separation of the media, and so on. However, because of the discrete representation of the continuous medium (the finite number of particles in a cell), calculational instability (fluctuations) often arises, the calculation of rarefied regions is difficult, and there are other problems. Limitations in the power of modern computers do not permit a considerable increase in the number of particles.

For problems in gas dynamics in the presence of a uniform medium, it seems more reasonable to use the concept of continuity by considering the mass flow across the boundaries of Eulerian cells rather than "particles."

Only numerical methods using high-speed computers and careful experiments allow us to find the complete solution to a complex gas-dynamics problem and to determine the necessary flow characteristics. Thus, the elaboration of numerical schemes, the calculation of different gas-dynamics problems, and the study of the analytical properties of the solutions and their asymptotic behavior are of significant interest at present.

1.3 Development of Numerical Algorithms

This chapter is, in essence, a review of the numerical methods used for the determination of the aerodynamic characteristics of high-speed vehicles with transonic and supersonic velocities. The numerical schemes were developed under our supervision and collaboration in the Moscow Physical Technical Institute and the Computing Center of the Academy of Sciences of the USSR. Here we discuss problems in the development and use of numerical algorithms for carrying out serial calculations in solving modern engineering problems arising in practice.

1.3.1 Steady-State Schemes

In determining the steady aerodynamic characteristics of bodies (especially when electronic computers of average power were employed) we made wide use of the following methods for solving steady gas-dynamics equations: the method of integral relations (MIR), the method of characteristics (MCh), and some finite-difference schemes (e.g., schemes with "artificial visocisity" and others). We wish to consider problems in which discontinuities and singularities are given beforehand, together with some associated boundary conditions; the solutions are to be carried out in regions where functions vary continuously.

As is known, three different MIR schemes were developed for the determination of flow in the region of a blunt nose, namely, using an approximation for the initial functions across the shock layer (scheme I), along it (scheme II), and in both directions (scheme III). As a result, the boundary-value problem was solved for an approximate system of ordinary differential (algebraic) equations. To solve the three-dimensional problem, some additional trigonometric approximations in the circumferential coordinate were introduced. For different flow conditions and different body shapes, one or another of the MIR schemes has been found applicable; they are widely used in our country as well as abroad [1-6].

The main advantage of these schemes is that, by means of different transformations, one succeeds eventually in approximating functions (or groups of functions) with comparatively weak variations. They allow us to obtain reliable results and a high degree of accuracy with a comparatively small number of interpolation nodes (usually three or four calculated points were used).

The choice of the independent variables, the form of the initial system of equations of motion (that is, the introduction of the integrals into the initial system, the use of the divergent form of the laws of conservation, and others), the use of conservation schemes, the approximation of the integrals, etc., are all of great importance in writing the numerical algorithm using MIR, and hence in producing results.

The main difficulty in carrying out MIR schemes is the solution of many parameter boundary problems for the approximating system of equations. This was overcome by means of different iteration schemes. Moreover, these schemes were used in transonic regions mainly for bodies of comparatively simple form, but for a supersonic zone one had to adopt another algorithm.

In calculating supersonic flow, the two- and three-dimensional MCh schemes of Chushkin, Magomedov, and their coworkers were used [7, 8]. As is known, once the initial form of the system is written in terms of characteristic variables, one requires the approximation of ordinary derivatives only. Using a fixed linear computational network, we get a system of finite-difference equations with its several advantages.

With the help of the above-mentioned approaches, a large number of gas-dynamics problems have been solved, namely, ideal-gas flows with chemical reactions and radiation, transonic and three-dimensional flows, and viscous flows. In most cases sufficiently steady and reliable results were obtained, which were in perfect agreement with experiment [6]. However, these approaches to the solution of the steady-state equation may be successfully used only for problems in which there are no singularities, discontinuities, intersections, and interactions. The application of these approaches is difficult for bodies of complex form with a large number of discontinuities. Besides, a single algorithm for the calculation of different types of flow is preferable.

1.3.2 Unsteady-State Schemes

The next step in the evolution of numerical methods, which was motivated by urgent practical needs and aided by the availability of electronic computers, was the development of nonsteady schemes and the use of the stability method for the solution of steady-state aerodynamic problems. We tried to keep to the general principles and ideas of the MIR and MCh in approximating the nonsteady equations with respect to space variables. The divergent or characteristic forms of the initial equations were used, the same calculational networks were employed, etc.

In this way the nonsteady schemes II and III of the method of integral relations and the network-characteristic method were developed [9, 10]. These allow us to consider rather complicated types of flow with a single algorithm. It is natural that the problems of computational stability and the attainment of steady-state solutions should become crucial. They require some specific technique such as the introduction of artificial viscosity into the initial system, and of dissipation terms into the difference equations. In a number of cases the accuracy of the results obtained is less than in the steady-state methods, but these approaches enabled us to consider new classes of problems, for example, the determination of the aerodynamic characteristics of three-dimensional flow for specific configurations, and the calculation of viscous transonic flows [10].

1.3.3 Large-Particles Method

Finally, in the third stage of development, it seemed reasonable and advantageous to introduce elements of the Harlow particle-in-cell method [11–13] into the algorithms. At first only the equation of continuity is represented as the mass flow across the Euler cell, using the simplest finite-difference or integral approximation along the coordinates.

Thus the modified method of large particles [14–15] came into existence, which (again by means of the stability process) allowed us to consider from one point of view such complicated tasks as, for example, the subsonic, transonic, and supersonic flow past a flat-nosed body in two dimensions or with axial symmetry. Such an approach is used in calculating viscous flows, and it may permit us to study the characteristics of separated flows.

It should be stressed that the development of the numerical schemes mentioned above benefits from the improvement and extension of the methods of solving the boundary-value problems for the corresponding approximating equations; the consideration of a new, wider class of problems;

and the development and improvement of electronic computers, machine languages, input and output arrangements, and so on.

1.4 Computational Experiments

In recent years the introduction of large computers has aroused a considerably greater interest in various numerical methods and algorithms whose realization borders on computational experiment. The need for such an approach to the solution of problems of mathematical physics is the result of ever-growing practical demands; in addition, it is connected to an attempt to construct more rational general theoretical models for the investigation of complex physical phenomena.

Let us outline the principal steps of a computational experiment. At first, one chooses a mathematical model of a physical object based on analytical study. Then one works out a tool for the investigation of the phenomenon in question, namely a difference scheme that permits the experiment itself to be carried out, i.e., the computational process. The next step comprises a detailed analysis of the results, leading to improvements and corrections in the mathematical model. This feedback procedure leads to modification and perfection of the methodology of numerical experiments.

The close analogy to physical experiments comprising similar steps is evident: analysis of a phenomenon under study; development of an experimental scheme; modification of design elements of the experimental installation; and measurements and their analysis.

In recent years the Computing Center of the USSR Academy of Sciences carried out a number of experiments associated with studies of complex gas-dynamic flows using a nonstationary method of large particles [14, 15]. Characteristic features of flows past bodies of different shapes were studied over a wide range of velocities, from subsonic through transonic and up to hypersonic. The results of a number of such experiments are presented in this chapter, but without the details of the computations.

Our approaches are based on the splitting of physical processes by a time step and on the stabilization of a process for the solution of stationary problems. The main purpose of this research is to consider mathematical models for more complex and general gas flows in the presence of large deformations. This approach finds application in the solution of both Euler [14, 15] and Navier-Stokes equations [16–18].

With the help of the large-particles method we succeeded in investigating complex gas-dynamics problems, including transonic overcritical phenomena, injected flows, and separation zones. It is important to note that the above class of problems was regarded from a single viewpoint: subsonic, transonic, and supersonic flows, transition through the sound velocity and the critical regime. The calculation of plane and axisymmetrical bodies was carried out by a single numerical algorithm as well [10, 14, 15, 19, 20].

It also seems promising to apply the main principles of the approach in question to the simulation of rarefied gas flows. The application of a statistical variant of such an approach to the solution of the Boltzmann equation is studied in references 21-23. We shall not dwell on the description of the techniques (it is given in detail in the references); instead, only the characteristic features of each approach will be given.

The basis of the technique of splitting a pattern into physical processes is as follows: The simulated medium may be replaced by a system of N particles (fluid particles for a continuous medium, and molecules for a discrete one) which, at the initial instant of time, are distributed in cells of the Eulerian mesh in a coordinate space in accordance with the initial data.

The evolution of such a system in time Δt may be split into two stages: (1) changes in the internal state of subsystems in cells that are assumed to be "frozen" or stable (Eulerian stage for a continuous medium, and collision relaxation for a discrete one) and (2) subsequent displacement of all the particles proportional to their velocity and Δt, without changing the internal state (Lagrangian stage for a continuous medium, and free motion of molecules for a discrete one).

The stationary distribution of all the medium parameters is calculated after the process is stabilized in time.

2 "LARGE-PARTICLES" METHOD FOR THE STUDY OF COMPLEX GAS FLOWS

In numerical models constructed by Davidov [14, 15] on the basis of Eulerian equations, the mass of a whole fluid (Eulerian) cell, i.e., a "large particle" (from which comes the name of the method), is considered instead of the ensemble of particles in cells. Furthermore, nonstationary (and continuous) flows of these "large particles" across the Eulerian mesh are studied by means of finite-difference or integral representations of conservation laws.

Actually, conservation laws are used in the form of balance equations for a cell of finite dimensions (which is a usual procedure in deriving gas-dynamics equations without further limit transition from cell to point). As a result, we obtain divergent-conservative and dissipative-steady numerical schemes that allow us to study a wide class of complex gas-dynamics problems (overcritical regimes, turbulent flows in the wake of a body, diffraction problems, transition through the sound velocity etc.) [10, 14, 15, 20].

2.1 Calculations

Consider the motion of an ideal compressible gas. Our starting point is provided by the Euler differential equations in the divergence form (the equations of continuity, momentum, and energy):

$$\frac{\partial \rho}{\partial t} + \mathrm{div}\,(\rho \mathbf{v}) = 0$$

$$\frac{\partial \rho U}{\partial t} + \mathrm{div}\,(\rho U \mathbf{v}) + \frac{\partial P}{\partial x} = 0$$

$$\frac{\partial \rho v}{\partial t} + \mathrm{div}\,(\rho v \mathbf{v}) + \frac{\partial P}{\partial y} = 0 \tag{1}$$

$$\frac{\partial \rho E}{\partial t} + \mathrm{div}\,(\rho E \mathbf{v}) + \mathrm{div}\,(P \mathbf{v}) = 0$$

It was shown in [14] that, in the large-particle method, the set of equations of gas dynamics, written as laws of conservation in integral form, may be used instead of (1). The important point is that the difference scheme approximating the initial set of equations should be homogeneous, so that "through" computation may be performed without isolating singularities.

Equations (1) are completed by the equation of state

$$P = P(\rho, E, \mathbf{v}) = 0 \tag{2}$$

The various stages of the computational cycle will be considered separately. Let us briefly describe the main principles of the large-particle method. The region of integration is covered by a fixed (over space) Euler mesh composed of rectangular cells with sides Δx, Δy (Δz, Δr are along a cylindrical coordinate system).

In the first (Eulerian) stage of the calculations, only those quantities change which are related to a cell as a whole, and the fluid is supposed to be instantaneously decelerated. Hence, the convective terms of the form div $(\phi \rho \mathbf{W})$, where $\phi = (1, U, V, E)$, corresponding to displacement effects, are eliminated from Eq. (1). Then it follows from the equation of continuity, in particular, that the density field will be "frozen" and the initial system of equations will be of the form

$$\rho \frac{\partial U}{\partial t} + \frac{\partial P}{\partial x} = 0 \qquad \rho \frac{\partial v}{\partial t} + \frac{\partial P}{\partial y} = 0 \qquad \rho \frac{\partial E}{\partial t} + \mathrm{div}\,(P \mathbf{v}) = 0 \tag{3}$$

Here we have used both the simplest finite-difference approximations and, to improve the calculation stability, the schemes of the method of integral relations, in which "sweeping-through" approximations of the integrands with respect to rays ($N = 3, 4, 5$) are used.

In the second (Lagrangian) stage we find mass flows across the cell boundaries Δt $t^n + \Delta t$. In this case we assume the total mass to be transferred only by a velocity component normal to the boundary. Thus, for instance,

$$\Delta M^n_{i+1/2, j} = <\rho^n_{i+1/2, j}> \; <U^n_{i+1/2, j}> \Delta y \; \Delta t \qquad (4)$$

The symbol $< >$ denotes the value (of ρ and U) across the cell boundary. The choice of these values is of great importance, since they substantially influence the stability and accuracy of the calculations. The consideration of the flow direction is characteristic of all possible ways of writing ΔM^n.

Here different kinds of representations for ΔM^n, of the first and second orders of accuracy, are considered. These are based on central differences, without account being taken of the flow direction and other conditions, as well as on the discrete model of a continuous medium comprising a combination of particles of a fixed mass in a cell [14, 15].

In the third (final) stage we estimate the final fields of the Euler flow parameters at the instant of time $t^{n+1} = t^n + \Delta t$ (all the errors in the solution of the equations are "removed"). As was pointed out, the equations at this stage are laws of conservation of mass M, impulse $\mathbf{P}$, and total energy written for a particular cell in the difference form

$$F^{n+1} = F^n + \Sigma \, \Delta F_b{}^n \qquad \text{where} \qquad F = (M, \mathbf{P}, E) \qquad (5)$$

According to these equations, inside the flow field there are no sources or sinks of M and $\mathbf{P}$, and their variations in time Δt are caused by interaction across the external boundary of the flow region.

2.2 Boundary Conditions

To retain the unified nature of the computations and to avoid special expressions for the boundary cells, layers of fictitious cells are introduced along all the boundaries, into which the parameters from the neighboring flow cells are written. The number of such layers depends on the order of the difference scheme (one layer for the first order of accuracy, etc.). Two kinds of boundaries then have to be distinguished: the rigid boundary (or axis of symmetry) and the "open" boundary of the computational region.

In the first case, the velocity component normal to the boundary changes sign, while the remaining flow parameters are taken unchanged (nonpenetration condition). The normal velocity component thus vanishes on the rigid walls, and the flow transmission conditions are thus realized. It will be shown below that another type of boundary condition is possible, namely, walls without slip (condition of sticking). In this case both velocity components change sign, and the entire velocity vector vanishes at the wall (condition of sticking).

The fluid can flow in and out through the open boundaries of the region, and some conditions on the continuity of the movement are required. Let the fluid flow into the rectangular mesh from the left; then the parameters of the entering flow will be specified here. On the remaining open boundaries of the region, we extrapolate the parameters of the flow "from within," i.e., we transfer to the fictitious layer the parameter values of the layer nearest to the boundary (zero-order extrapolation). A more complicated statement of the conditions is possible, as is more accurate extrapolation (linear, quadratic, etc.).

It is natural that the outer boundary of the region should be fairly remote from the source of the disturbance, in which case methods of "outward" flow extrapolation are possible. This topic will be discussed in more concrete terms below. It is merely mentioned here that the basic principle underlying the statement of the conditions is that no substantial disturbances should penetrate through the open boundaries of the region into the computational region.

2.3 Viscosity Effects

It has already been remarked that our approach employs homogeneous difference schemes, whereby "through" computation by a unified algorithm is possible both in the smooth-flow regions and at discontinuities. This is achieved by using finite-difference schemes with a viscosity approximation. Let us dwell briefly on this topic.

While the equations of gas dynamics for a nonviscous gas were taken as the initial equations, viscosity effects are in fact inherent in our difference scheme. They are produced, firstly, by the introduction into the scheme of an explicit term with artificial viscosity ("viscosity pressure") and secondly by the presence of an essentially schematic viscosity, dependent on the structure of the finite-difference equations.

The form of the approximation viscosity and estimates for the stability of the scheme can be obtained by writing, as Taylor series, the difference operators appearing in the equations in all three stages. The terms of zero (lowest) order should then represent the initial differential equations, while the structure of the approximation viscosity can be determined by retaining higher-order terms in the expansions ("expansion errors"). The resulting differential equations will be termed the *differential approximation* of the finite-difference scheme, while an expansion up to second-order terms in time and space is termed the *first differential approximation* [14, 15, 24].

The stability of the difference schemes may be investigated by means of the differential approximation. Such investigations were made by Yanenko and Shokin for one-dimensional quasilinear equations of the hyperbolic type [24]. While a strict mathematical groundwork has not yet been supplied for the case of nonlinear equations, the method of differential approximations has in fact been used here [13].

Taking the one-dimensional case for simplicity, let us describe the first differential approximation of our difference scheme. Take, say, U_{i+1}^n, write it as function $U(x + \Delta x, t)$, and expand each term of the finite-difference equations in Taylor series in the neighborhood of the point (x, t).

For instance, in computations of ΔM^n from the expressions (4) of the second order of accuracy, we obtain

$$\frac{\partial \rho}{\partial t} + \frac{\partial \rho U}{\partial x} = 0$$

$$\frac{\partial \rho U}{\partial t} + \frac{\partial (P + \rho U^2)}{\partial x} = -\frac{\partial q}{\partial x} + \frac{\partial}{\partial x}\left(\rho \epsilon \frac{\partial U}{\partial x}\right) \tag{6}$$

$$\frac{\partial \rho E}{\partial t} + \frac{\partial}{\partial x}\left[U(P + \rho E)\right] = -\frac{\partial q U}{\partial x} + \frac{\partial}{\partial x}\left(\rho \epsilon \frac{\partial E}{\partial x}\right)$$

or, using expressions (4) of the first order of accuracy,

$$\frac{\partial \rho}{\partial t} + \frac{\partial \rho U}{\partial x} = \frac{\partial}{\partial x}\left(\epsilon \frac{\partial \rho}{\partial x}\right)$$

$$\frac{\partial \rho U}{\partial t} + \frac{\partial (P + \rho U^2)}{\partial x} = -\frac{\partial q}{\partial x} + \frac{\partial}{\partial x}\left(\epsilon \frac{\partial \rho U}{\partial x}\right) + \rho \epsilon \frac{\partial^2 U}{\partial x^2} \tag{7}$$

$$\frac{\partial \rho E}{\partial t} + \frac{\partial}{\partial x}\left[U(P + \rho E)\right] = -\frac{\partial q U}{\partial x} + \epsilon \frac{\partial^2 (\rho E)}{\partial x^2} + \frac{\partial \epsilon}{\partial x}\frac{\partial \rho E}{\partial x}$$

where $\epsilon = |U|\,\Delta x/2$.

The differential approximations may be written similarly in the case of two-dimensional problems.

On the left-hand sides of (6) and (7), the exact expressions for the initial differential equations have been obtained, while on the right we have the terms that are a consequence of the presence of "viscosity" effects in the difference equations. The terms involving result from the explicit introduction of an artificial viscosity, while the terms involving are due to the schematic viscosity, which appears when the exact differential equations are replaced by finite-difference equations ("expansion errors").

It may easily be seen that, when the mesh is refined ($\Delta x \to 0$), we have $\epsilon \to 0$ and the equations of the differential approximation convert into the exact set of initial equations. In concrete computations (owing to Δx, Δt, ... being finite), terms containing ϵ always appear implicitly in the difference scheme even when $q = 0$; these terms are, in turn, analogous to the dissipative terms of the Navier-Stokes equations. The role of the coefficient of actual viscosity is here played by the coefficient of schematic viscosity, which depends on the local flow velocity and the size of the difference mesh.

In the two-dimensional case, it follows from the equations of momentum that the schematic viscosity (with $q = 0$) has the tensor form

$$\Delta = \frac{\rho}{2} \begin{vmatrix} U \Delta x \cdot \dfrac{\partial U}{\partial x} & v \Delta y \cdot \dfrac{\partial U}{\partial y} \\[2ex] U \Delta x \cdot \dfrac{\partial v}{\partial x} & v \Delta y \cdot \dfrac{\partial v}{\partial y} \end{vmatrix} = \frac{1}{2} \rho \mathbf{v} \, \Delta \mathbf{r} \cdot \nabla \mathbf{v} \qquad (6a)$$

where $\Delta \mathbf{r} = \Delta x \, \mathbf{i} + \Delta y \, \mathbf{j}$

It is clear from this that, owing to the presence of the vectors $\Delta \mathbf{r}$ and $\mathbf{v}$, the schematic viscosity does not possess invariance under Galileo transformations; in practice it only appears in zones where the gradient is large, e.g., in the shock wave, at the body surface, and at a breakaway of the flow. The coefficient of schematic viscosity ϵ (and hence the width of the "smeared" shock wave obtained) then depends on the size of the local flow velocity and the cell size. In regions of smooth flow, where the gradients of the flow parameters are relatively small, the influence of the schematic viscosity is negligible.

It will be shown below that in certain cases [when expressions (4) of the first order of accuracy are used for computing ΔM^n], the schematic viscosity ensures stable computation without introducing an explicit term involving the pseudoviscosity; whereas when the second-order expressions (4) are used in regions where the local velocity is small compared with the velocity of sound, the introduction of a term with q is necessary to obtain a stable solution.

2.4 Stability of the Scheme

While it is natural for different types of difference equations to be possible in different stages, the computations become strongly unstable on occasion, and rapidly increasing and oscillating solutions appear, which no longer reflect the behavior of the solutions of the initial differential equations.

The difference schemes quoted above are of the multilayer type, while the difference equations are strongly nonlinear with variable coefficients. This makes it impossible to employ Fourier's method, devised for linear equations with constant coefficients, for investigating the stability of the difference scheme as a whole. In essence, Fourier's method presupposes that the equations are linearized in the neighborhood of the flow with constant parameters, and it ignores nonlinear effects (influence of the flow gradients), which are sometimes the true sources of the instability.

A heuristic approach will therefore be employed here to analyze the stability of the difference schemes, based on a consideration of their differential approximations [13–15] and appropriate for nonlinear equations.

In this approach, we determine the signs of the coefficients α_i ("diffusion coefficients") in the dissipative terms of the differential approximation; these terms contain second partial derivatives in the space variables. For example, a linear equation can be indicated such that, when the value of the coefficient is negative, the equation of the differential approximation admits a solution that is exponentially increasing in time (unstable) [14].

In short, the necessary conditions for stability are obtained here from the condition $\alpha > 0$ (parabolicity condition). In the case of linear equations, the results of stability analyses obtained by means of the differential approximation, and obtained by Fourier's method, are exactly the same.

Let us examine how the different ways of writing the equation of continuity (second stage of the computations) contribute to the instability, assuming that the equations of momentum and energy are stable.

If ΔM^n is determined from expressions (4) of the second order of accuracy, we find, on expanding the relevant differential equations in Taylor series and retaining terms containing $\partial^2 \rho / \partial x^2$,

$$\frac{\partial \rho}{\partial t} + \frac{\partial \rho U}{\partial x} = \Delta_1 - \frac{\Delta t}{2}(U^2 + c^2)\frac{\partial^2 \rho}{\partial x^2} \tag{8}$$

If ΔM^n is evaluated from the expressions (4) of the first order of accuracy, we get

$$\frac{\partial \rho}{\partial t} + \frac{\partial \rho U}{\partial x} = \Delta_1^* + \left[\frac{\Delta x}{2}|U| - \frac{\Delta t}{2}(U^2 + c^2) - \frac{\Delta x^2}{4}\frac{\partial U}{\partial x}\right]\frac{\partial^2 \rho}{\partial x^2} \tag{9}$$

where Δ_1 and Δ_1^* are terms of the first differential approximation proportional to Δx and containing the first derivatives. In our case [14, 15],

$$\Delta x \approx 0.071 \quad \Delta t \approx 0.0071 \quad \rho_\infty = 1 \quad U_{-\infty} = 1 \tag{10}$$

In practical computations, when shock waves, contact discontinuities, and rarefaction waves appear,

$$\rho|U| \approx 1 \quad \left|\frac{\partial U}{\partial x}\right|\Delta x < 0.3 \quad \left|\frac{\partial \rho}{\partial x}\right|\Delta x < 2$$

It follows that the coefficient of $\partial^2 \rho / \partial x^2$ in (9) is positive, whereas it is negative in (8); i.e., scheme (8) has rapidly increasing solutions and is computationally unstable, while scheme (9) is stable.

2.5 Advantages

It follows from the very character of the construction of the calculation scheme that a complete system of nonstationary gas-dynamics equations is essentially solved here, while each calculation cycle represents a completed process in calculating a given time interval. Besides all initial nonstationary equations, the boundary conditions of the problem are satisfied, and the real fluid flow at the time in question is determined.

Thus, the large-particles method allows us to obtain the characteristics of nonstationary gas flows and, by means of the stability process, their steady magnitudes as well. Such an approach is especially applicable to problems in which a complete or partial development of physical phenomena with respect to time takes place. For example, in a study of transonic gas flows, the flows around finite bodies, flow in local supersonic zones, and separation regions develop comparatively slowly while the major part of the field develops rather rapidly.

In contrast to the FLIC method [25], our investigation is wholly devoted to systematic calculations of a wide class of compressible flows in gas-dynamics problems (transonic regimes, discontinuity, separation, and "injected" flows).

The divergent forms of the initial and difference equations are considered in the large-particle method; the energy is used; different kinds of approximations are used in the first and second stages; and additional density calculations are introduced in the final stage, which helps us to remove fluctuations and makes it possible to obtain satisfactory results with a relatively small network (usually 1000–2500 cells are used). All this results in completely conservative schemes; i.e., laws of conservation for the whole mesh region are an algebraic consequence of difference equations. Fractional cells are introduced for the calculation of bodies with a curvature in the slope of the contour.

The investigation of the schemes obtained (approximation problems, viscosity, stability, etc.) was carried out by successively considering the zeroth, first, and second differential approximations [13–15]. The investigation showed that the large-particle method yields divergent-conservative and dissipative-steady schemes for sweeping-through calculations. These enable us to carry out stable calculations for a wide class of gas-dynamics problems without introducing explicit terms with artificial viscosity. This may be of particular significance in studying flows around bodies with a curvature in the slope of the contour, since the methods of introducing explicit terms with artificial viscosity are different for whole and fractional cells. Moreover, by varying only the second stage of the calculation procedure, we can arrive at the conservative "particle-in-cell" method so that the calculational algorithm is of general use.

As for discontinuities, the stability of the calculations is provided here by the presence of approximate viscosity in the schemes (dissipative terms in difference equations), which results in a "smearing" of shock waves into several calculating cells and the formation of a wide boundary layer near the body. It should be stressed that the magnitude of the approximate viscosity is proportional to a local flow velocity and to the dimension of the difference mesh; therefore, its effect is practically evident only in zones with high gradients.

2.6 Results

We now give some results of the calculations of transonic and overcritical flows around profiles and plane and axisymmetrical bodies obtained by the large-particle method [20].

It is reasonable to charac terize the overcritical regimes of transonic flows around bodies with the value of the critical Mach number of the oncoming flow M_∞^* (when a sonic point develops on the body), as well as with the extent of a local supersonic zone (as compared to a characteristic dimension of the body) and with its intensity (maximum supersonic velocity realized in the zone).

Figure 1 presents flow-field patterns (lines M = constant) for a 24% circular arc profile ($\nu = 0$) extending from purely subsonic ($M_\infty = 0.6$) to supersonic ($M_\infty = 1.5$) regimes. The dynamics of the formation and development of a local supersonic zone, transitions through the critical Mach number (here $M_\infty^* = 0.65$), sound velocity, and other details are shown.

Figures 1*b*-1*g* illustrate a supercritical flow around a profile ($0.65 < M_\infty < 1$). One can distinctly see the position of the shock in the region of crowded lines M = constant that bounds the local supersonic line together with the sonic line ($M = 1$). The region of low velocities is located behind the shock wave. When the velocity of the flow increases, it reaches the parameters of an undisturbed flow at a large distance from the body. With $M_\infty > 0.9$, the zone becomes considerable both in size and in intensity (supersonic velocities are attainable up to $M = 1.7$-1.8), and in the case of a sonic flow (Fig. 1*g*), lines of the level $M = 1$ end at infinity.

The asymmetry of the whole flow pattern is noticeable (even at purely subsonic velocities in Fig. 1*a*); it results from nonpotentiality of the flow (supercritical regimes) and from the presence of viscous effects as well (subsonic regimes, formation of a wake behind the body, etc.).

In the case of a supersonic flow around a profile (Fig. 1*h*, where $M_\infty = 1.5$), a shock wave ahead of the body develops and bounds the disturbed region. Behind the wave, in the vicinity of the axis of symmetry, a region of subsonic velocities is realized. Afterward, the flow velocity along the contour of the body increases, and, as a result, an "ending" shock occurs near the stern of the body.

For comparison, the results of calculations by the above method of a flow around a 24% axisymmetrical spindlelike body ($\nu = 1$ and $0.8 \leqslant M_\infty \leqslant 2.5$) are given in Fig. 2. Here a critical regime occurs at $M_\infty^* \approx 0.86$. Local supersonic zones are less developed as compared to the plane case, and are of weaker intensity (for example, values of $M \approx 1.3$-1.4 are realized), although, naturally, the main singularities of transonic flow are seen here too.

Figure 3 compares the flow fields calculated by the above method (solid

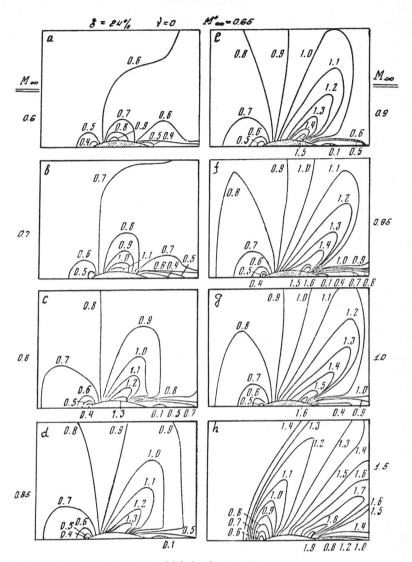

FIG. 1 Flow-field patterns for a 24% circular arc pattern.

line) and those of the Wood-Gooderum [26] experiment (dashed line). Subcritical (Fig. 3a, where $M_\infty = 0.725$) and supercritical (Fig. 3b, where $M_\infty = 0.761$) flows around a 12% profile are shown (in accordance with the calculations and the experiment, $M_\infty^* = 0.74$).

Analysis of the internal reference tests and the results of the comparisons reveal that the error in the calculations carried out by the large-particle method does not usually exceed several percent. The calculations were carried out using a Soviet BESM-6 computer; the time of the calculation in this case did not exceed an hour.

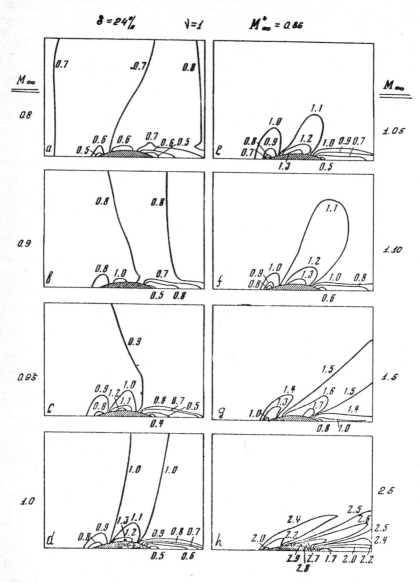

FIG. 2 Flow-field patterns for a 24% axisymmetrical body.

Figures 4-6 show results of calculations for some complicated flows past bodies of different shapes in the presence of discontinuities in the wake as well as under the influence of a fluid injected upstream from the front surface of the body. Such flows are of great practical interest in the study of wakes, turbulence, etc.

The results of the numerical experiments with fluid injection are given in

Figs. 4*a*, 5, and 6*b* and *c*. They include the case of the interaction of a supersonic flow around a finite thick circular disk (Fig. 4*a*, where $M_\infty = 3.5$), a 24% body of revolution (Fig. 5, where $M_\infty = 3.5$), and a sphere (Fig. 6*b*, where $M_\infty = 3.5$, and Fig. 6*c*, where $M_\infty = 6$) with a sonic axial stream (i.e., one where $M_c = 1.0$, $\rho_c = 2.0$, $U_c = 1.0$, and $v_c = 0$) issuing out of a nozzle situated on the axis of symmetry of the body. Figure 6*d* presents results for the case in which distributed injection of the flow takes place at the surface of a sphere. In all the figures, streamlines, shock waves, horizontal velocity lines (dots), and sonic lines (circles) are indicated; dashes denote lines separating the main flow from the injected stream. In Figs. 6*a* (sphere with $M_\infty = 3.5$) and 4*b* (cylinder with $M_\infty = 2.0$), results obtained for flows past the same bodies without injection ($M_c = 0$) are presented for comparison.

The action of the jet markedly complicates the flow pattern. For instance, in the flow past a cylinder, the head shock wave (*ABCD* in Fig. 4*a*) is pushed toward the oncoming flow, and its distance from the body increases significantly. The jet issues out of the body in the direction of the axis of symmetry at a sonic velocity and expands, forming a local supersonic region

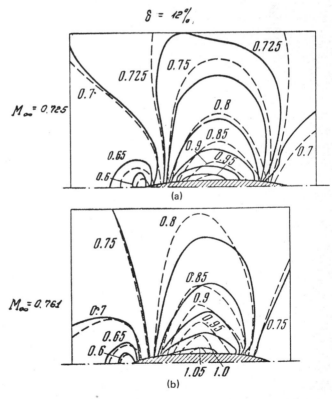

FIG. 3 Calculated (solid) and experimental (dashed) flow fields around a 12% profile: (*a*) subcritical; (*b*) supercritical.

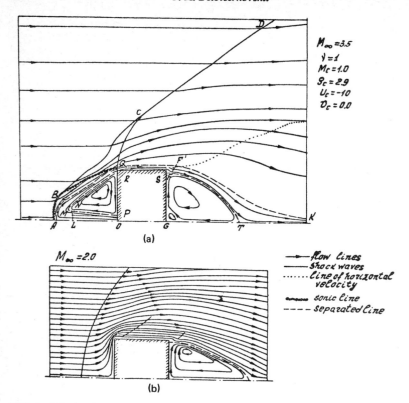

FIG. 4 Calculated results: (*a*) with fluid injection and (*b*) without injection.

OLMNPO that is closed by a triple λ-shock intersection (normal front *ML*, oblique front *MN*, transverse front *MP*), having a common point *M*. In front of the body, a stagnation zone with a complicated vortex structure develops; sonic line *BQ* is situated much lower than in a jetless flow. Behind the front stagnation zone, a secondary shock *QC* is formed, and at some distance from the body it merges with the head shock wave *ABCD* (at point *C*).

Behind the bodies in Figs. 4–6, both with and without injection, one can

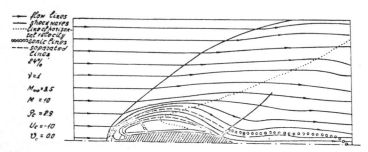

FIG. 5 Calculated results with fluid injection.

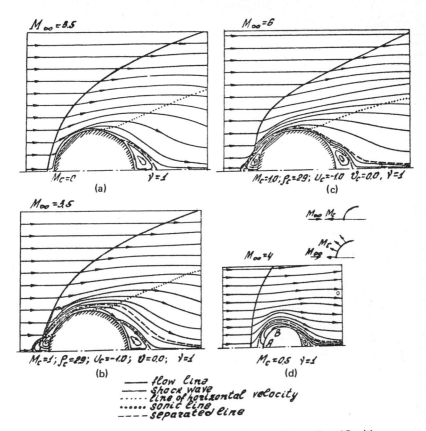

FIG. 6 Calculated results: (*a*) without injection; (*b*, *c*) with injection; (*d*) with injection at the body surface.

observe the development of separated zones of recirculating backward flows. In the cases considered, these zones are closed, localized in the wake of the body, and separated from the external flow by a "nonflow" line, i.e., a contact surface indicated by dashes in the figures. In the vicinity of the separation (it is interesting to note that in Fig. 4 the separation point is situated somewhat lower than the rear angular point of the body), a transverse shock wave *FF* develops. Backward recirculation flows are essentially subsonic and rarefied (gas density and pressure are low here), so that viscosity effects are negligible.

The large-particle method has also been applied to the study of internal gas flows, diffraction problems, and other problems.

Figure 7 presents some results of computations for flow through a straight channel ($\nu = 0$ in Fig. 7*a*) and a straight tube ($\nu = 1$ in Fig. 7*b*) in the presence of a central body ($M_\infty = 1.5$) for the case in which a triple shock intersection is formed as a result of the interaction of the flow with the upper

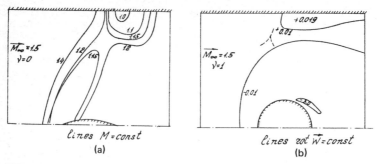

lines M = const
(a)

lines rot W = const
(b)

FIG. 7 Calculated results for (a) a straight channel and (b) a straight tube in the presence of a central body.

wall. (This can be seen by examining the behavior of the lines M = constant in Fig. 7a and rot W = constant in Fig. 7b.)

In calculating separated flows using various mesh sizes, the cell dimensions of the large particles were changed several times so that, in the wave of the body of size R, from 4 to 30 computational intervals were used (Fig. 8). In all cases there was an ample reserve of computational stability (over 100 Courant, where the Courant number represents the ratio of the time step to the space width of the cell).

Figure 8 shows base flows behind an axisymmetric cylinder ($M_\infty = 2.0$) for $R = 14\Delta y$. A gradual development of the flow in time is shown (in dimensionless units) from $t^n = 21$ to $t^n = 31$, when the zone is practically located. Streamlines are represented by solid lines; velocity vectors, by arrows. It follows from this diagram that at $t^n \approx 25$ the flow has already been formed but still continues to "breathe." It is interesting to note that similar flow patterns were obtained with denser meshes and (which is quite important) the zone "breathing," i.e., changes in its dimensions, internal structure, and other

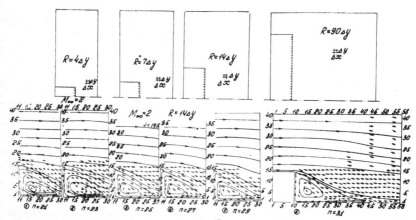

FIG. 8 Base flows behind an axisymmetric cylinder ($M_\infty = 2.0$).

features of the flow, occurred approximately at the same time intervals t^n in various approximations.

The formation of breakaway zones in the case of strong interaction seems to be explained here by the fact that, as a result of viscosity effects and the treatment of the boundary, conditions close to sticking conditions are realized on the body itself. A fairly wide boundary layer forms around the body surface (comparable to the thickness of the body at its tail), and this layer then moves away from the body surface and forms a breakaway zone with complicated vertical structure behind the tail part. It must be emphasized here that, while the boundary layer is in fact the result of viscosity effects in the scheme, in the breakaway zone itself the influence of the approximation viscosity ϵ (which is proportional to the local velocity and the size of the computational mesh; see above) is quite small, since in this zone only small values of the subsonic velocities are realized. Computations on different approximation meshes revealed only a slight change (within the limits of one step) in the zone contour.

The fact that the solution does not strongly depend on viscosity ($\epsilon \approx \rho|U|r$) shows, by the way, that flows corresponding to high Reynolds numbers can be treated by our methods of analysis. Thus, our calculations of separated zones might give quantitative information for the case of limiting flows (Re $\rightarrow \infty$) as, for example, the calculation of shock waves by a scheme including viscosity and other effects. Naturally, if necessary, the accuracy in determining the characteristic features of such zones can be further increased by using the results of preliminary calculations (e.g., the position of separation and closure points, or a zone contour) as initial data.

However, it should be pointed out that in the calculations the flow parameters on the front part of the body under study are determined comparatively quickly, while local supersonic zones and separation regions continue, as mentioned above, to "breathe." This may be due to a physical (nonstationary) character of the phenomenon itself. The difference scheme prescribed by the nonstationary large-particle method appears to be especially well suited for such a case.

3 COMPUTATION OF INCOMPRESSIBLE VISCOUS FLOWS

3.1 The Problem

At present, many numerical methods are known for the solution of Navier-Stokes equations describing viscous incompressible flows. Most of them were developed for equations relative to flow function Ψ and vortex ω.

A common disadvantage of these methods is the utilization in some form of a boundary condition (the Tom condition) for a vortex on a solid surface, which is omitted in the physical formulation of the problem. The rate of convergence of the numerical algorithms is limited by the presence of an additional iteration process due to this boundary condition for a solid-surface vortex.

Moreover, an apparent limitation of methods for the solution of the system (Ψ, ω) is connected with their inapplicability in cases of space viscous flows and compressible gas flows. This accounts for the recent interest in the numerical solution of Navier-Stokes equations represented in natural variables:

$$\frac{\partial v}{\partial t} + (\mathbf{v} \cdot \nabla)\mathbf{v} = -\nabla P + \nu \, \Delta \mathbf{v} \qquad \nabla \mathbf{v} = 0 \qquad (11)$$

where P = pressure
$\quad$ $\mathbf{v}$ = velocity vector
$\quad$ ν = coefficient of kinematic viscosity

Using the principles of the large-particle method, Gushchin and Shchennikov [16, 17, 29] studied viscous incompressible gas flows with velocity-pressure variables by means of a numerical scheme of splitting analogous to the SMAC method [27].

3.2 The Difference Scheme

The problem is to obtain, for Eqs. (11), a difference scheme with a high degree of accuracy that will enable us to carry out calculations for plane, axisymmetric, and space flows of a viscous incompressible fluid with a single algorithm.

Let us consider the following scheme for splitting a time cycle:

Stage I. Determination of the intermediate values of the velocities $\tilde{\mathbf{v}}(\tilde{U}, \tilde{v})$:

$$\frac{\tilde{\mathbf{v}} - \mathbf{v}^n}{\tau} = -(\mathbf{v}^n \nabla)\mathbf{v}^n + \nu \, \Delta \mathbf{v}^n \qquad (12a)$$

Stage II. Calculation of a pressure field:

$$\Delta P = \frac{\Delta \mathbf{v}}{\tau} \qquad \nabla \mathbf{v}^{n+1} = D^{n+1} = 0 \qquad (12b)$$

Stage III. Determination of the final values of the velocities:

$$\mathbf{v}^{n+1} = \tilde{\mathbf{v}} - \tau \cdot \nabla P \qquad (12c)$$

Stages I and III lead to the realization of the Navier-Stokes equation, and stages II and III are the conditions of solenoidality [second of Eqs. (11)]. In view of this, at stage I the transfer is accomplished only by convection and diffusion. The velocity field $\tilde{\mathbf{v}}$ so obtained does not satisfy a continuity equation ($\tilde{D} \neq 0$). Therefore, it is necessary to change ("correct") the field $\mathbf{v}$ at the expense of a pressure gradient P so that $D^{n+1} = 0$ (stage III); P is found by solving the Poisson equation (stage II).

For a proportional calculational mesh, a two-dimensional difference scheme of the second order of accuracy with respect to space is presented in [17]. The main difficulties of the numerical realization of the scheme involve the calculation of a pressure field and the formulation of boundary conditions.

It should be noted that in some works the projection of an equation of motion on the normal to the surface at boundary points is used as a boundary condition on a solid surface. This reduces the efficiency of the numerical method, since such conditions are not part of the physical formulation of the problem.

Easton [28] proposes an original modification of the boundary conditions, in the MAC method, that allows homogeneous boundary conditions to be provided for pressure.

Moreover, in the SMAC method [27] and the modified MAC method [28], owing to the difference schemes chosen, realization of the sticking condition necessarily results in the determination of a boundary vortex value on a solid surface satisfying the Tom condition of the first order of accuracy. In addition, in the SMAC method, the sticking condition does not provide a balance of forces on a solid surface.

An essential point of the proposed method is the choice of boundary conditions. For the solution of problems concerned with viscous incompressible flow around bodies of finite dimensions, we can distinguish two basic types of boundary conditions: conditions on a solid surface and those on a line sufficiently remote from a body. Let us dwell on each of these conditions.

For boundary conditions on a solid surface, we have

$$v_{i,-1/2}^n = 0 \qquad \text{nonpenetration condition}$$
$$U_{i+1/2,-1/2}^n = 0 \qquad \text{sticking condition} \tag{13}$$

From the latter, it follows that

$$\tilde{U}_{i+1/2,0} = \frac{U_{i+1/2,0}^n}{2} + \frac{U_{i+1/2,1}^n}{6} + O(h^3) \tag{14}$$

Condition (14) allows us to determine a boundary value for $\tilde{U}$ with the second order of accuracy with respect to internal field points. We thus avoid the necessity of introducing a layer of fictitious cells (inside a solid body), which in schemes of the MAC, SMAC, and modified MAC types [28] gives rise to the inaccurate calculation of a vortex value on a solid surface with the first order of accuracy. Note that, in the limits of the proposed approach, one need not calculate a vortex value on a solid surface. The latter can be determined from a calculated velocity field using some of the difference representations of the vortex expression

$$\omega = \frac{\partial U}{\partial y} - \frac{\partial v}{\partial x}$$

at the boundary points.

The boundary conditions on a line remote from a body are those for undisturbed flow; for the case $\bar{U}_\infty \parallel OX$, they have the form

$$v^n_{i, N+1/2} = 0 \qquad U^n_{i+1/2, N} = U_\infty$$

In the calculation of a pressure field, homogeneous boundary conditions are attained with the help of an approach in [28] that consists in the following. Supposing $v^{n+1}_{i,-1/2} = 0$ (for the case of a solid surface) and $v^{n+1}_{i,N+1/2} = 0$ (for the case of a line remote from a body), we have, from the finite-differences approximation (12c),

$$\tilde{v}_{i,-1/2} = \frac{\tau}{h}(P_{i,0} - P_{i,-1}) \qquad \tilde{v}_{i,N+1/2} = \frac{\tau}{h}(P_{i,N+1} - P_{i,N}) \qquad (15)$$

By taking account of (15), it is now not difficult to write a difference equation to calculate the pressure in the boundary cells [17].

The stationary solution of system of equations (12) is derived by repeating the above stages until the following criterion of establishment is fulfilled:

$$\max_{i,j} |U^{n+\lambda}_{i+1/2, j} - U^n_{i+1/2, j}| \leqslant \epsilon^*$$

Stability can be investigated in stages. The stability criterion at the first stage can be supplied by the first differential approximation (condition of α parabolicity). With regard to Eqs. (12a), the first differential approximation is [17]

$$\frac{\partial U}{\partial t} + \frac{\partial U^2}{\partial x} + \frac{\partial Uv}{\partial y} = \left(\nu - \frac{\tau}{2}U^2\right)\frac{\partial^2 U}{\partial x^2} + \left(\nu - \frac{\tau}{2}v^2 - \frac{h^2}{4}\frac{\partial v}{\partial y}\right)\frac{\partial^2 U}{\partial y^2}$$

$$\frac{\partial v}{\partial t} + \frac{\partial Uv}{\partial x} + \frac{\partial v^2}{\partial y} = \left(\nu - \frac{\tau}{2}U^2 - \frac{h^2}{4}\frac{\partial U}{\partial x}\right)\frac{\partial^2 v}{\partial x^2} + \left(\nu - \frac{\tau}{2}v^2\right)\frac{\partial^2 v}{\partial y^2} \qquad (16)$$

The stability criterion for the difference scheme employed follows from (16):

$$\tau \leqslant \frac{4\nu}{U^2 + v^2}$$

By eliminating P from (12b) and (12c), it is easy to show the absolute stability of the second and third stages by means of the Fourier method.

Thus, the difference scheme for the method enables us to calculate a flow without vortex and pressure values on a solid surface. This markedly increases the accuracy of the calculations, and the results attest to its effectiveness. The difference scheme (of the second order of accuracy) provides us with a single algorithm for calculating viscous incompressible flows around plane, axisymmetric, and three-dimensional bodies of complex configuration, as well as internal flows with a wide range of Reynolds numbers [17].

3.3 Results

A great number of external hydrodynamics problems were solved with this method. In a wide range of Reynolds numbers $(1 \leqslant Re \leqslant 10^3)$, viscous incompressible flows around different bodies of finite dimensions were studied: a rectangular slab and a cylinder of finite length with axis parallel to the velocity vector of the flow [29], a sphere and a cylinder with axis perpendicular to $\bar{U}_\infty$, a rectangular parallelepiped (three-dimensional flow) [7], as well as bodies of more complex form.

Figure 9 shows flow patterns around a cylinder (a plane problem) for Reynolds numbers 1, 10, 30, and 50 ($Re = 2Rv_\infty/\nu$, where R is the cylinder radius).

Figure 10 presents flow patterns around a cylinder for $Re = 10^3$ at times $T_1 = 162$, $T_2 = 166$, and $T_3 = 170$. In the latter case, a nonsteady flow pattern is observed (the stagnation zone seems to grow, and at some instant of time there is a "flopping" and overshooting of the fluid from the stagnation zone). The result probably should be verified.

Figures 11 and 12 present calculated results for the full Navier-Stokes equations for nonstationary three-dimensional flows. There, the problem is that of viscous incompressible flow around a cube (with linear dimension $2a$) when the oncoming flow velocity $\tilde{U}_\infty$ is parallel to an axis $0x$. Owing to the presence of two planes of symmetry ($0xy$ and $0zx$), the calculation is carried out in the positive quadrant $0xyz$ (Fig. 11).

As is known, there are difficulties in presenting the results of studies of three-dimensional flows. We give here the velocity profiles U (parallel to a vector $\bar{v}_\infty$) for various sections a with $x = $ constant.

Figure 11 shows, for $Re = 1$ ($Re = 2av_\infty/\nu$), a velocity profile U in an undisturbed flow ($x = -\infty$) and for a section $x = 3a$. Figure 12a shows, for $Re = 1, 8, 40$, and 100, the dynamics of the change in the velocity profile U along the cube (section a_1 coincides with the cube face $x = 2a$; the distance between sections is a constant $\Delta x = 0.5a$). Figure 12b illustrates the change

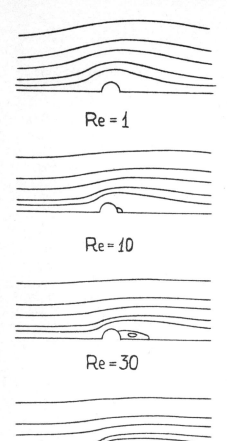

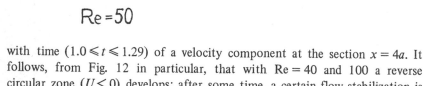

FIG. 9 Flow patterns around a cylinder.

with time $(1.0 \leqslant t \leqslant 1.29)$ of a velocity component at the section $x = 4a$. It follows, from Fig. 12 in particular, that with Re = 40 and 100 a reverse circular zone $(U < 0)$ develops; after some time, a certain flow stabilization is observed. For further details concerning these results, the reader is referred to references 16, 17, and 29.

4 COMPUTATION OF VISCOUS COMPRESSIBLE GAS FLOW (CONSERVATIVE FLOW METHOD)

4.1 The Method

The calculation of viscous compressible gas flows was performed by L. I. Severinov and A. I. Babakov with the help of an approximation of the conservation laws in integral form for each cell of the calculation scheme ("flow" method) [18]. The conservation laws for the mass, momentum, and energy of a finite volume have the form

$$\frac{\partial}{\partial t} \iiint_{\Omega} F \, d\Omega = - \oiint_{S_{\Omega}} \mathbf{Q}_F \, ds \qquad F = \{M, X, Y, Z, E\} \qquad (17)$$

where S_{Ω} = lateral surface of Ω

M = mass

X, Y, Z = momentum components

E = energy terms in a cell volume Ω

$\mathbf{Q}_F$ = flow density vector for each of the quantities

Equations (17) account for boundary conditions and are solved numerically for each cell of the calculation region.

If the values of $M^n = M(t^n)$, X^n, Y^n, Z^n, and E^n are known at an instant $t^n = \tau n$, where τ is a time integration step, then at the instant $t^{n+1} = (n+1)\tau$, these quantities can be calculated with error $O(\tau^2)$ as follows [18]:

$$F^{n+1} = F^n - \tau \oiint_{S_{\Omega}} \mathbf{Q}_F^{\,n} \cdot ds \qquad (18)$$

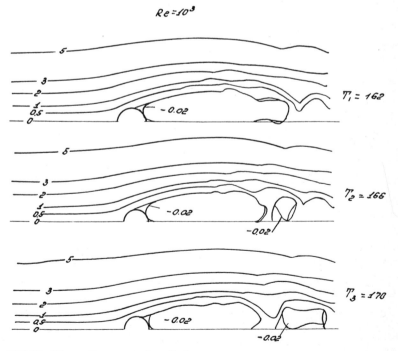

FIG. 10 Flow patterns around a cylinder.

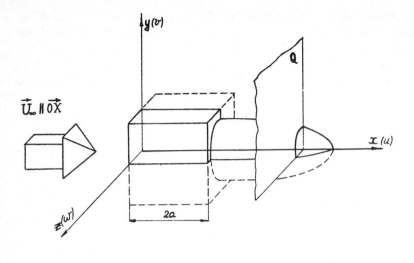

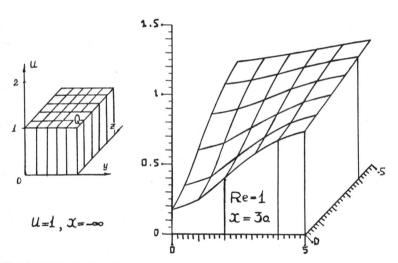

FIG. 11 Calculated results for nonstationary three-dimensional flow.

Supplementary conditions, the form of which depends on the particular problem, must make it possible to determine the flow-density vectors on the boundary of the domain in which the solution is sought. The three-dimensional coordinate system, the shapes of the elementary volumes Ω, and the methods of determining the field variables and their first derivatives on the surfaces S_Ω must be chosen to ensure stability and monotonicity of the difference method and a fairly simple approximation of the integrals in the system (18).

In the solution of a specific problem, the integrals in (18) are calculated on separate segments of the surface which are the boundaries between two adjacent volumes Ω. Depending on the directions of the flow density vectors, the values of M, X, Y, Z, and E vary (they increase in some cells and decrease in others) by quantities determined by the flows of mass, momentum, and

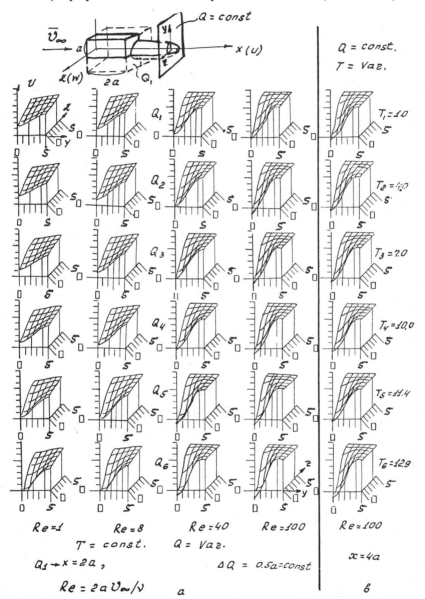

FIG. 12 Calculated results for nonstationary three-dimensional flow.

total energy through the corresponding segments of the boundary. Apart from
rounding errors, this calculation method cannot lead to the loss or generation
of the quantities M, X, Y, Z, and E due to computational errors. Therefore,
the flow method is conservative with respect to mass and momentum
components and total energy [18].

In the finite-differences equations (18), Stokes' assumption of the equality
of the mean values of the principal tensions (with reversed sign) and pressure
has been used. If the field variables are sufficiently smooth and the
assumptions used in calculating the mass, momentum, and energy flow-density
vectors are satisfied, the conservation laws (17) imply the complete Navier-
Stokes equations for a compressible gas, with the volume Ω arbitrary.

4.2 Analysis

We now consider some problems in the numerical investigation of Eqs. (17).

Knowing the values of the quantities M, X, Y, Z, and E, we can calculate,
for a given cell volume Ω fixed in space, the average values of the distribution
densities of the quantities ρ, ξ, η, ζ, and ϵ, where

$$\rho = \frac{M}{\Omega} \quad \xi = \frac{X}{\Omega} \quad \eta = \frac{Y}{\Omega} \quad \zeta = \frac{L}{\Omega} \quad \epsilon = \frac{E}{\Omega}$$

On the basis of these functions, it is easy to arrive at generally accepted
field variables—the components U, v, W of the velocity $\mathbf{v}$ and the specific
internal energy of the gas e:

$$U = \frac{\xi}{\rho} \quad v = \frac{\eta}{\rho} \quad W = \frac{\zeta}{\rho} \quad e = \frac{\epsilon}{\rho} - \frac{U^2 + v^2 + W^2}{2}$$

Using certain procedures of interpolation and numerical differentation, we
determine the values of the field variables and the first derivatives of U, v, W,
and e on the boundaries S of the cells Ω [18].

In determining the values of all the functions (except the distribution
densities ρ, ξ, η, ζ, and ϵ) and the first derivatives present, we have used
symmetric formulas; for example,

$$U_{m+1/2,\kappa} = \frac{U_{m+1,\kappa} + U_{m,\kappa}}{2}$$

$$\left(\frac{\partial U}{\partial x}\right)_{m+1/2,x} = \frac{U_{m+1,\kappa} - U_{m,\kappa}}{h_1}$$

$$\left(\frac{\partial U}{\partial y}\right)_{m+1/2,\kappa} = \frac{U_{m,\kappa+1} - U_{m,\kappa-1} + U_{m+1,\kappa+1} - U_{m+1,\kappa-1}}{4h_2}$$

To calculate the density values of the distributions ρ, ξ, η, ζ, and ϵ, we take asymmetric formulas:

$$\rho_{m+1/2,\kappa} = \begin{cases} 1.5\rho_{m,\kappa} - 0.5\rho_{m-1,\kappa} & \text{if } U_{m+1/2,n} > 0 \\ 1.5\rho_{m+1,\kappa} - 0.5\rho_{m+2,\kappa} & \text{if } U_{m+1/2,n} < 0 \end{cases}$$

These equations ensure second-order approximation accuracy.

In approximating flow-density vectors $\mathbf{Q}_F$, an essential element of the method is that the distribution densities of additive characteristics such as densities F are calculated on the boundary S_Ω of volume Ω in a nonsymmetrical way (extrapolation toward a gas flow). The other parameters, e.g., pressure, transfer velocities of additive characteristics, and derived axes U, v, and W are calculated according to symmetry formulas in the viscous-stress tensor and in the thermal-conduction law. We believe it allows us to take account of influence regions, which are an important factor in the investigation of complex physical flow patterns.

The presence of a "convective" transfer renders space directions unequivalent, and it is desirable to take account of this fact in constructing difference schemes.

The transition to integral conservation laws essentially requires the approximation of derivatives whose order is one lower than those approximated in methods for the numerical solution of Navier-Stokes equations.

It is not hard to see that in essence the flow method is conservative with respect to mass, momentum, and total energy; conservativeness occurs both locally (for each cell of a difference mesh) and integrally, i.e., for the whole calculational region [18]. As follows from (17), the conservative property results because the approach is based upon the difference approximation of conservation laws written for each cell of the calculation mesh in terms of surface integrals of vectors of flow densities $\mathbf{Q}_F$; i.e., the conservation law is used in a form that holds true for an arbitrary gas volume.

Indeed, in the solution of an actual problem, the surface integrals in (17) are calculated on separate surface segments S_Ω that constitute boundaries between two adjacent volumes Ω. Depending upon the direction of the flow vectors, the values of $F = \{M, X, Y, E\}$ vary (they increase in some cells and decrease in others), and the new values are determined by flows of mass, momentum, and total energy across coincident boundary zones. Within the accuracy of error rounding, calculations of this kind cannot result in the loss or formation of quantities F owing to the calculation procedures themselves, which testifies to its conservativeness.

A transfer (and, therefore, an approximation) of function "complexes"—distribution densities of the mass, momentum, and energy—is performed, which is consistent with the physics of the phenomenon. The approach is based upon the generality of a "transfer factor" (hence the name flow method). From the viewpoint of obeying conservation laws, the scheme analysis seems significant, as it is known that a calculational scheme provides more accurate results when it rigorously preserves the quantities that occur in the physical process involved.

The flow method is essentially a development of the large-particles method. The difference formulas of the flow method can be deduced by using a scheme for splitting (3)–(5) for a "transfer" of the components of the quantities F.

4.3 Results

A systematic study of the characteristics of viscous compressible gas flow around bodies of finite dimensions was carried out with the above approach, for a wide range of Reynolds numbers Re. The method formally "works" for large values of Re; however, the results are reliable when the boundary-layer thickness is much greater than a step of the calculational mesh. It should be emphasized that the division of the flow-density vector $\mathbf{Q}_F$ into convective and viscous components allows us to easily use the given algorithm for the calculation of ideal-gas flow.

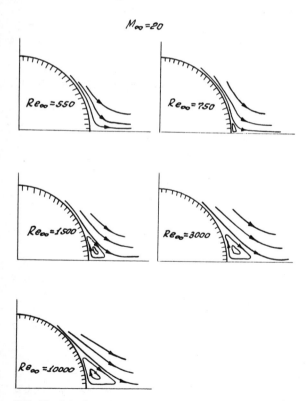

FIG. 13 Calculated flow patterns for a sphere.

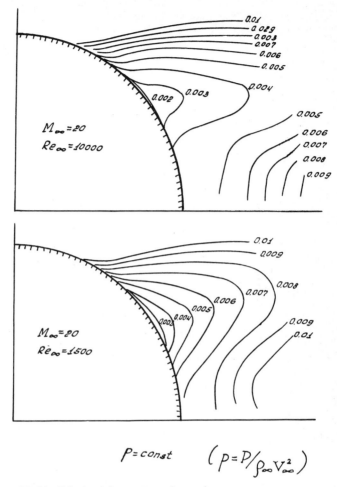

$P = const$ $\left(p = P \big/ \rho_\infty V_\infty^2 \right)$

FIG. 14 Calculated flow patterns for a sphere.

The results given below were obtained by means of a method of establishment for the solution of a stationary boundary-value problem. According to the investigation of a linear model and the calculations, the difference scheme of the second order of accuracy proves conventionally stable and conventionally monotonous [18]. The reliability of the results was experimentally examined for a general case by dividing a step of the calculation mesh, by using various forms of boundary conditions, and by comparing with the results of other calculations, with the results for an ideal gas, and with experiment [18, 30].

Figures 13 and 14 give the flow patterns for a sphere (separation zones of a reverse circular flow) with $M_\infty = 20$ and $550 \leqslant \mathrm{Re}_\infty \leqslant 10^4$. Figure 14 shows the behavior of lines $P = $ constant in a separation zone behind the sphere, with $\mathrm{Re}_\infty = 10^4$ and 1500 ($\mathrm{Re}_\infty = R\upsilon_\infty / \nu$).

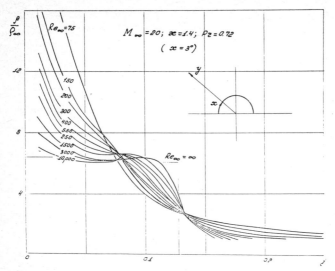

FIG. 15 Density behavior across a shock layer near the axis of symmetry.

Figure 15 illustrates the density behavior across a shock layer in the region of the axis of symmetry ($x = 3°$) for $75 \leqslant Re_\infty \leqslant 10^4$. A density graph for an ideal gas ($M_\infty = 20$, $x = 1.4$, $Re_\infty = \infty$) is also plotted. The behavior of the curves shows that, for $Re_\infty \to \infty$, the gas density tends to acquire its limit value in a viscous nonthermoconducting gas; a tendency toward the formation of a shock wave is distinctly seen.

Figure 16 shows a pressure distribution along a body (relative to the

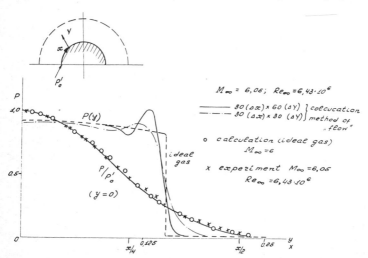

FIG. 16 Flow-method results compared with experiment and ideal-gas behavior.

pressure at a critical point). Solid lines designate the results obtained by the flow method ($M_\infty = 6.05$, $Re_\infty = 6.43 \times 10^6$); crosses, experimental data (G. M. Riabinkova); and circles, the results for an ideal gas (Belotserkovskii [31]). There is very good agreement among the data. Thus, a limit transition from the viscous equations (17) to an ideal gas is obtained.

The above approach therefore allows the study of viscous compressible gas flows in a wide range of flow regimes (separation zones inclusive) up to large Reynolds numbers.

5 STATISTICAL MODEL FOR THE INVESTIGATION OF RAREFIED GAS FLOWS

5.1 The Model

The applicability of a statistical variant of our general approach was investigated by Yanitsky [21-23] for the solution of the Boltzmann equation. The main problem in this field is the development and investigation of a model for the behavior of a gas medium consisting of a finite number of particles. The model is based upon a combination of the splitting idea of the large-particles method in terms of Bird's statistical treatment [32, 33] and Kats' ideas [34] about the existence of models asymptotically equivalent to the Boltzmann equation.

As is typical of particle-in-cell methods, the simulated medium is replaced by a system containing a finite number N of particles of fixed mass. At a given instant of time t_α in each cell j, there are $N(\alpha, j)$ particles endowed with certain velocities. The main calculation cycle is comprised of two stages:

At the first stage, particles collide only with their counterparts in a cell (collision relaxation).
At the second stage, they are only displaced and interact with the boundary of a reference volume and with the surface of a body (collisionless relaxation).

The main distinction between the model suggested in references 21-23 and Bird's model is that, at the first stage of the calculation, each group of N particles in a cell is regarded as Kats' statistical model for an ideal monoatomic gas consisting of a finite number of particles in a homogeneous coordinate space. In simulating collisions, our approach makes use of Monte Carlo methods for the numerical solution of the main equation of Kats' model; this enables us to correctly determine the time between particle collisions in accordance with collision statistics for an ideal gas.

In contrast to previously proposed Bird's methods [32, 33], the approach in references 21-23 is a rigorously Markovian process. The main equation of this approach is linear (unlike the Boltzmann equation), which substantially simplifies the numerical realization of the algorithm. The feature of the propagation of molecular chaotic motion implies that Kats' model is asymptotically equivalent to the Boltzmann equation without convective derivative. The integration of the main equation of Kats' model results (with accuracy up to the realization of the assumption of molecular chaotic motion) in the Boltzmann equation.

For the realization of the second stage of the calculation of the evolution of a simulated gas, it is suggested in references 21–23 that use be made of the numerical algorithms for the displacement of particles, utilizing incomplete information about the positions of the particles in a coordinate space. This reduces the required volume of processor memory, which significantly increases the method's effectiveness. The method can be realized in a two- or three-dimensional coordinate space as well.

5.2 The Method

Let us dwell here upon the principal aspects of the suggested statistical particle-in-cell method [21–23]. We suppose that the problem of rarefied-gas flow around a body can be solved by means of a distribution function and that the gas is monoatomic. Then any macroparameter of gas flow $\Psi(t, \overset{\bullet}{\mathbf{x}})$ related to a molecular feature $\Psi(\mathbf{c})$ is a functional of the form

$$\Psi(t, \mathbf{x}) = \frac{1}{n(t, \mathbf{x})} \int \Psi(\mathbf{c}) \cdot f(t, \mathbf{x}, \mathbf{c}) \, d\mathbf{c}$$

where $f(t, \mathbf{x}, \mathbf{c})$ is a molecular distribution function in a six-dimensional space $(\mathbf{x}, \mathbf{c})$ of the coordinates and velocities of the particles.

If Ω denotes the region of a control volume, and Γ the boundary Ω comprising a body surface as well, then the problem is reduced to obtaining the solution of the Boltzmann equation

$$\frac{\partial f}{\partial t} + \mathbf{c} \cdot \frac{\partial f}{\partial \mathbf{x}} = \int (f' \cdot f_1' - ff_1) g \, d\delta \cdot d\mathbf{c}_1 \tag{19}$$

satisfying the given initial parameters

$$f(t + 0, \mathbf{x}, \mathbf{c}) = f_0(\mathbf{x}, \mathbf{c}) \quad \mathbf{x} \in \Omega \quad -\infty < c_{X,Y,Z} < +\infty \tag{20}$$

and boundary conditions

$$f(t, \mathbf{x}_\Gamma, \mathbf{c}) = \int \kappa(\mathbf{c}, \mathbf{c}_1) f(t, \mathbf{x}_\Gamma, \mathbf{c}_1) d\mathbf{c}_1 \quad \mathbf{cn}(\mathbf{x}_\Gamma) > 0 \quad \mathbf{c}_1 \mathbf{n}(\mathbf{x}_\Gamma) < 0 \tag{20a}$$

Here $n(x_\Gamma)$ is normal to the surface Γ at the point $x_\Gamma \in \Gamma$ and directed into the volume Ω; the nucleus shape κ is derived from the interaction law "gas-surface."

In deducing the Boltzmann equation, the following suppositions are made:

1 The mechanics of the collisions are described in the classical way.
2 The force fields of molecules are spherically symmetric.
3 Only binary collisions are considered (two molecules take part in any collision).
4 The molecules move randomly [the hypothesis of molecular chaos is valid; i.e., the distribution function of molecular pairs $f_2(t, x, c_1, c_2) = f_1(t, x, c_1) \cdot f_1(t, x, c_2)$, which implies a statistical independence of particles].
5 The collision time is negligibly small.

The difficulty in constructing the solution of the Boltzmann equation in a nonlinear integrodifferential form results both from the great number of independent variables (there are seven of them in the general case: time, geometric coordinates, and molecular-velocity components) and from the complex structure of an integral of collisions.

Quadratic nonlinearity in the integrands, their dependence upon "dashed" functions of distribution (determined by the values of molecular velocities after a collision), the high level of multiplicity of integration (equal to five in a general case), and the complex formulation of boundary condition (20) are the main peculiarities which complicate the direct solution of the Boltzmann equation (19) and the application of ordinary numerical algorithms.

For an approximate solution of the problem formulated in this fashion, we shall construct a statistical model of an ideal monoatomic gas consisting of N particles* with coordinates r_i, and velocities c_i $(i = 1, 2, \ldots, N)$ so that the equation of evolution of the model approximates Eq. (19). The only additional assumption is that of molecular chaos:

$$f_2(t, x, c_1, c_2) = f_1(t, x, c_1) \cdot f_1(t, x, c_2) \tag{21}$$

where

$$f_S(t, x, c_1, \ldots, c_S) \equiv \frac{N}{(N-S)!} \mathfrak{F}_S(t, r_1, \ldots, r_S, c_1, \ldots, c_S)$$

and with $r_1 = r_2 = \cdots = r_S = x$, $\mathfrak{F}_S$ being an S-partial function of distribution in a phase space of $6N$ dimensions.

---

*A real gas is modeled by an ensemble of about 1000 rigid ball-like molecules that can be regarded as typical representatives of many trillions (10^{12}) of molecules, e.g., in the study of phenomena occurring in a real shock wave [22].

If $\{\mathbf{R}(t), \; \mathbf{c}(t)\} = \{\mathbf{r}_1(t), \; \mathbf{c}_1(t); \dots; \; \mathbf{r}_N(t), \; \mathbf{c}_N(t)\}$ designates the model state at time t, the solution of the problem is reduced to the numerical realization of a finite number of trajectories $\{\mathbf{R}(t), \; \mathbf{c}(t)\}$ with initial parameters corresponding to (20); the modeling of particle interaction with the boundary Γ is accomplished in accordance with the given nucleus K [Eq. (20*a*)]. Once a number of trajectories are realized, one can calculate any macroparameter using adequate estimates of the Monte Carlo method for integrals.

The synthesis of the basic ideas of splitting, the particle method, and Kats' statistical model enables us to construct the desired model $\{\mathbf{R}(t), \; \mathbf{c}(t)\}$ for a space-inhomogeneous case when $\partial f/\partial x \neq 0$.

Let us suppose that at time interval t_α ($\alpha = 0, 1, \dots$) in a cell with center x_j ($j = 1, 2, \dots, J$) there are $N(\alpha, j)$ particles with velocities $\{\mathbf{c}_1, \dots, \mathbf{c}_{N(\alpha, j)}\}$. The center x_j of a cell in which a particular particle is situated is taken as a coordinate r_i of a particle i. The state of such a modeling gas $\{\mathbf{R}, \; \mathbf{c}\}$ is uniquely defined by a sequence of J points of the form

$$\{\mathbf{R}(t), \; \mathbf{c}(t)\} \sim \{N(\alpha, j); \mathbf{c}_1, \dots, \mathbf{c}_{N(\alpha, j)}\} \quad j = 1, 2, \dots, J \quad N = \sum_{j=1}^{J} N(\alpha, j)$$

The principal cycle of calculation of the model evolution at time Δt is split into two stages.

At the first stage, for a gas at rest, we model the variation of the internal state of subsystems enclosed in the cells: collisions of particles (with their counterparts in a cell) in subsystems $\{\mathbf{c}_1, \dots, \mathbf{c}_N\}$ are simulated independently in each cell, and thus the particles acquire new velocities.

Vector $\mathbf{c} = \{\mathbf{c}_1, \dots, \mathbf{c}_N\}$ is regarded here as a state of Kats' model. Let $\varphi(t, \mathbf{c})$ be the density of the probabilistic distribution of the state $\mathbf{c}(t)$; then the governing equation of this model (Kats' "master equation" [24]) has the form

$$\frac{\partial \varphi(t, \mathbf{c})}{\partial t} = \frac{1}{V} \sum_{1 \leqslant l < m \leqslant N} q_{lm} \int [\varphi(t, \mathbf{c}_{lm}) - \varphi(t, \mathbf{c})] d\sigma_{lm} \equiv K\varphi(t, \mathbf{c}) \quad (22)$$

Here K is Kats' operator of collisions; $q_{lm} = |\mathbf{c}_l - \mathbf{c}_m|$, where $\mathbf{c}_l$ and $\mathbf{c}_m$ denote the velocities of the lth and mth particles upon their collision; $d\sigma_{lm}$ is a differential section of elastic dissipation of a pair of particles $(\mathbf{c}_l, \mathbf{c}_m)$; and a normalizing parameter V is determined by the choice of measurement units and can be interpreted as a cell volume.

If we introduce distribution functions

$$f_S(t, \mathbf{c}_1, \dots, \mathbf{c}_S) = \frac{N!}{(N-S)! \, V^S} \int \varphi(t, \mathbf{c}) \prod_{i=S+1}^{N} d\mathbf{c}_i$$

then by integrating (22) it is not difficult to obtain

$$\frac{\partial f_1(t, \mathbf{c}_1)}{\partial t} = \int [f_2(t, \mathbf{c}_1', \mathbf{c}_2') - f_2(t, \mathbf{c}_1, \mathbf{c}_2)] q_{12} d\sigma_{12} d\mathbf{c}_2$$

which coincides with the Boltzmann equation having a zero convective derivative when satisfying equality (21).

The algorithm for realization of the first calculation stage of the evolution of a space-inhomogeneous model corresponds to the Monte Carlo method of numerical solution of Kats' basic equation (22), which (unlike the Boltzmann equation) is linear.

At the second stage, we model a collisionless transfer of particles from a particular cell to any neighboring cells without changing the internal state of the subsystems; their interaction with a control-volume boundary and a body surface is also considered. This stage corresponds to the Monte Carlo method of numerical solution of the Boltzmann free molecular equation in the form

$$\frac{\partial f}{\partial t} + \mathbf{c} L f = 0 \tag{23}$$

Here L is a finite-difference operator approximating a derivative $\partial/\partial x$; its introduction is closely related to an incomplete description of the system state in a coordinate space.

The simplest numerical algorithms for this approach [21–23] correspond to the solution of time-explicit, conventionally stable finite-difference schemes of the first order of accuracy for Kats' equations and the Boltzmann free molecular equation, respectively. Here, the equation of evolution of a modeling gas $\{R(t_\alpha), \mathbf{c}(t_\alpha)\}$ with enough accuracy to satisfy the equality defining molecular chaos has the form (a one-dimensional flow)

$$\frac{\Delta_\alpha f}{\Delta t} + c_x \frac{\Delta_\alpha f}{\Delta x} = J[ff_1] - \Delta t \, c_x \frac{\Delta_\alpha J[ff_1]}{\Delta x} \tag{24}$$

Here $\Delta_\alpha/\Delta t$ and $\Delta_\alpha/\Delta x$ are finite-difference first-order operators approximating derivatives $\partial/\partial t$ and $\partial/\partial x$, respectively. $J[ff_1]$ designates the right-hand side of the Boltzmann equation. The finite-difference scheme given is conventionally stable, and it approximates the Boltzmann equation within the accuracy of $O(\Delta t)$ and $O(\Delta x)$. As mentioned above, incomplete information concerning the space positions of the particles is used here.

The above approach can naturally be extended to the cases of two- and three-dimensional space. The extension to plane and space flows is trivial, and it consists of a sequence of one-dimensional displacements along coordinate axes. This corresponds to the splitting of a multidimensional transfer equation

$$\frac{\partial f}{\partial t} + c\frac{\partial f}{\partial x} = 0$$

into a sequence of one-dimensional finite-difference schemes.

The Boltzmann equation is known to imply a molecular chaos or a statistical independence of particles.* Our model includes the same premises as the Boltzmann equation, but without the molecular-chaos (or statistical independence) assumption. Consequently, in the model, there exists a statistical particle independence giving rise to a molecular-chaos disturbance. It should be noted that the inherent statistical independence rests upon theoretical and physical premises and does not depend upon the mesh dimension (it exists at $\Delta x \to 0$ as well).

The results of calculations of rarefied-gas flows reveal that:

1   Results for various quantities of particles in a cell (e.g., with $N = 3$ and $N = 20$) practically coincide.

*The molecular-chaos hypothesis implies that particle velocities are statistically independent. (M. N. Kogan, "Rarefied Gas Dynamics," M. Nauka, 1967.)

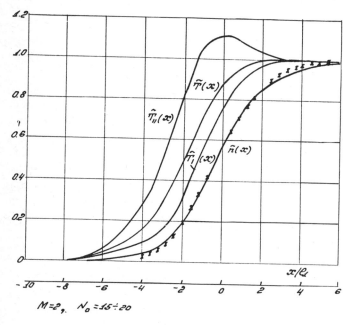

$M = 2, \quad N_0 = 15 \div 20$

——— calculation by the present statistical method

ɪ ɪ   numerical solution of the Boltzmann equation.

(Cheremisin's data [26])

FIG. 17   Statistical method versus direct numerical integration.

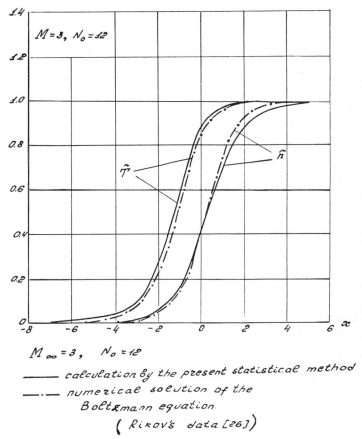

$M_\infty = 3$, $N_0 = 12$

_____ calculation by the present statistical method

_.___ numerical solution of the
            Boltzmann equation

(Rikov's data [26])

FIG. 18 Statistical method versus direct numerical integration.

2   The results are in good agreement with the solution of the Boltzmann equation (Cheremisin's and Rykov's data). Therefore, the molecular-chaos disturbance is small in the problems involved here (though statistical particle independence exists, it is weakly manifested in rarefied-gas problems and, apparently, it can be neglected here).

For turbulence, statistical independence is of crucial significance, and we can suppose that the statistical independence of the suggested method will become apparent when turbulent flows are considered.

5.3 Results

The model was tested with a problem dealing with the structure of a direct shock in a gas consisting of elastic balls in the Mach-number range $M_\infty = 1.25 \div 4$. Figures 17 and 18 show graphs of the density $\tilde{n}(x)$, longitudinal temperature $\tilde{T}_\parallel(x)$, transverse temperature $\tilde{T}_1(x)$, and total

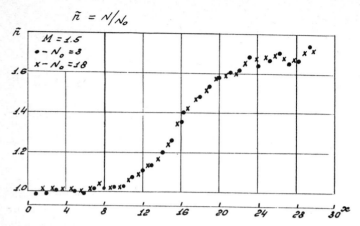

FIG. 19　Comparison of results for $N_0 = 3$ and $N_0 = 18$.

temperature $\tilde{T}(x)$ for Mach numbers of 2 and 3. The unit of length is a free mean path of the molecules in the flow. The relation $\Delta t/\Delta x$ is chosen to satisfy stability conditions. The average number of particles in the cells corresponding to the oncoming flow is $N_0 = 15 \div 20$ $(M = 2)$ and $N_0 = 12$ $(M = 3)$. For comparison, the figures give the density $\tilde{n}(x)$ and temperature

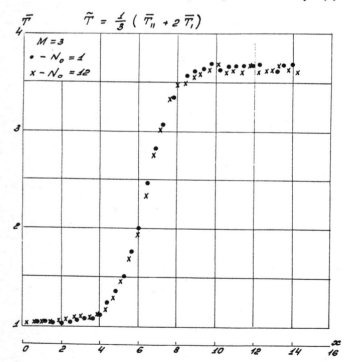

FIG. 20　Comparison of results for $N_0 = 1$ and $N_0 = 12$.

$\tilde{T}(x)$ obtained by direct numerical integration of the Boltzmann equation [35, 36] on a network Δx similar to the one used in our calculations $(\Delta x = 0.2 \div 0.3)$.

Figures 19 and 20 show the dependence of the results upon the average number N_0 of particles in the cells. In gas-dynamics problems concerned with a rarefied gas, this dependence is obviously rather weak.

This approach is probably also suitable for the investigation of turbulent gas flows.

6 CONCLUSION

In conclusion, we should say that a series of numerical algorithms (the computational experiment) allows us to effectively investigate a wide class of complex gas-dynamics phenomena from a single viewpoint. These methods may also be used for obtaining the aerodynamic characteristics of bodies, vehicles, and aircraft.

While individual fine and local details of the flow may be missed by this approach, it hardly seems likely that the basic properties of a flow will not be significantly determined by an exact description of small structures. The computational experiment is the only way of obtaining the general characteristics of a complex phenomenon and a picture of the flow as a whole.

REFERENCES

1 Dorodnitsyn, A. A.: On One Method of the Equation Solution of a Laminar Boundary Layer, *Zh. Prikl. Mekh. Tekh. Fiz.*, vol. 1, no. 3, pp. 111–118, 1960.

2 Belotserkovskii, O. M.: A Flow with a Detached Shock Wave around a Circular Cylinder, *Dokl. Akad. Nauk SSSR*, vol. 113, no. 3, pp. 509–512, 1975.

3 Belotserkovskii, O. M.: A Flow with a Detached Shock Wave around a Symmetrical Profile, *Prikl. Mat. Mekh.*, vol. 22, no. 2, pp. 206–219, 1958.

4 Belotserkovskii, O. M., and P. I. Chushkin: A Numerical Method of Integral Relation, *Zh. Vychisl. Mat. Mat. Fiz.*,[*] vol. 2, no. 5, pp. 731–759, 1962.

5 Belotserkovskii, O. M., A. Bulekbayev, and V. G. Grudnitskii: Algorithms for Schemes of the Method of Integral Relations Applied to the Calculations of Mixed Gas Flows, *Zh. Vychisl. Mat. Mat. Fiz.*, vol. 6, no. 6, pp. 1064–1081, 1966.

6 Belotserkovskii, O. M. (ed.): "Flow Past Blunt Bodies in Supersonic Flow: Theoretical and Experimental Results," *Tr. Vychisl. Tsentr. Akad. Nauk SSSR*, published by Computing Center AN SSSR, Moscow, 1966 (1st edition), 1967 (2nd edition, revised and extended).

7 Chushkin, P. I.: Blunt Bodies of Simple Form in Supersonic Gas Flow, *Prikl. Mat. Mekh.*, vol. 24, no. 5, pp. 927–930, 1960.

8 Chushkin, P. I.: Method of Characteristics for Three-dimensional Supersonic Flow, *Tr. Vychisl. Tsentr. Akad. Nauk SSSR*, Moscow, 1968.

9 Magomedov, K. M., and A. S. Kholodov: On the Construction of Difference Schemes for Equations of Hyperbolic Type Based on Characteristic Coordinates, *Zh. Vychisl. Mat. Mat. Fiz.*, vol. 9, no. 2, pp. 373–386, 1969.

[*]This journal is translated by Pergamon Press as *U.S.S.R. Computational Mathematics and Mathematical Physics.*

10 Belotserkovskii, O. M. (ed.): "Numerical Investigation of Modern Problems in Gas Dynamics," Izd. Nauka, Moscow, 1974.

11 Harlow, F. H.: The Particle-in-Cell Computing Method for Fluid Dynamics, in Berni Alder, Sidney Fernbach, and Manuel Rotenberg (eds.), "Methods in Computational Physics," vol. 3, Academic, New York, 1964.

12 Rich, M.: A Method for Eulerian Fluid Dynamics, report LAMS-2826, Los Alamos Scientific Laboratory, New Mexico, 1963.

13 Hirt, C. W.: Heuristic Stability Theory for Finite-Difference Equation, J. Comput. Phys., vol. 2, no. 4, pp. 339–355, 1968.

14 Belotserkovskii, O. M., and Yu. M. Davidov: The Use of Unsteady Methods of "Large Particle" for Problems of External Aerodynamics, preprint Vychisl. Tsentr. Akad. Nauk SSR, 1970.

15 Belotserkovskii, O. M., and Yu. M. Davidov: A Non-stationary "Coarse Particle" Method for Gas Dynamical Computations, Zh. Vychisl. Mat. Mat. Fiz., vol. 11, no. 1, pp. 182–207, 1971.

16 Gushchin, V. A., and V. V. Shchennikov: On One Numerical Method for the Solution of the Navier-Stokes Equation, Zh. Vychisl. Mat. Mat. Fiz., vol. 14, no. 2, pp. 512–520, 1974.

17 Belotserkovskii, O. M., V. A. Gushchin, and V. V. Shchennikov: Method of Splitting Applied to the Solution of Problems of Viscous Incompressible Fluid Dynamics, Zh. Vychisl. Mat. Mat. Fiz., vol. 15, no. 1, pp. 197–207, 1975.

18 Belotserkovskii, O. M., and L. I. Severinov: The Conservative "Flow" Method and the Calculation of the Flow of a Viscous Heat-conducting Gas Past a Body of Finite Size, Zh. Vychisl. Mat. Mat. Fiz., vol. 13, no. 2, pp. 385–397, 1973.

19 Belotserkovskii, O. M., and E. G. Shifrin: Transonic Flow behind a Detached Shock Wave, Zh. Vychisl. Mat. Mat. Fiz., vol. 9, no. 4, pp. 908–931, 1969.

20 Belotserkovskii, O. M., and Yu. M. Davidov: Computation of Transonic "Super-critical" Flows by the "Coarse Particle" Method, Zh. Vychisl. Mat. Mat. Fiz., vol. 13, no. 1, pp. 147–171, 1973.

21 Yanitsky, V. E.: Use of Poisson's Stochastic Process to Calculate the Collision Relaxation of a Non-Equilibrium Gas, Zh. Vychisl. Mat. Mat. Fiz., vol. 13, no. 2, pp. 505–510, 1973.

22 Yanitsky, V. E.: Application of Random Motion Processes for Modeling Free Molecular Gas Motion, Zh. Vychisl. Mat. Mat. Fiz., vol. 14, no. 1, pp. 259–262, 1974.

23 Belotserkovskii, O. M., and V. E. Yanitsky: Statistical "Particle-in-Cell" Method for the Solution of the Problem of Rarefied Gas Dynamics, Zh. Vychisl. Mat. Mat. Fiz., vol. 15, no. 5, pp. 1195–1208, 1975 (part I); vol. 15, no. 6, pp. 1553–1567, 1975 (part II).

24 Yanenko, N. N., and Y. I. Shokin: On the First Differential Approximation of Difference Schemes for Hyperbolic Sets of Equations, Sib. Mat. Zh., vol. 10, no. 5, pp. 1173–1187, 1969.

25 Gentry, R. A., R. E. Martin, and J. Daly: An Eulerian Differencing Method for Unsteady Compressible Flow Problems, J. Comput. Phys., vol. 1, pp. 87–118, 1966.

26 Ferrari, C., and F. G. Tricomi: "Transsonic Aerodynamics," Academic, New York, 1968.

27 Amsden, A. A., and F. H. Harlow: The SMAC Method, report LA-4370, Los Alamos Scientific Laboratory, New Mexico, 1970.

28 Easton, C. R.: Homogeneous Boundary Conditions for Pressure in MAC Method, J. Comput. Phys., vol. 9, no. 2, pp. 375–379, 1972.

29 Gushchin, V. A., and V. V. Shchennikov: Solution of Problems of Viscous Incompressible Fluid Dynamics by the Method of Splitting, Sb. Vychisl. Mat. Mat. Fiz., no. 2, 1974.

30 Babakov, A. V., O. M. Belotserkovskii, and L. I. Severinov: Numerical Investigation of a Viscous Heat-conducting Gas Flow Past a Blunt Body of Finite Size, *Izv. Acad. Nauk SSSR*, Mech. jidkosti i gaza, no. 3, pp. 112–123, 1975.

31 Belotserkovskii, O. M.: Calculation of the Flow around Axially Symmetric Bodies with a Detached Shock Wave, preprint, Computing Center AN SSSR, 1961.

32 Bird, G. A.: The Velocity Distribution Function within a Shock Wave, *J. Fluid Mech.*, vol. 30, no. 3, pp. 479–487, 1967.

33 Bird, G. A.: Direct Simulation and the Boltzmann Equation, *Phys. Fluids*, vol. 13, no. 11, pp. 2677–2681, 1970.

34 Kats, M.: "Probability and Related Topics in Physical Sciences," Izd. "Mir," 1965.

35 Cheremisin, F. G.: Numerical Solution of the Boltzmann Kinetic Equation for One-Dimensional Stationary Gas Motion, *Zh. Vychisl. Mat. Mat. Fiz.*, vol. 10, no. 3, pp. 654–665, 1970.

36 Rikov, V. A.: On Averaging the Boltzmann Kinetic Equation with Respect to a Transverse Velocity for the Case of One-dimensional Gas Motion, *Izv. Acad. Nauk SSSR*, Mech. jidkosti i gaza, no. 4, pp. 120–127, 1969.

Author Index

Adamson, T. C., 55, 86
Agrell, Mrs., 299n
Agrell, N., 338
Ahmed, S. R., 297
Albano, E., 188, 233
Albone, C. M., 30, 85, 164, 174, 176–178, 301, 338
Alfrey, C. P., 111, 112, 131
Ambrosiani, J., 61, 86
Amsden, A. A., 364, 365, 386
Andersen, M. N., 96, 151
Arlinger, B. G., 30, 55, 85, 223, 235
Ashley, H., 159, 232

Babakov, A. I., 368, 375, 387
Bailey, F. R., 30, 85, 159, 161, 162, 168, 175, 176, 180, 233, 301, 305, 309, 310, 337, 338
Baker, T. J., 30, 55, 85
Baldwin, B. S., 3, 83
Ballhaus, W. F., 29, 30, 84, 85, 159, 161, 162, 168, 170, 175, 176, 200, 201, 203, 208–209, 234, 337
Bateman, H., 3, 83
Bauer, F., 46, 52–54, 85, 86, 223–235
Bavitz, P. C., 52, 86
Beam, R. M., 77, 87, 190, 200, 203, 209, 234
Belotserkovskii, O. M., 345, 346, 347, 348, 349, 350, 351, 353, 354, 355, 356, 364, 365, 366, 367, 368, 369, 372, 373, 375, 377, 378, 381, 385, 387
Belotserkovskiy, S. M., 264–266, 297
Bentz, J. C., 142
Ben-Zui, F., 111, 112
Berger, S. A., 150
Bernstein, E. F., 112, 153
Besigk, G., 297
Binion, T. W., 180

Bird, G. A., 377, 387
Blackshear, P. L., 112, 153
Blakemore, W. S., 111, 152
Bollay, W., 264, 266, 297
Boxwell, D. A., 205, 234
Brais, M. P., 111, 153
Brandt, A., 29, 85
Braunwald, N. S., 111, 153
Buckley, F. T., Jr., 108, 109, 123
Bulekbayev, A., 345, 385
Buneman, O., 27, 28, 49, 50, 84
Buzbee, B. L., 27, 84

Callaghan, J. C., 111, 112, 115, 153
Campbell, D. R., 122, 124
Cannon, J. N., 135, 154
Caradonna, F. X., 203, 205, 234
Carlson, L. A., 74, 86
Carmichael, R. L., 249, 250, 296
Caro, C. G., 97, 151
Caughey, D. A., 30, 55, 64, 73, 85
Chafee, E. E., 91, 150
Chandravatna, P. A., 111, 112, 152
Charm, S. E., 150
Charpin, F., 71, 86, 166, 233
Cheng, H. K., 27, 75, 84, 86
Cheng, L. C., 108–110, 147, 149, 152
Cheng, R. T., 109
Cheng, S. I., 103, 151
Cheremisin, F. G., 383, 385, 387
Chow, R., 55, 86
Chushkin, P. I., 345, 385
Clapworthy, P. J., 30, 85
Clark, M. E., 108–110, 141, 145–150
Coder, D. W., 108, 109, 123
Cokelet, G. R., 89, 150
Cole, J. D., 1, 4, 8, 15, 29, 83, 84, 157, 163, 301, 310, 337, 338

389

Subject Index